Nissan Almera
Service and Repair Manual

John S. Mead

Models covered

(4053 - 8AJ1 - 240)

Nissan Almera Saloon and Hatchback models, including special/limited editions
1.4 litre (1392cc) and 1.6 litre (1597cc) petrol engines

Does not cover 2.0 litre GTi or Diesel models

© J H Haynes & Co. Ltd. 2005

A book in the **Haynes Service and Repair Manual Series**

ISBN **978 1 78521 522 3**

British Library Cataloguing in Publication Data
A catalogue record for this book is available from the British Library.

J H Haynes & Co. Ltd.
Haynes North America, Inc

www.haynes.com

Disclaimer

There are risks associated with automotive repairs. The ability to make repairs depends on the individual's skill, experience and proper tools. Individuals should act with due care and acknowledge and assume the risk of performing automotive repairs.

The purpose of this manual is to provide comprehensive, useful and accessible automotive repair information, to help you get the best value from your vehicle. However, this manual is not a substitute for a professional certified technician or mechanic.

This repair manual is produced by a third party and is not associated with an individual vehicle manufacturer. If there is any doubt or discrepancy between this manual and the owner's manual or the factory service manual, please refer to the factory service manual or seek assistance from a professional certified technician or mechanic.

Even though we have prepared this manual with extreme care and every attempt is made to ensure that the information in this manual is correct, neither the publisher nor the author can accept responsibility for loss, damage or injury caused by any errors in, or omissions from, the information given.

Contents

LIVING WITH YOUR NISSAN ALMERA

MAINTENANCE

Routine maintenance and servicing

Contents

REPAIRS AND OVERHAUL

Engine and associated systems

Transmission

Brakes and suspension

Body equipment

Wiring diagrams

REFERENCE

Index

Advanced driving

Many people see the words 'advanced driving' and believe that it won't interest them or that it is a style of driving beyond their own abilities. Nothing could be further from the truth. Advanced driving is straightforward safe, sensible driving - the sort of driving we should all do every time we get behind the wheel.

An average of 10 people are killed every day on UK roads and 870 more are injured, some seriously. Lives are ruined daily, usually because somebody did something stupid. Something like 95% of all accidents are due to human error, mostly driver failure. Sometimes we make genuine mistakes - everyone does. Sometimes we have lapses of concentration. Sometimes we deliberately take risks.

For many people, the process of 'learning to drive' doesn't go much further than learning how to pass the driving test because of a common belief that good drivers are made by 'experience'.

Learning to drive by 'experience' teaches three driving skills:

- [ ] Quick reactions. (Whoops, that was close!)
- [ ] Good handling skills. (Horn, swerve, brake, horn).
- [ ] Reliance on vehicle technology. (Great stuff this ABS, stop in no distance even in the wet...)

Drivers whose skills are 'experience based' generally have a lot of near misses and the odd accident. The results can be seen every day in our courts and our hospital casualty departments.

Advanced drivers have learnt to control the risks by controlling the position and speed of their vehicle. They avoid accidents and near misses, even if the drivers around them make mistakes.

The key skills of advanced driving are **concentration,** effective all-round **observation, anticipation** and **planning.** When **good vehicle handling** is added to these skills, all driving situations can be approached and negotiated in a safe, methodical way, leaving nothing to chance.

Concentration means applying your mind to safe driving, completely excluding anything that's not relevant. Driving is usually the most dangerous activity that most of us undertake in our daily routines. It deserves our full attention.

Observation means not just looking, but seeing and seeking out the information found in the driving environment.

Anticipation means asking yourself what is happening, what you can reasonably expect to happen and what could happen unexpectedly. (One of the commonest words used in compiling accident reports is 'suddenly'.)

Planning is the link between seeing something and taking the appropriate action. For many drivers, planning is the missing link.

If you want to become a safer and more skilful driver and you want to enjoy your driving more, contact the Institute of Advanced Motorists at www.iam.org.uk, phone 0208 996 9600, or write to IAM House, 510 Chiswick High Road, London W4 5RG for an information pack.

Working on your car can be dangerous. This page shows just some of the potential risks and hazards, with the aim of creating a safety-conscious attitude.

General hazards

Scalding

• Don't remove the radiator or expansion tank cap while the engine is hot.
• Engine oil, automatic transmission fluid or power steering fluid may also be dangerously hot if the engine has recently been running.

Burning

• Beware of burns from the exhaust system and from any part of the engine. Brake discs and drums can also be extremely hot immediately after use.

Crushing

• When working under or near a raised vehicle, always supplement the jack with axle stands, or use drive-on ramps. *Never venture under a car which is only supported by a jack.*

• Take care if loosening or tightening high-torque nuts when the vehicle is on stands. Initial loosening and final tightening should be done with the wheels on the ground.

Fire

• Fuel is highly flammable; fuel vapour is explosive.
• Don't let fuel spill onto a hot engine.
• Do not smoke or allow naked lights (including pilot lights) anywhere near a vehicle being worked on. Also beware of creating sparks (electrically or by use of tools).
• Fuel vapour is heavier than air, so don't work on the fuel system with the vehicle over an inspection pit.
• Another cause of fire is an electrical overload or short-circuit. Take care when repairing or modifying the vehicle wiring.
• Keep a fire extinguisher handy, of a type suitable for use on fuel and electrical fires.

Electric shock

• Ignition HT voltage can be dangerous, especially to people with heart problems or a pacemaker. Don't work on or near the ignition system with the engine running or the ignition switched on.

• Mains voltage is also dangerous. Make sure that any mains-operated equipment is correctly earthed. Mains power points should be protected by a residual current device (RCD) circuit breaker.

Fume or gas intoxication

• Exhaust fumes are poisonous; they often contain carbon monoxide, which is rapidly fatal if inhaled. Never run the engine in a confined space such as a garage with the doors shut.
• Fuel vapour is also poisonous, as are the vapours from some cleaning solvents and paint thinners.

Poisonous or irritant substances

• Avoid skin contact with battery acid and with any fuel, fluid or lubricant, especially antifreeze, brake hydraulic fluid and Diesel fuel. Don't syphon them by mouth. If such a substance is swallowed or gets into the eyes, seek medical advice.
• Prolonged contact with used engine oil can cause skin cancer. Wear gloves or use a barrier cream if necessary. Change out of oil-soaked clothes and do not keep oily rags in your pocket.
• Air conditioning refrigerant forms a poisonous gas if exposed to a naked flame (including a cigarette). It can also cause skin burns on contact.

Asbestos

• Asbestos dust can cause cancer if inhaled or swallowed. Asbestos may be found in gaskets and in brake and clutch linings. When dealing with such components it is safest to assume that they contain asbestos.

Special hazards

Hydrofluoric acid

• This extremely corrosive acid is formed when certain types of synthetic rubber, found in some O-rings, oil seals, fuel hoses etc, are exposed to temperatures above 400°C. The rubber changes into a charred or sticky substance containing the acid. *Once formed, the acid remains dangerous for years. If it gets onto the skin, it may be necessary to amputate the limb concerned.*
• When dealing with a vehicle which has suffered a fire, or with components salvaged from such a vehicle, wear protective gloves and discard them after use.

The battery

• Batteries contain sulphuric acid, which attacks clothing, eyes and skin. Take care when topping-up or carrying the battery.
• The hydrogen gas given off by the battery is highly explosive. Never cause a spark or allow a naked light nearby. Be careful when connecting and disconnecting battery chargers or jump leads.

Air bags

• Air bags can cause injury if they go off accidentally. Take care when removing the steering wheel and/or facia. Special storage instructions may apply.

Diesel injection equipment

• Diesel injection pumps supply fuel at very high pressure. Take care when working on the fuel injectors and fuel pipes.

⚠️ *Warning: Never expose the hands, face or any other part of the body to injector spray; the fuel can penetrate the skin with potentially fatal results.*

Remember...

DO

• Do use eye protection when using power tools, and when working under the vehicle.

• Do wear gloves or use barrier cream to protect your hands when necessary.

• Do get someone to check periodically that all is well when working alone on the vehicle.

• Do keep loose clothing and long hair well out of the way of moving mechanical parts.

• Do remove rings, wristwatch etc, before working on the vehicle – especially the electrical system.

• Do ensure that any lifting or jacking equipment has a safe working load rating adequate for the job.

DON'T

• Don't attempt to lift a heavy component which may be beyond your capability – get assistance.

• Don't rush to finish a job, or take unverified short cuts.

• Don't use ill-fitting tools which may slip and cause injury.

• Don't leave tools or parts lying around where someone can trip over them. Mop up oil and fuel spills at once.

• Don't allow children or pets to play in or near a vehicle being worked on.

The Nissan Almera range was introduced in the UK in October 1995 with the 3-door and 5-door Hatchback models. The 4-door Saloon model was added to the range in May 1996, with a 3-door GTi model (not covered by this manual) following in July. During the course of production, various minor changes and refinements were carried out and numerous special editions were added to the model range.

The Almera range is available with three sizes of petrol engines and one diesel engine. Covered in this manual are the 1.4 litre and 1.6 litre petrol engines, both being of double overhead camshaft (DOHC) 16-valve design. Both engines feature multi-point fuel injection and are equipped with an extensive range of emissions control systems. The engines are of a well-proven design and have been used extensively in a wide range of Nissan vehicles.

Fully-independent front suspension is fitted, with semi-independent multi-link suspension used at the rear.

A five-speed manual transmission is fitted as standard across the range, with a four-speed electronically-controlled automatic transmission optionally available on 1.6 litre models.

A wide range of standard and optional equipment is available within the Almera range including central locking, electric windows, an electric sunroof, an anti-lock braking system and supplementary restraint system.

For the home mechanic, the Almera is a straightforward vehicle to maintain, and most of the items requiring frequent attention are easily accessible.

Your Nissan Almera Manual

The aim of this manual is to help you get the best value from your vehicle. It can do so in several ways. It can help you decide what work must be done (even should you choose to get it done by a garage), provide information on routine maintenance and servicing, and give a logical course of action and diagnosis when random faults occur. However, it is hoped that you will use the manual by tackling the work yourself. On simpler jobs, it may even be quicker than booking the car into a garage and going there twice, to leave and collect it. Perhaps most important, a lot of money can be saved by avoiding the costs a garage must charge to cover its labour and overheads.

The manual has drawings and descriptions to show the function of the various components, so that their layout can be understood. Then the tasks are described and photographed in a clear step-by-step sequence.

References to the 'left' or 'right' are in the sense of a person in the driver's seat, facing forward.

Acknowledgements

Thanks are due to Draper Tools Limited, who provided some of the workshop tools, and to all those people at Sparkford who helped in the production of this Manual.

We take great pride in the accuracy of information given in this manual, but vehicle manufacturers make alterations and design changes during the production run of a particular vehicle of which they do not inform us. No liability can be accepted by the authors or publishers for loss, damage or injury caused by any errors in, or omissions from, the information given.

The following pages are intended to help in dealing with common roadside emergencies and breakdowns. You will find more detailed fault finding information at the back of the manual, and repair information in the main chapters.

If your car won't start and the starter motor doesn't turn

☐ If it's a model with automatic transmission, make sure the selector is in P or N.
☐ Open the bonnet and make sure that the battery terminals are clean and tight.
☐ Switch on the headlights and try to start the engine. If the headlights go very dim when you're trying to start, the battery is probably flat. Get out of trouble by jump starting (see next page) using a friend's car.

If your car won't start even though the starter motor turns as normal

☐ Is there fuel in the tank?
☐ Is there moisture on electrical components under the bonnet? Switch off the ignition, then wipe off any obvious dampness with a dry cloth. Spray a water-repellent aerosol product (WD 40 or equivalent) on ignition and fuel system electrical connectors like those shown in the photos. Pay special attention to the ignition coil wiring connector and HT leads.

A Check that the HT leads are securely connected by pushing them onto the spark plugs . . .

B . . . and onto the distributor cap.

C Check that the wiring connector is securely connected to the distributor.

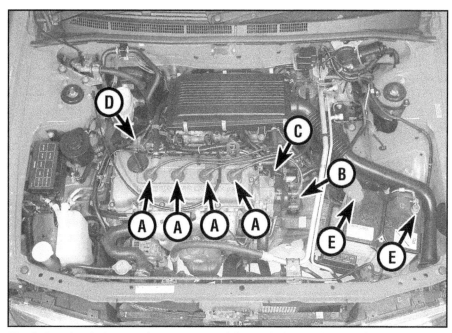

Check that electrical connections are secure (with the ignition switched off) and spray them with a water-dispersant spray like WD40 if you suspect a problem due to damp.

D Check all engine related earth connections and wiring multiplugs for security.

E Check the security and condition of the battery terminals.

Jump starting

When jump-starting a car using a booster battery, observe the following precautions:

✔ Before connecting the booster battery, make sure that the ignition is switched off.

✔ Ensure that all electrical equipment (lights, heater, wipers, etc) is switched off.

✔ Take note of any special precautions printed on the battery case.

✔ Make sure that the booster battery is the same voltage as the discharged one in the vehicle.

✔ If the battery is being jump-started from the battery in another vehicle, the two vehicles MUST NOT TOUCH each other.

✔ Make sure that the transmission is in neutral (or PARK, in the case of automatic transmission).

 Jump starting will get you out of trouble, but you must correct whatever made the battery go flat in the first place. There are three possibilities:

1 The battery has been drained by repeated attempts to start, or by leaving the lights on.

2 The charging system is not working properly (alternator drivebelt slack or broken, alternator wiring fault or alternator itself faulty).

3 The battery itself is at fault (electrolyte low, or battery worn out).

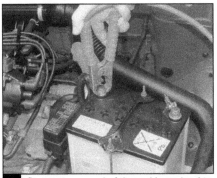

1 Connect one end of the red jump lead to the positive (+) terminal of the flat battery

2 Connect the other end of the red lead to the positive (+) terminal of the booster battery.

3 Connect one end of the black jump lead to the negative (-) terminal of the booster battery

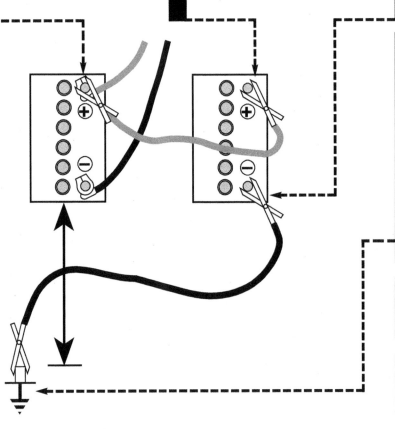

4 Connect the other end of the black jump lead to a bolt or bracket on the engine block, well away from the battery, on the vehicle to be started.

5 Make sure that the jump leads will not come into contact with the fan, drive-belts or other moving parts of the engine.

6 Start the engine using the booster battery and run it at idle speed. Switch on the lights, rear window demister and heater blower motor, then disconnect the jump leads in the reverse order of connection. Turn off the lights etc.

Wheel changing

Some of the details shown here will vary according to model. However, the basic principles apply to all vehicles.

⚠️ *Warning: Do not change a wheel in a situation where you risk being hit by another vehicle. On busy roads, try to stop in a lay-by or a gateway. Be wary of passing traffic while changing the wheel - it is easy to become distracted by the job in hand.*

Preparation

☐ When a puncture occurs, stop as soon as it is safe to do so.
☐ Park on firm level ground, if possible, and well out of the way of other traffic.
☐ Use hazard warning lights if necessary.

☐ If you have one, use a warning triangle to alert other drivers of your presence.
☐ Apply the handbrake and engage first or reverse gear (or Park on models with automatic transmission).

☐ Chock the wheel diagonally opposite the one being removed – a couple of large stones will do for this.
☐ If the ground is soft, use a flat piece of wood to spread the load under the jack.

Changing the wheel

1 The spare wheel and tools are stored in the luggage compartment beneath the floor covering. Unscrew the retaining bolt to release the spare wheel.

2 Lift out the jack and tool tray, located adjacent to the spare wheel, then remove the spare wheel from its location.

3 Where applicable, prise off the wheel trim from the wheel with the flat tyre, using the jack handle.

4 Loosen each wheel nut by half a turn. Where anti-theft wheel nuts are used, a special adapter will be required.

5 Locate the jack head between the two notches at the reinforced jacking point nearest the wheel to be changed. Ensure that the groove in the jack head is properly engaged with the sill. Use the wheelbrace engaged with the jack handle to raise the car until the wheel is clear of the ground.

6 Unscrew the wheel nuts, lift off the wheel and position it under the car for safety. Fit the spare wheel and wheel nuts and tighten moderately with the wheelbrace.

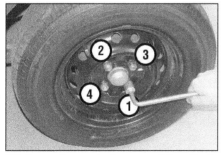

7 Lower the car to the ground then finally tighten the wheel nuts in a diagonal sequence. Stow the punctured wheel and tools back in the luggage compartment and secure them in position.

Finally...

☐ Remove the wheel chocks.

☐ Stow the jack and tools in the correct locations in the car.

☐ Check the tyre pressure on the wheel just fitted. If it is low, or if you don't have a pressure gauge with you, drive slowly to the nearest garage and inflate the tyre to the right pressure.

☐ Have the damaged tyre or wheel repaired as soon as possible.

☐ The wheel nuts should be slackened and retightened to the specified torque at the earliest possible opportunity.

Identifying leaks

Puddles on the garage floor or drive, or obvious wetness under the bonnet or underneath the car, suggest a leak that needs investigating. It can sometimes be difficult to decide where the leak is coming from, especially if the engine bay is very dirty already. Leaking oil or fluid can also be blown rearwards by the passage of air under the car, giving a false impression of where the problem lies.

 Warning: Most automotive oils and fluids are poisonous. Wash them off skin, and change out of contaminated clothing, without delay.

 The smell of a fluid leaking from the car may provide a clue to what's leaking. Some fluids are distinctively coloured. It may help to clean the car carefully and to park it over some clean paper overnight as an aid to locating the source of the leak.
Remember that some leaks may only occur while the engine is running.

Sump oil

Engine oil may leak from the drain plug...

Oil from filter

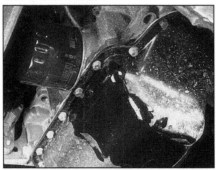

...or from the base of the oil filter.

Gearbox oil

Gearbox oil can leak from the seals at the inboard ends of the driveshafts.

Antifreeze

Leaking antifreeze often leaves a crystalline deposit like this.

Brake fluid

A leak occurring at a wheel is almost certainly brake fluid.

Power steering fluid

Power steering fluid may leak from the pipe connectors on the steering rack.

Towing

When all else fails, you may find yourself having to get a tow home – or of course you may be helping somebody else. Long-distance recovery should only be done by a garage or breakdown service. For shorter distances, DIY towing using another car is easy enough, but observe the following points:

☐ Towing eyes are fitted at the front and rear of the vehicle for attachment of the tow rope. The towing eyes can be accessed through slots, or from underneath the bumpers.

☐ Use a proper tow-rope – they are not expensive. The vehicle being towed must display an ON TOW sign in its rear window.

☐ Always turn the ignition key to the 'on' position when the vehicle is being towed, so that the steering lock is released, and that the direction indicator and brake lights will work.

☐ Only attach the tow-rope to the towing eyes provided.

☐ Before being towed, release the handbrake and select neutral on the transmission. **Note:** *On models with automatic transmission, special precautions apply. If in doubt, do not tow, or transmission damage may result.*

☐ Note that greater-than-usual pedal pressure will be required to operate the brakes, since the vacuum servo unit is only operational with the engine running.

☐ On models with power steering, greater-than-usual steering effort will also be required.

☐ The driver of the car being towed must keep the tow-rope taut at all times to avoid snatching.

☐ Make sure that both drivers know the route before setting off.

☐ Only drive at moderate speeds and keep the distance towed to a minimum. Drive smoothly and allow plenty of time for slowing down at junctions.

⚠ *Warning: To prevent damage to the catalytic converter, a vehicle must not be push-started, or started by towing, when the engine is at operating temperature. Use jump leads (see 'Jump starting').*

Introduction

There are some very simple checks which need only take a few minutes to carry out, but which could save you a lot of inconvenience and expense.

These *Weekly checks* require no great skill or special tools, and the small amount of time they take to perform could prove to be very well spent, for example;

☐ Keeping an eye on tyre condition and pressures, will not only help to stop them wearing out prematurely, but could also save your life.

☐ Many breakdowns are caused by electrical problems. Battery-related faults are particularly common, and a quick check on a regular basis will often prevent the majority of these.

☐ If your car develops a brake fluid leak, the first time you might know about it is when your brakes don't work properly. Checking the level regularly will give advance warning of this kind of problem.

☐ If the oil or coolant levels run low, the cost of repairing any engine damage will be far greater than fixing the leak, for example.

Underbonnet check points

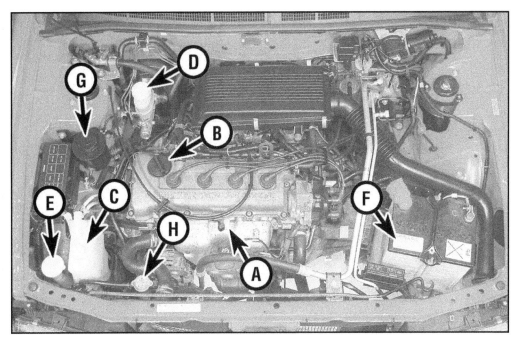

A Engine oil level dipstick

B Engine oil filler cap

C Coolant expansion tank

D Brake fluid reservoir

E Screen washer fluid reservoir

F Battery

G Power steering fluid reservoir

H Radiator filler cap

Engine oil level

Before you start
✔ Make sure that your car is on level ground.
✔ Check the oil level before the car is driven, or at least 5 minutes after the engine has been switched off.

 HAYNES HINT *If the oil is checked immediately after driving the vehicle, some of the oil will remain in the upper engine components, resulting in an inaccurate reading on the dipstick.*

The correct oil
Modern engines place great demands on their oil. It is very important that the correct oil for your car is used (See 'Lubricants and fluids').

Car Care
● If you have to add oil frequently, you should check whether you have any oil leaks. Place some clean paper under the car overnight, and check for stains in the morning. If there are no leaks, the engine may be burning oil.

● Always maintain the level between the upper and lower dipstick marks (see photo 3). If the level is too low severe engine damage may occur. Oil seal failure may result if the engine is overfilled by adding too much oil.

1 The dipstick is located at the front of the engine and is brightly-coloured for easy identification.

2 Withdraw the dipstick. Using a clean rag or paper towel, remove all oil from the dipstick. Insert the clean dipstick into the tube as far as it will go, then withdraw it again.

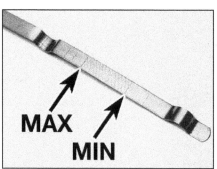
3 Note the oil level on the end of the dipstick, which should be within the cross-hatched area.

4 Oil is added through the filler cap aperture. Add the oil a little at a time, checking the level on the dipstick often. Using a funnel will help to reduce spillage. Don't overfill (see *Car Care*).

Coolant level

⚠ **Warning: DO NOT attempt to remove the expansion tank pressure cap when the engine is hot, as there is a very great risk of scalding. Do not leave open containers of coolant about, as it is poisonous.**

Car Care
● With a sealed-type cooling system, adding coolant should not be necessary on a regular basis. If frequent topping-up is required, it is likely there is a leak. Check the radiator, all hoses and joint faces for signs of staining or wetness, and rectify as necessary.

● It is important that antifreeze is used in the cooling system all year round, not just during the winter months. Don't top-up with water alone, as the antifreeze will become too diluted.

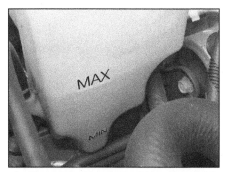
1 The coolant level should be checked in the expansion tank, which is located at the right-hand side of the engine compartment. The expansion tank has MAX and MIN level markings. When the engine is cold, the level should be between the two marks. When the engine is hot, the level may rise slightly above the MAX mark.

2 If topping-up is necessary, wait until the engine is cold. Slowly unscrew the expansion tank cap to release any pressure present in the cooling system, and remove it. Add a mixture of water and antifreeze to the expansion tank until the coolant level is up to the MAX mark, then refit the cap.

3 If the expansion tank is empty, check the coolant level in the radiator. Slowly unscrew the expansion tank cap to release any pressure present in the cooling system, and remove it. Top-up the radiator to the level of the filler opening then refit the cap. When the radiator has been topped-up, top-up the expansion tank, as described previously.

Brake fluid level

Warning:
● Brake fluid can harm your eyes and damage painted surfaces, so use extreme caution when handling and pouring it.
● Do not use fluid that has been standing open for some time, as it absorbs moisture from the air, which can cause a dangerous loss of braking effectiveness.

 HAYNES HiNT
● Make sure that your car is on level ground.
● The fluid level in the reservoir will drop slightly as the brake pads wear down, but the fluid level must never be allowed to drop below the MIN mark.

Safety First!
● If the reservoir requires repeated topping-up this is an indication of a fluid leak somewhere in the system, which should be investigated immediately.

● If a leak is suspected, the car should not be driven until the braking system has been checked. Never take any risks where brakes are concerned.

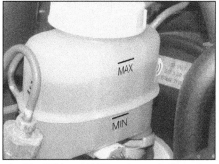

1 The MIN and MAX marks are indicated on the side of the reservoir which is located at the rear of the engine compartment on the driver's side. The fluid level must be kept between the marks at all times.

2 If topping-up is necessary, first wipe clean the area around the filler cap to prevent dirt entering the hydraulic system, then unscrew the cap.

3 Carefully add fluid, taking care not to spill it onto the surrounding components. Use only the specified fluid; mixing different types can cause damage to the system. After topping-up to the correct level, securely refit the cap and wipe off any spilt fluid.

Power steering fluid level

Before you start
✔ Park the vehicle on level ground.
✔ Set the steering wheel straight-ahead.
✔ The engine should be turned off.

HAYNES HiNT
For the check to be accurate, the steering must not be turned once the engine has been stopped.

Safety First!
● The need for frequent topping-up indicates a leak, which should be investigated immediately.

1 The reservoir is located on the right-hand side of the engine compartment, in front of the suspension turret. The reservoir has a filler cap incorporating a dipstick for level checking. Wipe the area around the filler cap, then unscrew the cap, and read off the level on the dipstick.

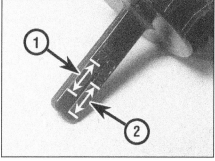

2 If the vehicle has recently been driven on a reasonable run, the level should be read from the HOT range (1) on the dipstick. If the vehicle has not recently been driven, and the engine is cold, the level should be read from the COLD range (2).

3 If necessary, top-up to the relevant MAX mark using the specified type of fluid, then refit the filler cap. Take great care not to allow any dirt or foreign matter to enter the hydraulic system, and do not overfill the reservoir.

Tyre condition and pressure

It is very important that tyres are in good condition, and at the correct pressure - having a tyre failure at any speed is highly dangerous. Tyre wear is influenced by driving style - harsh braking and acceleration, or fast cornering, will all produce more rapid tyre wear. As a general rule, the front tyres wear out faster than the rears. Interchanging the tyres from front to rear ("rotating" the tyres) may result in more even wear. However, if this is completely effective, you may have the expense of replacing all four tyres at once! Remove any nails or stones embedded in the tread before they penetrate the tyre to cause deflation. If removal of a nail does reveal that the tyre has been punctured, refit the nail so that its point of penetration is marked. Then immediately change the wheel, and have the tyre repaired by a tyre dealer.

Regularly check the tyres for damage in the form of cuts or bulges, especially in the sidewalls. Periodically remove the wheels, and clean any dirt or mud from the inside and outside surfaces. Examine the wheel rims for signs of rusting, corrosion or other damage. Light alloy wheels are easily damaged by "kerbing" whilst parking; steel wheels may also become dented or buckled. A new wheel is very often the only way to overcome severe damage.

New tyres should be balanced when they are fitted, but it may become necessary to re-balance them as they wear, or if the balance weights fitted to the wheel rim should fall off. Unbalanced tyres will wear more quickly, as will the steering and suspension components. Wheel imbalance is normally signified by vibration, particularly at a certain speed (typically around 50 mph). If this vibration is felt only through the steering, then it is likely that just the front wheels need balancing. If, however, the vibration is felt through the whole car, the rear wheels could be out of balance. Wheel balancing should be carried out by a tyre dealer or garage.

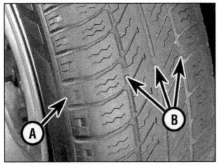

1 Tread Depth - visual check
The original tyres have tread wear safety bands (B), which will appear when the tread depth reaches approximately 1.6 mm. The band positions are indicated by a triangular mark on the tyre sidewall (A).

2 Tread Depth - manual check
Alternatively, tread wear can be monitored with a simple, inexpensive device known as a tread depth indicator gauge.

3 Tyre Pressure Check
Check the tyre pressures regularly with the tyres cold. Do not adjust the tyre pressures immediately after the vehicle has been used, or an inaccurate setting will result.

Tyre tread wear patterns

Shoulder Wear

Underinflation (wear on both sides)
Under-inflation will cause overheating of the tyre, because the tyre will flex too much, and the tread will not sit correctly on the road surface. This will cause a loss of grip and excessive wear, not to mention the danger of sudden tyre failure due to heat build-up.
Check and adjust pressures
Incorrect wheel camber (wear on one side)
Repair or renew suspension parts
Hard cornering
Reduce speed!

Centre Wear

Overinflation
Over-inflation will cause rapid wear of the centre part of the tyre tread, coupled with reduced grip, harsher ride, and the danger of shock damage occurring in the tyre casing.
Check and adjust pressures

If you sometimes have to inflate your car's tyres to the higher pressures specified for maximum load or sustained high speed, don't forget to reduce the pressures to normal afterwards.

Uneven Wear

Front tyres may wear unevenly as a result of wheel misalignment. Most tyre dealers and garages can check and adjust the wheel alignment (or "tracking") for a modest charge.
Incorrect camber or castor
Repair or renew suspension parts
Malfunctioning suspension
Repair or renew suspension parts
Unbalanced wheel
Balance tyres
Incorrect toe setting
Adjust front wheel alignment
Note: *The feathered edge of the tread which typifies toe wear is best checked by feel.*

Battery

Caution: Before carrying out any work on the vehicle battery, read the precautions given in 'Safety first' at the start of this manual.

✔ Make sure that the battery tray is in good condition, and that the clamp is tight. Corrosion on the tray, retaining clamp and the battery itself can be removed with a solution of water and baking soda. Thoroughly rinse all cleaned areas with water. Any metal parts damaged by corrosion should be covered with a zinc-based primer, then painted.

✔ Periodically (approximately every three months), check the charge condition of the battery as described in Chapter 5A.

✔ If the battery is flat, and you need to jump start your vehicle, see *Roadside Repairs*.

HAYNES HiNT

Battery corrosion can be kept to a minimum by applying a layer of petroleum jelly to the clamps and terminals after they are reconnected.

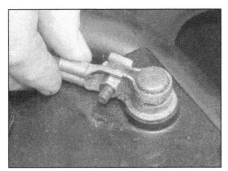

1 The battery is located on the left-hand side of the engine compartment. The exterior of the battery should be inspected periodically for damage such as a cracked case or cover, and the battery cable clamps should be checked for tightness to ensure good electrical connections.

2 If corrosion (white, fluffy deposits) is evident, remove the cables from the battery terminals, clean them with a small wire brush, then refit them. Automotive stores sell a tool for cleaning the battery post . . .

3 . . . as well as the battery cable clamps

Electrical systems

✔ Check all external lights and the horn. Refer to the appropriate Sections of Chapter 12 for details if any of the circuits are found to be inoperative.

✔ Visually check all accessible wiring connectors, harnesses and retaining clips for security, and for signs of chafing or damage.

HAYNES HiNT *If you need to check your brake lights and indicators unaided, back up to a wall or garage door and operate the lights. The reflected light should show if they are working properly.*

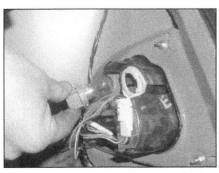

1 If a single indicator light, stop-light or headlight has failed, it is likely that a bulb has blown and will need to be renewed. Refer to Chapter 12 for details. If both stop-lights have failed, it is possible that the switch has failed (see Chapter 9).

2 If more than one indicator light or tail light has failed it is likely that either a fuse has blown or that there is a fault in the circuit (see Chapter 12). The main fuses are located in the fusebox on the driver's side of the facia, with supplementary fuses located in the engine compartment. For access to the main fuses, pull open the cover flap.

3 To renew a blown fuse, simply pull it out using the special plastic tool and fit a new fuse of the correct rating (see Chapter 12). If the fuse blows again, it is important that you find out why – a complete checking procedure is given in Chapter 12.

Screen washer fluid level

Screenwash additives not only keep the windscreen clean during foul weather, they also prevent the washer system freezing in cold weather – which is when you are likely to need it most. Don't top-up using plain water as the screenwash will become too diluted, and will freeze during cold weather.

On no account use coolant antifreeze in the washer system – this could discolour or damage paintwork.

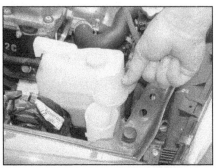

1 The windscreen/tailgate washer fluid reservoir filler is located at the front right-hand corner of the engine compartment, behind the headlight. Certain models are fitted with a facia-mounted warning light, to indicate when topping-up is required. On models without a warning light, a level indicator tube is fitted to the reservoir filler cap. To check the level, use a finger to cover the breather hole in the top of the filler cap, then pull the cap from the reservoir. Fluid will be retained in the tube to indicate the level in the reservoir.

2 When topping-up the reservoir, add a screenwash additive in the quantities recommended on the bottle. Use of a funnel will prevent spillage.

Wiper blades

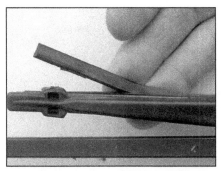

1 Check the condition of the wiper blades; if they are cracked or show any signs of deterioration, or if the glass swept area is smeared, renew them. Wiper blades should be renewed annually. Don't forget to check the tailgate wiper as well.

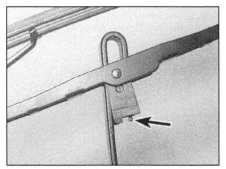

2 To remove a wiper blade, pull the arm fully away from the screen until it locks. Swivel the blade through 90°, press the locking tab with your fingers and slide the blade out of the arm's hooked end.

Lubricants and fluids

Engine	Multigrade engine oil, viscosity SAE 10W/30 to 15W/50, to API SG/SH or better
Cooling system	Ethylene glycol-based antifreeze and soft water
Manual gearbox	Hypoid gear oil, viscosity SAE 80W to API GL4
Automatic transmission	Dexron type II automatic transmission fluid (ATF)
Braking system	Hydraulic fluid to SAE J1703F, DOT 4
Power steering	Dexron type II automatic transmission fluid (ATF)

Choosing your engine oil

Engines need oil, not only to lubricate moving parts and minimise wear, but also to maximise power output and to improve fuel economy.

HOW ENGINE OIL WORKS

• *Beating friction*

Without oil, the moving surfaces inside your engine will rub together, heat up and melt, quickly causing the engine to seize. Engine oil creates a film which separates these moving parts, preventing wear and heat build-up.

• *Cooling hot-spots*

Temperatures inside the engine can exceed 1000° C. The engine oil circulates and acts as a coolant, transferring heat from the hot-spots to the sump.

• *Cleaning the engine internally*

Good quality engine oils clean the inside of your engine, collecting and dispersing combustion deposits and controlling them until they are trapped by the oil filter or flushed out at oil change.

OIL CARE - FOLLOW THE CODE

To handle and dispose of used engine oil safely, always:

0800 66 33 66
www.oilbankline.org.uk

• *Avoid skin contact with used engine oil. Repeated or prolonged contact can be harmful.*
• *Dispose of used oil and empty packs in a responsible manner in an authorised disposal site. Call 0800 663366 to find the one nearest to you. Never tip oil down drains or onto the ground.*

Tyre pressures

Refer to the manufacturer's tyre specification plate fitted to the driver's door rear pillar (visible when the door is open) for the correct tyre pressures for your particular vehicle. Pressures apply only to original-equipment tyres, and may vary if any other make or type is fitted; check with the tyre manufacturer or supplier for correct pressures if necessary.

Notes

Chapter 1
Routine maintenance and servicing

Contents

Degrees of difficulty

Easy, suitable for novice with little experience	**Fairly easy,** suitable for beginner with some experience	**Fairly difficult,** suitable for competent DIY mechanic 	**Difficult,** suitable for experienced DIY mechanic	**Very difficult,** suitable for expert DIY or professional 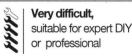

Lubricants and fluids

Refer to end of *Weekly Checks* on page 0•17

Capacities

Engine oil	3.2 litres
Cooling system:	
Manual transmission models	5.9 litres
Automatic transmission models	6.4 litres
Transmission:	
1.4 litre manual transmission models	2.8 litres
1.6 litre manual transmission models	2.9 litres
1.6 litre automatic transmission models	7.0 litres
Fuel tank	50.0 litres

Cooling system

	Antifreeze	Water
Antifreeze mixture (ethylene glycol antifreeze):		
Protection down to –15ºC	30%	70%
Protection down to –35ºC	50%	50%

Note: *Refer to antifreeze manufacturer for latest recommendations.*

Fuel system

Idle speed (not adjustable)*	700 ± 50 rpm (controlled by ECU)
Base idle speed	625 ± 50 rpm
Idle mixture CO content	Less than 1.0 % (not adjustable – controlled by ECU)

Although the idle speed is not adjustable, the base idle speed can be set as described (see Section 4).

Ignition system

	Type	Electrode gap
Spark plugs:		
1.4 litre engine	Bosch FR 7 D+	0.9 mm
	NGK BKR6E	0.8 to 0.9 mm
1.6 litre engine	Bosch FR 7 D+X	1.1 mm
	NGK BKR6E	0.8 to 0.9 mm

Auxiliary drivebelts

	Setting	Limit
Drivebelt deflection:		
Alternator:		
With power steering	7 to 9 mm	11 mm
Without power steering	7 to 9 mm	10 mm
Air conditioning compressor	6 to 8 mm	9.5 mm
Power steering pump	4 to 6 mm	7.5 mm

Note: *In all cases, the drivebelt deflection is measured by applying a force of 98 N (10 kg) as described in the text. All figures are quoted for a 'used' drivebelt – if a new belt has been fitted, the setting deflection should be decreased by 1.0 mm.*

Brakes

Minimum front brake pad friction material thickness	2.0 mm
Minimum rear brake pad friction material thickness	1.5 mm
Minimum rear brake shoe lining thickness	1.5 mm
Number of clicks required to fully apply handbrake:	
Models with rear drum brakes	7 to 8 clicks
Models with rear disc brakes	8 to 9 clicks
Number of clicks required to operate handbrake 'on' warning light . . .	0 to 1 click

Suspension and steering

Front wheel toe setting	2.0 mm ± 2.0 mm toe-in

Torque wrench settings

	Nm	lbf ft
Air conditioning compressor drivebelt tensioning pulley nut	28	21
Automatic transmission drain plug	34	25
Cylinder block coolant drain plug	40	30
Engine sump drain plug	35	26
Manual transmission:		
Filler/level plug	29	21
Drain plug	29	21
Roadwheel nuts	110	81
Seat belt mounting bolts	50	37
Spark plugs	25	18

The maintenance intervals in this manual are provided with the assumption that you, not the dealer, will be carrying out the work. These are the minimum maintenance intervals based on the schedule recommended by the manufacturer for vehicles driven daily. If you wish to keep your vehicle in peak condition at all times, you may wish to perform some of these procedures more often. We encourage frequent maintenance because it enhances the efficiency, performance and resale value of your vehicle. If the vehicle is driven in dusty areas, used to tow a trailer, or driven frequently at slow speeds (idling in traffic) or on short journeys, more frequent maintenance intervals are recommended. Nissan recommend that many service intervals are halved for vehicles which are used under these conditions.

Every 250 miles (400 km) or weekly

☐ Refer to *Weekly checks*.

Every 9000 miles (15 000 km) or 12 months – whichever comes first

☐ Renew the engine oil and filter (Section 3)

Note: *Frequent oil and filter changes are good for the engine. We recommend changing the oil more frequently than the mileage specified here, or at least twice a year.*

Every 18 000 miles (30 000 km) or 12 months – whichever comes first

☐ Check and if necessary adjust the idle speed (Section 4)
☐ Check all underbonnet components and hoses for fluid leaks (Section 5)
☐ Check the manual transmission oil level (Section 6)
☐ Check the brake pads and renew if necessary (Section 7)
☐ Check and adjust the handbrake (Section 8)
☐ Check and adjust the clutch (Section 9)
☐ Check the condition of the air conditioning system components (see Section 10)
☐ Renew the spark plugs (Section 11)
☐ Renew the fuel filter (Section 12)
☐ Check the condition of the emissions control system hoses and components (Section 13)
☐ Check the operation of the exhaust gas sensor (Section 14)
☐ Check the condition of the auxiliary drivebelts, and renew if necessary (Section 15)

Every 18 000 miles (30 000 km) or 12 months – whichever comes first (continued)

☐ Check the rear brake shoes (where fitted) and renew if necessary (Section 16)
☐ Renew the brake fluid (Section 17)
☐ Check the automatic transmission fluid level (Section 18)
☐ Check the steering and suspension components for condition and security (Section 19)
☐ Check the condition of the driveshaft rubber gaiters (Section 20)
☐ Check the wheel alignment (Section 21)
☐ Check the balance of each roadwheel (Section 22)
☐ Check the operation and security of all seat belts (Section 23)
☐ Lubricate all hinges and locks (Section 24)
☐ Carry out a road test (Section 25)

Every 36 000 miles (60 000 km) or 2 years – whichever comes first

In addition to all the items listed above, carry out the following:

☐ Renew the air filter (Section 26)
☐ Check the ignition system components (Section 27)
☐ Renew the PCV filter (Section 28)
☐ Check the operation of the braking system servo unit and check valve (Section 29)
☐ Renew the coolant (Section 30)
☐ Renew the manual transmission oil (Section 31)
☐ Renew the automatic transmission fluid (Section 32)

Underbonnet view

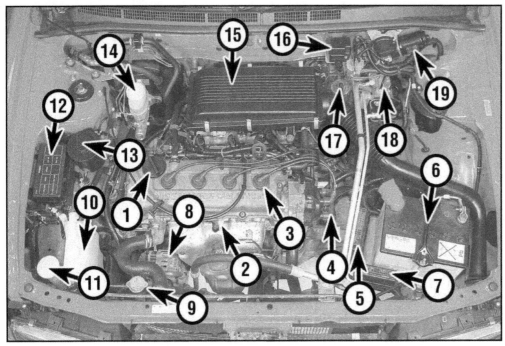

1 Engine oil filler cap
2 Engine oil dipstick
3 Spark plug HT lead
4 Distributor
5 Fusible link and fusebox
6 Battery
7 Fusible link and relay box
8 Alternator
9 Radiator filler cap
10 Coolant expansion tank
11 Windscreen washer reservoir
12 Relay box
13 Power steering fluid reservoir
14 Brake master cylinder fluid reservoir
15 Air cleaner housing
16 ABS relay box
17 Fuel filter
18 ABS hydraulic unit
19 Windscreen wiper motor

Front underbody view

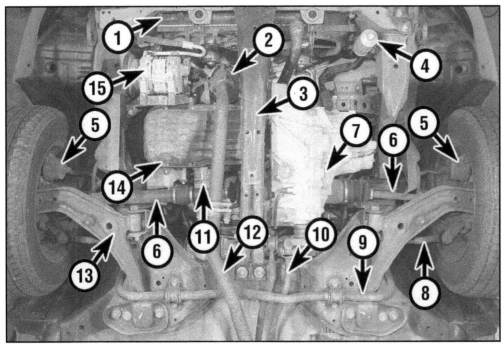

1 Radiator drain plug
2 Catalytic converter
3 Centre member
4 Air conditioning receiver/dryer
5 Brake caliper
6 Driveshaft
7 Manual transmission drain plug
8 Track rod end
9 Anti-roll bar
10 Gearchange linkage selector rod
11 Oil filter
12 Exhaust system front pipe
13 Suspension lower arm
14 Sump drain plug
15 Alternator

Rear underbody view

1 Fuel tank
2 Exhaust system intermediate pipe
3 Trailing arm
4 Shock absorber lower mounting
5 Coil spring
6 Axle beam
7 Lateral link
8 Control rod
9 Exhaust system rear silencer
10 Brake caliper
11 Fuel tank filler pipe

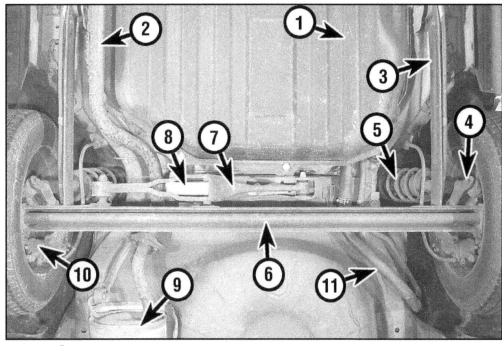

Maintenance procedures

1 General information

This Chapter is designed to help the home mechanic maintain his/her vehicle for safety, economy, long life and peak performance.

The Chapter contains a master maintenance schedule, followed by sections dealing specifically with each task on the schedule. Visual checks, adjustments, component renewal and other helpful items are included. Refer to the accompanying illustrations of the engine compartment and the underside of the vehicle for the locations of the various components.

Servicing of your vehicle in accordance with the mileage/time maintenance schedule and the following sections will provide a planned maintenance programme, which should result in a long and reliable service life. This is a comprehensive plan, so maintaining some items but not others at the specified service intervals will not produce the same results.

As you service your vehicle, you will discover that many of the procedures can – and should – be grouped together, because of the particular procedure being performed, or because of the close proximity of two otherwise-unrelated components to one another. For example, if the vehicle is raised for any reason, the exhaust can be inspected at the same time as the suspension and steering components.

The first step in this maintenance programme is to prepare yourself before the actual work begins. Read through all the sections relevant to the work to be carried out, then make a list and gather all the parts and tools required. If a problem is encountered, seek advice from a parts specialist, or a dealer service department.

2 Regular maintenance

1 If, from the time the vehicle is new, the routine maintenance schedule is followed closely, and frequent checks are made of fluid levels and high-wear items, as suggested throughout this manual, the engine will be kept in relatively good running condition, and the need for additional work will be minimised.
2 It is possible that there will be times when the engine is running poorly, due to lack of regular maintenance. This is even more likely if a used vehicle, which has not received regular and frequent maintenance checks, is purchased. In such cases, additional work may need to be carried out, outside of the regular maintenance intervals.
3 If engine wear is suspected, a compression test (Chapter 2A) will provide valuable information regarding the overall performance of the main internal components. Such a test can be used as a basis to decide on the extent of the work to be carried out. If for example a compression test indicates serious internal engine wear, conventional maintenance as described in this Chapter will not greatly improve the performance of the engine, and may prove a waste of time and money, unless extensive overhaul work (Chapter 2B) is carried out first.
4 The following series of operations are those most often required to improve the performance of a generally poor-running engine:

Primary operations

a) Check all the engine-related fluids (See 'Weekly checks').
b) Check the condition of all hoses, and check for fluid leaks (Section 5).
c) Clean, inspect and test the battery (See 'Weekly checks' and Chapter 5A).
d) Renew the spark plugs (Section 11).
e) Check the condition and tension of the auxiliary drivebelt(s) (Section 15).
f) Check the condition of the air filter, and renew if necessary (Section 26).
g) Inspect the distributor cap and rotor arm (Section 27).
h) Inspect the ignition HT leads (Section 27).

5 If the above operations do not prove fully effective, carry out the following secondary operations:

Secondary operations

All items listed under Primary operations, plus the following:

a) Check the charging system (Chapter 5A).
b) Check the ignition system (Chapter 5B).
c) Check the fuel, exhaust and emission control systems (Chapter 4A or 4B).
d) Renew the distributor cap and rotor arm (Section 27).
e) Renew the ignition HT leads (Section 27).

Every 9000 miles or 12 months – whichever comes first

3 Engine oil and filter renewal

HAYNES HINT
Frequent oil and filter changes are the most important preventative maintenance procedures which can be undertaken by the DIY owner. As engine oil ages, it becomes diluted and contaminated, which leads to premature engine wear.

1 Before starting this procedure, gather together all the necessary tools and materials. Also make sure that you have plenty of clean rags and newspapers handy, to mop up any spills. Ideally, the engine oil should be warm, as it will drain more easily, and more built-up sludge will be removed with it. Take care not to touch the exhaust or any other hot parts of the engine when working under the vehicle. To avoid any possibility of scalding, and to protect yourself from possible skin irritants and other harmful contaminants in used engine oils, it is advisable to wear gloves when carrying out this work.

2 Access to the underside of the vehicle will be greatly improved if it can be raised on a lift,

HAYNES HINT

As the drain plug releases from the threads, move it away sharply so the stream of oil issuing from the sump runs into the container, not up your sleeve.

driven onto ramps, or jacked up and supported on axle stands (see *Jacking and vehicle support*). Whichever method is chosen, make sure that the vehicle remains level, or if it is at an angle, that the drain plug is at the lowest point. The drain plug is located at the rear of the sump.

3 Remove the oil filler cap from the cylinder head cover (twist it anti-clockwise and withdraw it).

4 Using a spanner, or preferably a suitable socket and bar, slacken the drain plug about half a turn. Position the draining container under the drain plug, then remove the plug completely. If possible, try to keep the plug pressed into the sump while unscrewing it by hand the last couple of turns **(see Haynes Hint)**.

5 Allow some time for the oil to drain, noting that it may be necessary to reposition the container as the oil flow slows to a trickle.

6 After all the oil has drained, wipe the drain plug and the sealing washer with a clean rag. Examine the condition of the sealing washer – renew it if it shows signs of scoring or other damage which may prevent an oil-tight seal. Clean the area around the drain plug opening, and refit the plug complete with the washer. Tighten the plug securely – preferably to the specified torque, using a torque wrench.

7 The oil filter is located at the rear right-hand side of the cylinder block – access is most easily obtained from underneath the vehicle.

8 Move the container into position under the oil filter.

9 Use an oil filter removal tool (if required) to slacken the filter initially, then unscrew it by

3.9 Removing the oil filter from underneath the vehicle

hand the rest of the way **(see illustration)**. Empty the oil from the old filter into the container. To ensure that the old filter is completely empty before disposal, puncture the filter dome in at least two places and allow any remaining oil to drain through the punctures and into the container.

10 Use a clean rag to remove all oil, dirt and sludge from the filter sealing area on the engine. Check the old filter to make sure that the rubber sealing ring has not stuck to the engine. If it has, carefully remove it.

11 Apply a light coating of clean engine oil to the sealing ring on the new filter, then screw the filter into position on the engine. Lightly tighten the filter until its sealing ring contacts the block, then tighten it through a further two-thirds of a turn.

12 Remove the old oil and all tools from under the vehicle then, if applicable, lower the vehicle to the ground.

13 Fill the engine through the filler hole in the cylinder head cover, using the correct grade and type of oil (see *Weekly checks*). Pour in half the specified quantity of oil first, then wait a few minutes for the oil to drain into the sump. Continue to add oil, a small quantity at a time, until the level is up to the lower mark on the dipstick. Adding a further 1.0 litre will bring the level up to the upper mark on the dipstick.

14 Start the engine and run it for a few minutes, while checking for leaks around the oil filter seal and the sump drain plug. Note that there may be a delay of a few seconds before the low oil pressure warning light goes out when the engine is first started, as the oil circulates through the new oil filter and the engine oil galleries before the pressure builds-up. Do not run the engine above idle speed while the warning light is on.

15 Stop the engine, and wait a few minutes for the oil to settle in the sump once more. With the new oil circulated and the filter now completely full, recheck the level on the dipstick, and add more oil as necessary.

16 Dispose of the used engine oil and filter safely, referring to *General repair procedures* in the Reference Chapter. Do not discard the old filter with domestic household waste. The facility for waste oil disposal provided by many local council refuse tips generally has a filter receptacle alongside.

Every 18 000 miles or 12 months – whichever comes first

4 Idle speed check and adjustment

1 Before checking the idle speed, always check first the following:
 a) *Check that the ignition timing is accurate (Chapter 5B).*
 b) *Check that the spark plugs are in good condition and correctly gapped (Section 11).*
 c) *Check that the accelerator cable is correctly adjusted (Chapter 4A).*
 d) *Check that the crankcase breather hoses are secure, with no leaks or kinks (Section 13).*
 e) *Check that the air cleaner filter element is clean (Section 26).*
 f) *Check that the exhaust system is in good condition (Chapter 4A).*
 g) *If the engine is running very roughly, check the compression pressures as described in Chapter 2A.*

2 Take the car on a journey of sufficient length to warm it up to normal operating temperature.
3 Ensure that all electrical loads are switched off; if the car is not equipped with a tachometer, connect one following its manufacturer's instructions. Note the idle speed, comparing it with that specified. If adjustment is necessary, proceed as follows.
4 Stop the engine and turn off the ignition. Disconnect the wiring connector from the throttle potentiometer, which is mounted onto the side of the throttle housing (see Chapter 4A for further information).
5 Start the engine, clear excess fuel from the inlet manifold by racing the engine two or three times to between 2000 and 3000 rpm, then allow it to idle again. With the throttle potentiometer disconnected, the engine will now be idling at the slightly slower base idle speed. Check that the base idle speed is within

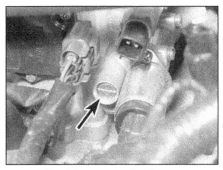

4.5 The idle speed adjusting screw (arrowed) is on the right-hand side of the throttle housing

the limits given in the Specifications. If adjustment is necessary, the idle speed adjusting screw is situated on the right-hand side of the throttle housing (see illustration). Screw it in or out as necessary to obtain the specified base idle speed. When the base idle speed is correctly set, switch off the engine and reconnect the throttle potentiometer wiring connector.

5 Hose and fluid leak check

1 Visually inspect the engine joint faces, gaskets and seals for any signs of water or oil leaks. Pay particular attention to the areas around the cylinder head cover, cylinder head, oil filter and sump joint faces. Over a period of time, some very slight seepage from these areas is to be expected – what you are really looking for is any indication of a serious leak. Should a leak be found, renew the offending gasket or oil seal by referring to the appropriate Chapters in this manual.
2 Also check the security and condition of all the engine-related pipes and hoses. Ensure that all cable ties or securing clips are in place and in good condition. Clips which are broken or missing can lead to chafing of the hoses pipes or wiring which could cause more serious problems in the future.
3 Carefully check the radiator hoses and heater hoses along their entire length. Renew any hose which is cracked, swollen or deteriorated. Cracks will show up better if the hose is squeezed. Pay close attention to the hose clips that secure the hoses to the cooling system components. Hose clips can pinch and puncture hoses, resulting in cooling system leaks. If the crimped-type hose clips are used, it may be a good idea to fit standard worm-drive clips.
4 Inspect all the cooling system components (hoses, joint faces, etc) for leaks. Where any problems of this nature are found on system components, renew the component or gasket with reference to Chapter 3 (see Haynes Hint).
5 Where applicable, inspect the automatic transmission fluid cooler hoses for leaks or deterioration.
6 With the vehicle raised, inspect the petrol tank and filler neck for punctures, cracks and other damage. The connection between the filler neck and tank is especially critical. Sometimes a rubber filler neck or connecting hose will leak due to loose retaining clamps or deteriorated rubber.
7 Carefully check all rubber hoses and metal fuel lines leading away from the petrol tank. Check for loose connections, deteriorated

hoses, crimped lines and other damage. Pay particular attention to the vent pipes and hoses, which often loop up around the filler neck and can become blocked or crimped. Follow the lines to the front of the vehicle, carefully inspecting them all the way. Renew damaged sections as necessary.
8 Check the condition of all brake fluid hoses.
9 From within the engine compartment, check the security of all fuel hose attachments and pipe unions, and inspect the fuel hoses and vacuum hoses for kinks, chafing and deterioration.
10 Where applicable, check the condition of the power steering fluid hoses and pipes.

6 Manual transmission oil level check

1 Park the car on a level surface. The oil level must be checked before the car is driven, or at least 5 minutes after the engine has been switched off. If the oil is checked immediately after driving the car, some of the oil will remain distributed around the transmission components, resulting in an inaccurate level reading. To improve access, position the car over an inspection pit, or raise the car off the ground and position it on axle stands, (see *Jacking and vehicle support*) making sure the vehicle remains level to the ground.
2 Wipe clean the area around the filler/level plug, which is on the front face of the transmission. Unscrew the plug and clean it.
3 The oil level should reach the lower edge of the filler/level hole. A certain amount of oil will have gathered behind the filler/level plug and will trickle out when it is removed; this does **not** necessarily indicate that the level is correct. To ensure that a true level is established, wait until the initial trickle has

A leak in the cooling system will usually show up as white- or rust-coloured deposits on the area adjoining the leak.

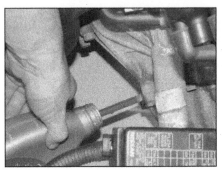

6.3 Topping-up the manual transmission oil

stopped, then add oil as necessary until a trickle of new oil can be seen emerging **(see illustration)**. The level will be correct when the flow ceases; use only good-quality oil of the specified type.

4 Refilling the transmission is an extremely awkward operation; above all, allow plenty of time for the oil level to settle properly before checking it. If a large amount had to be added to the transmission and a large amount flows out on checking the level, refit the filler/level plug, and take the vehicle on a short journey. This will allow the new oil to be distributed

HAYNES HINT

For a quick check, the thickness of friction material on each brake pad can be measured through the aperture in the caliper body.

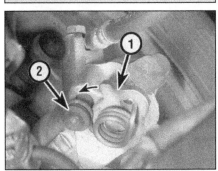

8.5 Ensure that, when released, the handbrake lever rests against the stopper bolt – models with rear disc brakes

1 Handbrake lever (shown here with handbrake applied)
2 Stopper bolt

fully around the transmission components. On returning, recheck the level when the oil has settled again.

5 If the transmission has been overfilled so that oil flows out as soon as the filler/level plug is removed, check that the car is completely level (front-to-rear and side-to-side). If necessary, allow the surplus to drain off into a suitable container.

6 When the level is correct, refit the filler/level plug, tightening it to the specified torque wrench setting. Wash off any spilt oil.

7 Brake pad condition check

Front brake pads

1 Firmly apply the handbrake, then jack up the front of the vehicle and support it securely on axle stands (see *Jacking and vehicle support*). Remove the front roadwheels.

2 If any pad's friction material is worn to the specified thickness or less, *all four pads must be renewed as a set* **(see Haynes Hint)**. **Note:** *If any pad is approaching the minimum thickness, consider renewal as a precautionary measure in case the pads wear out before the next service.*

3 For a comprehensive check, the brake pads should be removed and cleaned. This will permit the operation of the caliper to be checked, and the condition of the brake disc itself to be fully examined on both sides. Refer to Chapter 9 for further information.

Rear brake pads

4 Chock the front wheels then jack up the rear of the vehicle and support it securely on axle stands (see *Jacking and vehicle support*). Remove the rear roadwheels.

5 Proceed as described for the front brake pads in paragraphs 2 and 3.

8 Handbrake check and adjustment

1 The rear brakes are of the self-adjusting type, and the only adjustments required are to the operating cables.

2 The handbrake should be capable of holding the parked vehicle stationary, even on steep slopes, when applied with moderate force. The mechanism should be firm and positive in feel, with no trace of stiffness or sponginess from the cables, and the mechanism should release immediately the handbrake lever is released. If the mechanism does not operate satisfactorily, it should be checked immediately.

3 To check the operation of the handbrake, chock the front wheels then jack up the rear of the vehicle and support it securely on axle stands (see *Jacking and vehicle support*).

4 Fully release the handbrake, and check that the rear roadwheels can be rotated by hand – slight dragging is acceptable, but it should be possible to turn each wheel easily without undue force.

5 On models with rear disc brakes, check that the handbrake levers on the calipers return to rest against the stopper bolts with the handbrake fully released **(see illustration)**. This will prove easier if the rear roadwheels are removed.

6 Depress the brake pedal several times to establish the correct shoe-to-drum, or pad-to-disc clearance, as applicable.

7 With the pedal released, again, check that the rear roadwheels can be rotated.

8 Apply normal moderate pressure to operate the handbrake lever, and count the number of clicks necessary to bring the lever to the fully-applied position (check that the roadwheels are locked with the lever fully applied). The number of clicks should be as specified (see Specifications).

9 If the number of clicks required to fully apply the handbrake is not as specified, proceed as follows.

10 Working inside the vehicle, remove the centre console as described in Chapter 11.

11 The handbrake adjuster nut is located under the lever on the threaded end of the front cable.

12 Turn the adjuster nut as required **(see illustration)**, and recheck the adjustment as described in paragraph 8, until the handbrake operates correctly over the specified number of clicks.

13 Check that the handbrake 'on' warning light illuminates after the specified number of handbrake clicks (see Specifications). If necessary, bend the switch bracket to give the correct adjustment.

14 On completion, where applicable refit the roadwheels, then lower the vehicle to the ground.

9 Clutch adjustment check and control mechanism lubrication

1 Check that the clutch pedal moves smoothly and easily through its full travel.

8.12 Using a spanner to turn the handbrake adjuster nut

11.2 Disconnecting the HT lead from the spark plug

11.4 Removing the spark plug

Check also that the clutch itself functions correctly, with no trace of slip or drag, then adjust the clutch as described in Chapter 6. If excessive effort is required to operate the clutch, check first that the cable is correctly routed and undamaged, then remove the pedal to ensure that its pivot is properly greased. Refer to Chapter 6 for further information.

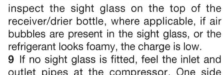

10 Air conditioning system check

Note: *Before proceeding, refer to the precautions given in Chapter 3 regarding work on the air conditioning system.*

1 Check the tension and condition of the auxiliary drivebelt which drives the air conditioning compressor, as described in Section 15.
2 Check the condition of the condenser fins, and clean if necessary (where applicable, remove the front grille panel for access. Clean dirt and insects, etc, from the fins using compressed air, or a soft brush. Be careful not to damage the condenser.
3 Operate the air conditioning system for at least 10 minutes each month, even during cold weather, to keep the seals, etc, in good condition.
4 Regularly inspect the refrigerant pipes, hoses and unions for security and condition.
5 The most common cause of poor cooling is simply a low system refrigerant charge. If a noticeable drop in cool air output occurs, one of the following checks will help to determine if the refrigerant level is low.
6 Warm-up the engine to normal operating temperature.
7 Move the temperature control knob to the coldest setting, and move the blower motor control knob to the highest setting. Open the doors (to ensure that the air conditioning system does not shut off as soon as it cools the passenger compartment).

8 With the compressor engaged – the compressor clutch will make an audible click, and the centre of the clutch will rotate – inspect the sight glass on the top of the receiver/drier bottle, where applicable, if air bubbles are present in the sight glass, or the refrigerant looks foamy, the charge is low.
9 If no sight glass is fitted, feel the inlet and outlet pipes at the compressor. One side should be cold, and the other hot. If there is no perceptible difference in temperature between the two pipes, this indicates a fault with the compressor, a low refrigerant charge, or some other system fault – consult a Nissan dealer or air conditioning specialist for advice.
10 The air conditioning system will lose a proportion of its charge through normal seepage – so it is as well to regard periodic recharging as a maintenance operation. Recharging must be done by a Nissan dealer or an air conditioning specialist.
11 Do not under any circumstances attempt to open any of the refrigerant lines, or renew any or the components.

11 Spark plug renewal

1 The correct functioning of the spark plugs is vital for the correct running and efficiency of the engine. It is essential that the plugs fitted are appropriate for the engine (the suitable type is specified at the beginning of this Chapter). If this type is used and the engine is in good condition, the spark plugs should not need attention between scheduled replacement intervals. Spark plug cleaning is rarely necessary, and should not be attempted unless specialised equipment is available, as damage can easily be caused to the firing ends.
2 If the marks on the original-equipment spark plug (HT) leads cannot be seen, mark the leads one to four to correspond to the cylinder the lead serves (No 1 cylinder is at the

timing chain end of the engine). Pull the leads from the plugs by gripping the end fitting, not the lead, otherwise the lead connection may be fractured. Note that the spark plugs are deeply recessed, the HT lead end fittings are extended **(see illustration)**.
3 It is advisable to remove the dirt from the spark plug recesses using a clean brush, vacuum cleaner or compressed air before removing the plugs, to prevent dirt dropping into the cylinders.
4 Unscrew the plugs using a spark plug spanner, suitable box spanner or a deep socket and extension bar **(see illustration)**. Keep the socket aligned with the spark plug, otherwise if it is forcibly moved to one side, the ceramic insulator may be broken off. As each plug is removed, examine it as follows.
5 Examination of the spark plugs will give a good indication of the condition of the engine. If the insulator nose of the spark plug is clean and white, with no deposits, this is indicative of too hot a plug (a hot plug transfers heat away from the electrode slowly, a cold plug transfers heat away quickly) or a possible engine management system fault.
6 If the tip and insulator nose are covered with hard black-looking deposits, then this is also indicative of a possible problem in the engine management system. Should the plug be black and oily, then it is likely that the engine is fairly worn.
7 It is normal for the insulator nose to be covered with light tan to greyish-brown deposits, indicating that both the spark plug, and the engine are in good condition.
8 The spark plug electrode gap is of considerable importance as, if it is too large or too small, the size of the spark and its efficiency will be seriously impaired. The gap should be set to the value given in the Specifications at the beginning of this Chapter.
9 To set it, measure the gap with a feeler blade and then bend open, or closed, the outer plug electrode until the correct gap is achieved **(see illustration)**. The centre

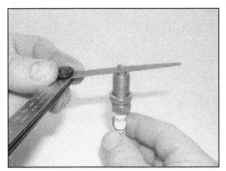

11.9 Measure the spark plug gap with a feeler blade . . .

electrode should never be bent, as this may crack the insulator and cause plug failure, if nothing worse.

10 Special spark plug electrode gap adjusting tools are available from most motor accessory shops, or from some spark plug manufacturers **(see illustration)**.

11 Before fitting the spark plugs, check that the threaded connector sleeves are tight, and that the plug exterior surfaces and threads are clean **(see Haynes Hint)**.

12 Tighten the plug to the specified torque using the spark plug socket and a torque wrench. Refit the remaining spark plugs in the same manner.

13 Connect the HT leads in their correct order, ensuring that they are correctly fitted into their retaining clips.

12 Fuel filter renewal

⚠️ *Warning: Before carrying out the following operation, refer to the precautions given in 'Safety first!' at the beginning of this manual, and follow them implicitly. Petrol is a highly-dangerous and volatile liquid, and the precautions necessary when handling it cannot be overstressed.*

1 The fuel filter is situated in the engine compartment, mounted on the engine compartment bulkhead on the left-hand side **(see illustration)**. For improved access, remove the air cleaner assembly as described in Chapter 4A.

2 Slide the filter out of the retaining clamp, noting which way round it is fitted to ensure correct fitting of the new filter.

3 Bearing in mind the information given in Chapter 4A on depressurising the fuel system, release the retaining clips and disconnect the fuel hoses from the filter.

4 Dispose safely of the old filter; it will be highly inflammable, and may explode if thrown on a fire.

5 Connect the fuel hoses to the new filter, and secure them in position with their retaining clips. Where applicable, the arrow on the filter must point towards the hose connected to the engine.

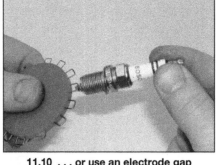

11.10 . . . or use an electrode gap adjusting tool

6 Slide the filter into position in the clamp then, where applicable, refit the air cleaner assembly with reference to Chapter 4A.

7 Start the engine, and check the filter hoses connections for leaks.

13 Emissions control systems check

1 Details of the emissions control system components and testing are given in Chapter 4B.

2 Checking consists simply of a visual check for obvious signs of damaged or leaking hoses and joints.

14 Exhaust gas sensor check

1 The exhaust gas sensor can be tested using the engine management ECCS control unit self-diagnosis facility as described in Chapter 4A.

15 Auxiliary drivebelt checking and renewal

1 There are either one, two or three auxiliary drivebelts fitted, depending on specification. The drivebelt arrangements are as follows.

Models without power steering or air

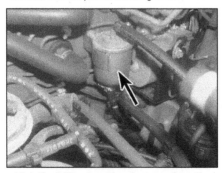

12.1 Fuel filter location (arrowed) on the engine compartment bulkhead

It's often difficult to insert spark plugs into their holes without cross-threading them. To avoid this possibility, fit a short piece of rubber hose over the end of the spark plug. The flexible hose acts as a universal joint, to help align the plug with the plug hole. Should the plug begin to cross-thread, the hose will slip on the spark plug, preventing thread damage.

conditioning – one drivebelt (drives the alternator and coolant pump).

Models with power steering but without air conditioning – two drivebelts (one drives the alternator, and the other the power steering pump and coolant pump).

Models with air conditioning but without power steering – two drivebelts (one drives the alternator and coolant pump, and the other the air conditioning compressor).

Models with power steering and air conditioning – three drivebelts (as for models with power steering, with an additional belt for the air conditioning compressor).

Checking drivebelt condition

2 Firmly apply the handbrake, then jack up the front of the vehicle and support it securely on axle stands (see *Jacking and vehicle support*). Remove the right-hand front roadwheel.

3 From underneath the front of the car, undo the retaining screws and remove the wheel arch liner from underneath the wing to gain access to the crankshaft pulley. If necessary, also undo the retaining screws and remove the engine undershield to improve access.

4 Using a suitable socket and extension bar fitted to the crankshaft pulley bolt, rotate the crankshaft so that the entire length of the drivebelt(s) can be examined. Examine the drivebelt(s) for cracks, splitting, fraying or damage. Check also for signs of glazing (shiny patches) and for separation of the belt plies. Renew the belt if worn or damaged.

5 If the condition of the belt is satisfactory, check the drivebelt tension as described below under the relevant sub-heading.

Air conditioning drivebelt

Removal

6 If not already done, proceed as described in paragraphs 2 and 3.

7 Disconnect the battery negative terminal (refer to *Disconnecting the battery* in the Reference Chapter).

8 Slacken the nut securing the tensioning pulley assembly to the engine.

9 Rotate the adjuster bolt to move the tensioner pulley away from the drivebelt until there is sufficient slack for the drivebelt to be removed from the pulleys

Refitting

10 Fit the belt around the pulleys, ensuring that the belt is of the correct type if it is being renewed, and take up the slack in the belt by tightening the adjuster bolt.

11 Tension the drivebelt as described in the following paragraphs.

Tensioning

12 If not already done, proceed as described in paragraphs 2 and 3.

13 Correct tensioning of the drivebelt will ensure that it has a long life. Beware, however, of overtightening, as this can cause wear in the alternator bearings.

14 The belt tension is checked at the mid-point between the pulleys on the top belt run. Referring to the Specifications given at the start of this Chapter, apply the specified force and check that the belt deflection is within the specified range.

15 To adjust the tension, with the tensioner pulley assembly retaining nut slackened, rotate the adjuster bolt until the correct tension is achieved. Once the belt is correctly tensioned, rotate the crankshaft a couple of times and recheck the tension.

16 When the belt is correctly tensioned, tighten the tensioner pulley assembly retaining nut to the specified torque setting, and reconnect the battery negative terminal.

17 Refit the wheel arch liner and engine undershield, securely tightening their fasteners, then refit the roadwheel and lower the vehicle to the ground.

Power steering drivebelt

Removal

18 If not already done, proceed as described in paragraphs 2 and 3.

19 Disconnect the battery negative terminal (refer to *Disconnecting the battery* in the Reference Chapter). On models with air conditioning, remove the air conditioning compressor drivebelt as described in paragraphs 8 and 9.

20 Slacken the power steering pump mounting nuts/bolts (as applicable).

21 Back off the adjuster bolt to relieve the tension in the drivebelt, then slip the drivebelt from the pulleys.

Refitting

22 Fit the belt around the pulleys, ensuring

that the belt is of the correct type if it is being renewed, and take up the slack in the belt by tightening the adjuster bolt.

23 Tension the drivebelt as described in the following paragraphs. Where necessary, refit and tension the air conditioning compressor drivebelt as described in paragraphs 10 to 16.

Tensioning

24 If not already done, proceed as described in paragraphs 2 and 3.

25 Correct tensioning of the drivebelt will ensure that it has a long life. Beware, however, of overtightening, as this can cause wear in the alternator bearings.

26 The belt tension is checked at the mid-point between the pulleys on the upper belt run. Referring to the Specifications, apply the specified force and check that the belt deflection is within the specified range.

27 To adjust, with the upper mounting nut/bolt just holding the power steering pump firm, and the other mounting nut/bolt loosened, turn the adjuster bolt until the correct tension is achieved. Rotate the crankshaft a couple of times, recheck the tension, then securely tighten both the power steering pump mounting nuts/bolts.

28 Refit the wheel arch liner and engine undershield, securely tightening their fasteners, then refit the roadwheel and lower the vehicle to the ground. Reconnect the battery negative terminal.

Alternator drivebelt

Removal

29 If not already done, proceed as described in paragraphs 2 and 3.

30 Disconnect the battery negative terminal (refer to *Disconnecting the battery* in the Reference Chapter). On models with air conditioning, remove the air conditioning drivebelt as described in paragraphs 8 and 9. On models with power steering, remove the power steering drivebelt as described in paragraphs 20 and 21.

31 Slacken both the alternator upper and lower mounting nuts/bolts (as applicable) **(see illustration)**.

32 Back off the adjuster bolt to relieve the tension in the drivebelt, then slip the drivebelt from the pulleys **(see illustrations)**.

15.31 Slacken the alternator mounting bolts . . .

Refitting

33 Fit the belt around the pulleys, ensuring that the belt is of the correct type if it is being renewed, and take up the slack in the belt by tightening the adjuster bolt.

34 Tension the drivebelt as described in the following paragraphs. Where necessary, refit the air conditioning compressor drivebelt as described in paragraphs 10 to 16.

Tensioning

35 If not already done, proceed as described in paragraphs 2 and 3.

36 Correct tensioning of the drivebelt will ensure that it has a long life. Beware, however, of overtightening, as this can cause wear in the alternator bearings.

37 The belt should be tensioned is checked at the mid-point between the pulleys on the upper belt run. Referring to the Specifications given at the start of this Chapter, apply the specified force and check that the belt deflection is within the specified range.

38 To adjust, with the upper mounting nut/bolt just holding the alternator firm, and the lower mounting nut/bolt loosened, turn the adjuster bolt until the correct tension is achieved. Rotate the crankshaft a couple of times, recheck the tension, then securely tighten both the alternator mounting nuts/bolts.

39 Refit the wheel arch liner and engine undershield, securely tightening their fasteners, then refit the roadwheel and lower the vehicle to the ground. Reconnect the battery negative terminal.

15.32a . . . then loosen the adjuster bolt . . .

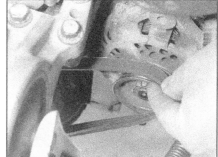

15.32b . . . and slip the drivebelt off the alternator pulley

16 Brake shoe check

1 Chock the front wheels then jack up the rear of the vehicle and support it securely on axle stands (see *Jacking and vehicle support*).
2 For a quick check, the thickness of friction material remaining on one of the brake shoes can be observed through the hole in the brake backplate which is exposed by prising out the sealing grommet **(see Haynes Hint)**. If a rod of the same diameter as the specified minimum friction material thickness is placed against the shoe friction material, the amount of wear can be assessed. A torch or inspection light will probably be required. If the friction material on any shoe is worn down to the specified minimum thickness or less, all four shoes must be renewed as a set. **Note:** *If any shoe is approaching the minimum thickness, consider renewal as a precautionary measure in case the shoes wear out before the next service.*
3 For a comprehensive check, the brake drums should be removed and cleaned. This will allow the wheel cylinders to be checked and the condition of the brake drum itself to be fully examined. Refer to the relevant Sections of Chapter 9 for further information.

17 Brake fluid renewal

⚠️ **Warning: Brake hydraulic fluid can harm your eyes and damage painted surfaces, so use extreme caution when handling and pouring it. Do not use fluid that has been standing open for some time, as it absorbs moisture from the air. Excess moisture content can cause a dangerous loss of braking effectiveness.**

1 The procedure is similar to that for the bleeding of the hydraulic system as described in Chapter 9. The brake fluid reservoir should be emptied by syphoning, using a clean poultry baster or similar before starting, then refilled with fresh fluid. Allowance should be made for the old fluid to be expelled when bleeding a section of the circuit.
2 Working as described in Chapter 9, open the first bleed screw in the sequence and pump the brake pedal gently until nearly all the fluid has been emptied from the master cylinder reservoir. Top-up to the MAX level with more fresh fluid, and continue pumping until new fluid can be seen emerging from the bleed screw. Tighten the screw and top the reservoir level up to the MAX level line.
3 Old hydraulic fluid is invariably much darker in colour than the new, making it easy to distinguish the two.
4 Work through all the remaining bleed screws in the sequence until new fluid can be seen at all of them. Be careful to keep the

HAYNES HiNT

For a quick check, the thickness of friction material remaining on one of the brake shoes can be observed through the hole in the brake backplate (arrowed).

master cylinder reservoir topped-up to above the MIN level at all times, or air may enter the system and greatly increase the length of the task.
5 When the operation is complete, check that all bleed screws are securely tightened, and that their dust caps are refitted. Wash off all traces of spilt fluid, and recheck the master cylinder reservoir fluid level.
6 Check the operation of the brakes before taking the car on the road.

18 Automatic transmission fluid level check

1 Take the vehicle on a short journey to warm the transmission up to normal operating temperature, then park the vehicle on level ground. The fluid level is checked using the dipstick located at the front of the engine compartment, directly below the distributor, which is mounted onto the left-hand end of the cylinder head.
2 With the engine idling and the selector lever in the P (Park) position, withdraw the dipstick from the tube, and wipe all the fluid from its end with a clean rag or paper towel. Insert the clean dipstick back into the tube as far as it will go, then withdraw it once more. Note the level on the end of the dipstick, noting that there are

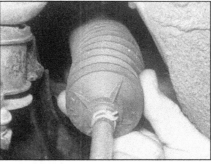

19.2 Checking a steering gear gaiter

two sets of level markings. On one side of the dipstick are the COLD upper and lower marks (which are in the form of cut-outs on the edge of the dipstick – for use when the fluid temperature is between 30°C and 50°C); on the other side are the HOT upper and lower marks (which are in the form of lines marked on the dipstick – for use when the fluid temperature is between 50°C and 80°C). If the vehicle is fully warmed-up, the HOT marks should be used.
3 If topping-up is necessary, add the required quantity of the specified fluid to the transmission via the dipstick tube. Use a funnel with a fine mesh gauze, to avoid spillage and to ensure that no foreign matter enters the transmission. Add fluid as necessary until the level is between the relevant set of upper and lower marks on the dipstick. **Note:** *Never overfill the transmission so that the fluid level is above the upper mark.*
4 After topping-up, take the vehicle on a short run to distribute the fresh fluid, then recheck the level again, topping-up if necessary.
5 Always maintain the level between the two dipstick marks. If the level is allowed to fall below the lower mark, various problems, and even severe transmission damage, could result.

19 Steering and suspension check

Suspension & steering check

1 Firmly apply the handbrake, then jack up the front of the vehicle and support it securely on axle stands (see *Jacking and vehicle support*).
2 Visually inspect the balljoint dust covers and the steering rack and pinion gaiters for splits, chafing or deterioration **(see illustration)**. Any wear of these components will cause loss of lubricant, together with dirt and water entry, resulting in rapid deterioration of the balljoints or steering gear.
3 On vehicles with power steering, check the fluid hoses for chafing or deterioration, and the pipe and hose unions for fluid leaks. Also check for signs of fluid leakage under pressure from the steering gear rubber gaiters, which would indicate failed fluid seals within the steering gear.
4 Grasp the roadwheel at the 12 o'clock and 6 o'clock positions, and try to rock it **(see illustration)**. Very slight free play may be felt, but if the movement is appreciable, further investigation is necessary to determine the source. Continue rocking the wheel while an assistant depresses the footbrake. If the movement is now eliminated or significantly reduced, it is likely that the hub bearings are at fault. If the free play is still evident with the footbrake depressed, then there is wear in the suspension joints or mountings.
5 Now grasp the wheel at the 9 o'clock and 3 o'clock positions, and try to rock it as

before. Any movement felt now may again be caused by wear in the hub bearings or the steering track rod balljoints. If the inner or outer balljoint is worn, the visual movement will be obvious.

6 Using a large screwdriver or flat bar, check for wear in the suspension mounting bushes by levering between the relevant suspension component and its attachment point. Some movement is to be expected as the mountings are made of rubber, but excessive wear should be obvious. Also check the condition of any visible rubber bushes, looking for splits, cracks or contamination of the rubber.

7 With the car standing on its wheels, have an assistant turn the steering wheel back-and-forth about an eighth of a turn each way. There should be very little, if any, lost movement between the steering wheel and roadwheels. If this is not the case, closely observe the joints and mountings previously described, but in addition check the steering column universal joints for wear, and also check the rack-and-pinion steering gear itself.

Strut/shock absorber check

8 Check for any signs of fluid leakage around the suspension strut/shock absorber body, or from the rubber gaiter around the piston rod. Should any fluid be noticed, the suspension strut/shock absorber is defective internally, and should be renewed. **Note:** *Suspension struts/shock absorbers should always be renewed in pairs on the same axle.*

9 The efficiency of the suspension strut/shock absorber may be checked by bouncing the vehicle at each corner. Generally speaking, the body will return to its normal position and stop after being depressed. If it rises and returns on a rebound, the suspension strut/shock absorber is probably suspect. Examine the suspension strut/shock absorber upper and lower mountings for any signs of wear.

20 Driveshaft gaiter check

1 With the vehicle raised and securely supported on stands (see *Jacking and vehicle support*), turn the steering onto full lock, then slowly rotate the roadwheel. Inspect the condition of the outer constant velocity (CV) joint rubber gaiters, squeezing the gaiters to open out the folds **(see illustration)**. Check for signs of cracking, splits or deterioration of the rubber, which may allow the grease to escape, and lead to water and grit entry into the joint. Also check the security and condition of the retaining clips. Repeat these checks on the inner CV joints. If any damage or deterioration is found, the gaiters should be renewed as described in Chapter 8.

2 At the same time, check the general condition of the CV joints themselves by first holding the driveshaft and attempting to rotate the wheel. Repeat this check by holding the

inner joint and attempting to rotate the drive-shaft. Any appreciable movement indicates wear in the joints, wear in the driveshaft splines, or a loose driveshaft retaining nut.

21 Wheel alignment check

Definitions

1 A vehicle's steering and suspension geometry is defined in four basic settings – all angles are expressed in degrees (toe settings are also expressed as a measurement); the relevant settings are camber, castor, steering axis inclination and toe setting. With the exception of front wheel toe setting, none of these settings are adjustable, and in all cases, special equipment is necessary to check them. Note that front wheel toe setting is often referred to as 'tracking' or 'front wheel alignment'.

Checking

2 Due to the special measuring equipment necessary to check the wheel alignment, and the skill required to use it properly, the checking and adjustment of these settings is best left to a Nissan dealer or similar expert. Note that most tyre-fitting shops now possess sophisticated checking equipment.

22 Roadwheel balance check

1 Accurate wheel balancing requires access to specialised test equipment and as such should be entrusted to a suitably-equipped Nissan dealer or tyre specialist.

23 Seat belt check

1 All models are fitted with three-point lap and diagonal inertia reel seat belts at both front and at the rear outer seats. The rear centre seat has either a two-point lap-type

belt of static type (ie, not inertia reel), or a three-point lap and diagonal inertia reel belt.

2 Inspect the belts for signs of fraying or other damage. Also check the operation of the buckles and retractor mechanisms, and ensure that all mounting bolts are securely tightened. Note that the bolts are shouldered so that the belt anchor points are free to rotate.

3 If there is any sign of damage, or any doubt about the condition of a belt, it must be renewed. If the vehicle has been involved in a collision, any belts in use at the time should be renewed as a matter of course, and all other belts should be checked carefully.

4 Use only warm water and non-detergent soap to clean the belts. Never use any chemical cleaners, strong detergents, dyes or bleaches. Keep the belts fully extended until they have dried naturally – do not apply heat to dry them.

24 Hinge and lock lubrication

1 Work around the vehicle, and lubricate the hinges of the bonnet, doors, and tailgate, or boot lid, with a light machine oil.

2 Lightly lubricate the bonnet release mechanism and the exposed sections of the inner cable with a smear of grease. Similarly, lubricate the tailgate/boot lid/fuel filler flap release mechanisms, where accessible.

3 Check carefully the security and operation of all hinges, latches and locks, adjusting them where required (see Chapter 11). Check the operation of the central locking system.

4 Check the condition and operation of the tailgate struts, renewing them if either is leaking or no longer able to support the tailgate securely when raised.

25 Road test

Instruments & electrical equipment

1 Check the operation of all instruments and electrical equipment.

19.4 Rocking the roadwheel to check steering/suspension components

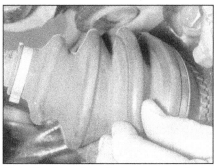

20.1 Checking driveshaft constant velocity joint (CV) joint gaiter

2 Make sure that all instruments read correctly, and switch on all electrical equipment in turn to check that it functions properly.

Steering & suspension

3 Check for any abnormalities in the steering, suspension, handling or road feel.
4 Drive the vehicle, and check that there are no unusual vibrations or noises.
5 Check that the steering feels positive, with no excessive 'sloppiness', or roughness, and check for any suspension noises when cornering and driving over bumps.

Drivetrain

6 Check the performance of the engine, clutch (where applicable), transmission and driveshafts.
7 Listen for any unusual noises from the engine, clutch and transmission.
8 Make sure that the engine runs smoothly when idling, and that there is no hesitation when accelerating.

9 Check that, where applicable, the clutch action is smooth and progressive, that the drive is taken up smoothly, and that the pedal travel is not excessive. Also listen for any noises when the clutch pedal is depressed.
10 On manual transmission models, check that all gears can be engaged smoothly without noise, and that the gear lever action is smooth and not abnormally vague or 'notchy'.
11 On automatic transmission models, make sure that all the gearchanges occur smoothly, without snatching, and without an increase in engine speed between changes. Check that all the gear positions can be selected with the vehicle at rest. If any problems are found, they should be referred to a Nissan dealer.
12 Listen for a metallic clicking sound from the front of the vehicle as the vehicle is driven slowly in a circle with the steering on full lock. Carry out this check in both directions. If a clicking noise is heard, this indicates wear in a driveshaft joint, in which case, the complete driveshaft must be renewed (see Chapter 8).

Braking system

13 Make sure that the vehicle does not pull to one side when braking, and that the wheels do not lock when braking hard.
14 Check that there is no vibration through the steering when braking.
15 Check that the handbrake operates correctly without excessive movement of the lever, and that it holds the vehicle stationary on a slope.
16 Test the operation of the brake servo unit as follows. Depress the footbrake four or five times to exhaust the vacuum, then start the engine. As the engine starts, there should be a noticeable 'give' in the brake pedal as vacuum builds-up. Allow the engine to run for at least two minutes and then switch it off. If the brake pedal is depressed again, it should be possible to detect a hiss from the servo as the pedal is depressed. After about four or five applications, no further hissing should be heard, and the pedal should feel considerably harder.

Every 36 000 miles or 2 years – whichever comes first

26 Air filter renewal

1 Release the retaining clips, then lift off the air cleaner cover (see illustrations).
2 Lift the air cleaner filter element out of the

26.1a Release the retaining clips . . .

housing, noting which way round it is fitted (see illustration).
3 Wipe the inside of the air cleaner housing and cover with a clean cloth to remove all traces of dirt and debris.
4 Install the new filter element, ensuring that it is the right way up and is correctly seated in the housing.
5 Refit the air cleaner cover, and secure it in position with its retaining clips.

27 Ignition system check

> **Warning: Due to the high voltages produced by the electronic ignition system, extreme care must be taken when working on the system with the ignition switched on. Persons with surgically-implanted cardiac pacemaker devices should keep well clear of the ignition circuits, components and test equipment.**

1 The ignition system components should be checked for damage or deterioration as described under the relevant sub-heading.

General

2 The spark plug (HT) leads should be checked whenever new spark plugs are installed in the engine. Where necessary, unclip the cover from the distributor.
3 Ensure that the leads are numbered before removing them, to avoid confusion when refitting. Pull the leads from the plugs by gripping the end fitting, not the lead, otherwise the lead connection may be fractured (see illustration).
4 Check inside the end fitting of each lead for signs of corrosion, which will look like a white crusty powder. Push the end fitting back onto the spark plug, ensuring that it is a tight fit on the plug. If not, remove the lead again and use pliers to carefully crimp the metal connector inside the end fitting until it fits securely on the end of the spark plug.
5 Using a clean rag, wipe the entire length of the lead, to remove any built-up dirt and

26.1b . . . then lift off the air cleaner cover

26.2 Lift the filter element from the housing, noting which way round it is fitted

27.3 Ensure that the HT leads are clearly numbered before removing them

grease. Once the lead is clean, check for burns, cracks and other damage. Do not bend the lead excessively, or pull the lead lengthwise – the conductor inside might break.

6 Disconnect the other end of the lead from the distributor cap. Again, pull only on the end fitting. Check for corrosion and a tight fit in the same manner as the spark plug end. Refit the lead securely on completion.

7 Check the remaining leads one at a time, in the same way.

8 If new spark plug (HT) leads are required, purchase a set for your specific car and engine.

9 Slacken and remove the distributor cap retaining screws. Remove the cap, and recover the cap seal. Wipe the cap clean, and carefully inspect it inside and out for signs of cracks, carbon tracks (tracking) and worn, burned or loose contacts. Check that the cap's carbon brush is unworn, free to move against spring pressure, and making good contact with the rotor arm. Also inspect the cap seal for signs of wear or damage, and renew if necessary. Slacken the retaining bolt, remove the rotor arm from the distributor shaft, and inspect the rotor arm **(see illustrations)**. It is common practice to renew the cap and rotor arm whenever new spark plug (HT) leads are fitted. When fitting a new cap, remove the leads from the old cap one at a time, and fit them to the new cap in the exact same location – do not simultaneously remove all the leads from the old cap, or firing order confusion may occur. On refitting, ensure that the rotor arm is pressed securely onto the distributor shaft, and securely tighten its retaining bolt. Ensure that the cap seal is in position, then fit the cap and securely tighten its retaining screws. Where necessary, refit the cover to the distributor.

10 Even with the ignition system in first-class condition, some engines may still occasionally experience poor starting, attributable to damp ignition components. To disperse moisture, suitable aerosol products can be very effective. Products are also available to provide a sealing coat to exclude moisture from the ignition system, and in extreme difficulty will help to start a car when only a very poor spark occurs.

Ignition timing

11 Check the ignition timing as described in Chapter 5B.

28 PCV filter renewal

1 Release the retaining clips, and lift off the air cleaner housing cover.

2 Lift the filter element out from the housing to gain access to the PCV filter.

3 Remove the filter from the air cleaner housing, and wipe clean the area around the filter **(see illustration)**.

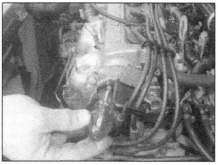

27.9a Undo the two screws and remove the distributor cap

4 Fit a new PCV filter to the housing, and refit the air cleaner filter element.

5 Seat the air cleaner cover on the housing, and secure it in position with the retaining clips.

29 Braking system vacuum servo unit check

1 To test the operation of the servo unit, depress the footbrake several times to exhaust the vacuum, and check that there is no change in the pedal stroke.

2 Depress the brake pedal, then start the engine whilst keeping the pedal firmly depressed. As the engine starts, there should be a noticeable 'give' in the brake pedal as the vacuum builds-up. Allow the engine to run for at least two minutes, then switch it off. If the brake pedal is now depressed, it should feel normal, but further applications should result in the pedal feeling firmer, with the pedal stroke decreasing with each application.

3 With the engine running, depress the brake pedal then, with the pedal still depressed, stop the engine. The pedal should not 'give' for at least 30 seconds.

4 If the servo does not operate as described, first inspect the servo unit check valve as described in Chapter 9.

5 If the servo unit still fails to operate satisfactorily, the fault lies within the unit itself. Repairs to the unit are not possible – if faulty, the servo unit must be renewed (see Chapter 9).

28.3 The PCV filter is fitted to the air cleaner housing base

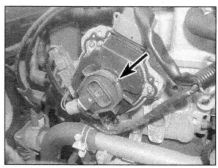

27.9b Slacken and remove the retaining bolt (arrowed) and pull the rotor arm off the shaft

30 Coolant renewal

⚠ *Warning: Wait until the engine is cold before starting this procedure. Do not allow antifreeze to come in contact with your skin, or with the painted surfaces of the vehicle. Rinse off spills immediately with plenty of water. Never leave antifreeze lying around in an open container, or in a puddle in the driveway or garage floor. Children and pets are attracted by its sweet smell, but antifreeze can be fatal if ingested.*

Cooling system draining

1 To drain the cooling system, first cover the radiator pressure tank cap with a wad of rag, and slowly turn the cap anti-clockwise to relieve the pressure in the cooling system (a hissing sound will normally be heard). Wait until any pressure remaining in the system is released, then continue to turn the cap until it can be removed.

2 Inside the car, move the heater temperature control lever fully to the HOT position.

3 Position a suitable container beneath the radiator drain plug. Loosen the drain plug, and allow the coolant to drain into the container.

4 If necessary, remove the coolant expansion tank, drain out the coolant, then refit the tank, ensuring that the hoses are securely reconnected. Take care not to spill coolant on the surrounding components.

5 Reposition the container under the cylinder block drain plug. The drain plug is located at the front of the cylinder block, at the transmission end **(see illustration)**.

6 Remove the cylinder block drain plug, then unscrew the air bleed screws. The bleed screws are located at the transmission end of the cylinder head, next to the distributor; at the rear timing chain end of the cylinder head; and/or in the heater hose at the rear of the engine **(see illustration)**.

7 If the coolant has been drained for a reason other than renewal, then provided it is clean and less than two years old, it can be re-used.

Cooling system flushing

8 If coolant renewal has been neglected, or if the antifreeze mixture has become diluted, then in time, the cooling system may gradually lose efficiency, as the coolant passages become restricted due to rust, scale deposits, and other sediment. The cooling system efficiency can be restored by flushing the system clean.

9 The radiator should be flushed independently of the engine, to avoid unnecessary contamination.

10 To flush the radiator, fit and tighten the radiator pressure cap, and if the radiator is fitted to the vehicle, clamp the hose running from the top of the radiator to the coolant expansion tank.

11 Disconnect the top and bottom hoses at the radiator, then insert a garden hose into the radiator top inlet. Direct a flow of clean water through the radiator, and continue flushing until clean water emerges from the radiator bottom outlet. If after a reasonable period, the water still does not run clear, the radiator can be flushed with a good proprietary cleaning agent. It is important that the cleaning agent manufacturer's instructions are followed carefully. If the contamination is particularly bad, insert the hose in the radiator bottom outlet, and flush the radiator in reverse ('reverse-flushing').

12 Remove the thermostat as described in Chapter 3, then temporarily refit the thermostat cover. Close the cooling system bleed screws if they have been opened.

13 With the radiator top and bottom hoses disconnected from the radiator, insert a hose into the radiator bottom hose. Direct a clean flow of water through the engine, and continue flushing until clean water emerges from the radiator top hose.

14 On completion of flushing, refit the thermostat with reference to Chapter 3, and reconnect the hoses.

Cooling system filling

15 Before attempting to fill the cooling system, make sure that all hoses and clips are in good condition, and that the clips are tight. Note that an antifreeze mixture must be used all year round, to prevent corrosion of the alloy engine components.

16 Ensure that the air bleed screws have been unscrewed (see paragraph 7), and reconnect the radiator bottom hose.

17 Position the container under the cylinder block drain plug, then refill the cooling system through the radiator filler neck, until coolant runs from the cylinder block drain plug aperture. Coat the threads of the drain plug with suitable sealant, then refit and tighten the plug.

18 Continue to fill the system through the radiator until coolant free from air bubbles emerges from the air bleed screws. Close the bleed screws once the coolant escaping is free from bubbles.

19 Continue to fill the radiator until the level reaches the filler opening, then fill the expansion tank until the coolant level reaches the MAX mark. Refit the radiator pressure cap, and the expansion tank cap.

20 Start the engine, and warm it up until it reaches normal operating temperature. Race the engine two or three times under no load, and check the coolant temperature gauge for signs of overheating.

21 Stop the engine, allow it to cool completely, then check for leaks, particularly around the disturbed components.

22 With the system cold (the system must be cold for an accurate coolant level indication), remove the radiator pressure cap (turn the pressure cap on the radiator anti-clockwise until it reaches the first stop; wait until any pressure remaining in the system is released, then push the cap down, turn it anti-clockwise to the second stop and lift off). The level should be up to the filler opening.

23 If necessary, top-up the coolant level in the radiator, then top-up the expansion tank to the MAX level mark. On completion, refit the radiator pressure cap (turn the cap clockwise as far as it will go to secure), and refit the expansion tank cap. Where applicable, refit the engine undershield.

Antifreeze mixture

24 Always use an ethylene-glycol based antifreeze which is suitable for use in mixed-metal cooling systems. The quantity of antifreeze and levels of protection are indicated in the Specifications.

25 Before adding antifreeze, the cooling system should be completely drained, preferably flushed, and all hoses and clips checked for condition and security.

26 After filling with antifreeze, a label should be attached to the radiator or expansion tank stating the type and concentration of antifreeze used, and the date installed. Any subsequent topping-up should be made with the same type and concentration of antifreeze.

27 Do not use engine antifreeze in the windscreen/tailgate/headlight washer system, as it will cause damage to the vehicle paintwork. A screenwash additive should be added to the washer system in the quantities recommended on the bottle.

31 Manual transmission oil renewal

1 This operation is much quicker and more efficient if the car is first taken on a journey of sufficient length to warm the engine/transmission up to normal operating temperature.

2 Park the car on level ground, switch off the ignition and apply the handbrake firmly. For

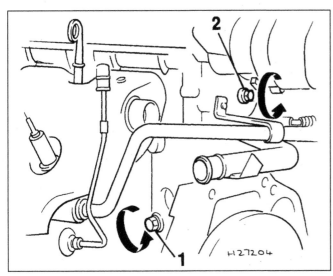

30.5 Coolant drain plug (1) and cylinder head air bleed screw (2)

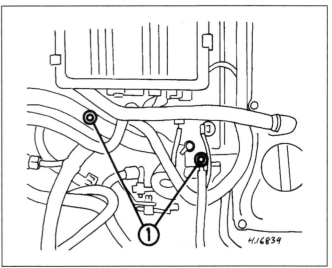

30.6 Cooling system air bleed screws (1)

improved access, jack up the front of the car and support it securely on axle stands (see *Jacking and vehicle support*). Note that the car must be lowered to the ground and level, to ensure accuracy, when refilling and checking the oil level.

3 Wipe clean the area around the filler/level plug, which is on the front face of the transmission. Unscrew the plug and clean it.

4 Position a suitable container under the drain plug situated on the left-hand side of the transmission differential housing **(see illustration)**.

5 Allow the oil to drain completely into the container. If the oil is hot, take precautions against scalding. Clean both the filler/level and the drain plugs, being especially careful to wipe any metallic particles off the magnetic inserts.

6 When the oil has finished draining, clean the drain plug threads and those of the transmission casing, then refit the drain plug, tightening it to the specified torque wrench setting. If the car was raised for the draining operation, lower it to the ground.

7 Refilling the transmission is an awkward operation. Above all, allow plenty of time for the oil level to settle properly before checking it. Note that the car must be parked on flat level ground when checking the oil level.

8 Refill the transmission with the exact amount of the specified type of oil, then check the oil level as described in Section 6; if the correct amount was poured into the transmission, and a large amount flows out on checking the level, refit the filler/level plug and take the car on a short journey so that the new

oil is distributed fully around the transmission components, then check the level again on your return.

9 When the level is correct, refit the filler/level plug, tightening it to the specified torque wrench setting. Wash off any spilt oil.

32 Automatic transmission fluid renewal

1 Take the vehicle on a short run to warm the transmission up to normal operating temperature.

2 Park the car on level ground, switch off the ignition and apply the handbrake firmly. For improved access, jack up the front of the car and support it securely on axle stands (see *Jacking and vehicle support*). Note that the car must be lowered to the ground and level, to ensure accuracy, when refilling and checking the fluid level.

3 Remove the dipstick, then position a suitable container under the transmission.

4 Unscrew the drain plug from the transmission sump, and allow the fluid to drain completely into the container. If the fluid is hot, take precautions against scalding. Clean the drain plug, being especially careful to wipe any metallic particles off the magnetic insert. Discard the original sealing washer; it should be renewed whenever it is disturbed.

5 When the fluid has finished draining, clean the drain plug threads and those of the transmission. Fit a new sealing washer to the drain plug and refit it to the transmission,

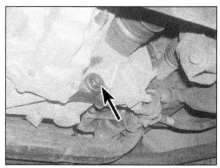

31.4 Manual transmission drain plug location (arrowed)

tightening it to the specified torque setting. If the car was raised for the draining operation, lower it to the ground.

6 Refill the transmission by adding the specified type of fluid to the transmission a little at a time via the dipstick tube. Use a funnel with a fine mesh gauze, to avoid spillage and to ensure that no foreign matter enters the transmission. Allow plenty of time for the level to settle properly before checking it as described in Section 18. Note that the car must be parked on flat level ground when checking the level.

7 Once the level is up to the MAX mark on the dipstick, refit the dipstick then start the engine and allow it to idle for a few minutes. Switch the engine off and recheck the fluid level, topping-up if necessary. Take the car on a short run to fully distribute the new fluid around the transmission and recheck the fluid level as described in Section 18.

Chapter 2 Part A:
Engine in-car repair procedures

Contents

Degrees of difficulty

Easy, suitable for novice with little experience		**Fairly easy,** suitable for beginner with some experience		**Fairly difficult,** suitable for competent DIY mechanic		**Difficult,** suitable for experienced DIY mechanic		**Very difficult,** suitable for expert DIY or professional	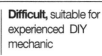

Specifications

Engine (general)

Designation	GA
Engine code:	
1.4 litre engines	GA14DE
1.6 litre engines	GA16DE
Capacity:	
1.4 litre engines	1392cc
1.6 litre engines	1597cc
Bore:	
1.4 litre engines	73.6 mm
1.6 litre engines	76.0 mm
Stroke:	
1.4 litre engines	81.8 mm
1.6 litre engines	88.0 mm
Direction of crankshaft rotation	Clockwise (viewed from right-hand side of vehicle)
No 1 cylinder location	At timing chain end of block
Firing order	1–3–4–2
Compression ratio:	
1.4 litre engines	9.5 : 1
1.6 litre engines	9.8 : 1
Cylinder compression pressures:	
Standard:	
1.4 litre engines	13.2 bars
1.6 litre engines	13.5 bars
Minimum:	
1.4 litre engines	11.2 bars
1.6 litre engines	11.5 bars
Maximum difference between cylinders (all engines)	1.0 bar

Valve clearances

Cold engine*:
Inlet	0.25 to 0.33 mm
Exhaust	0.32 to 0.40 mm

Hot engine:

For checking:
Inlet	0.21 to 0.49 mm
Exhaust	0.30 to 0.58 mm

For adjusting:
Inlet	0.32 to 0.40 mm
Exhaust	0.37 to 0.45 mm

* The valve clearances must always be checked with the engine hot. Although Nissan quote valve clearances for a hot and cold engine, the valve clearances should only be checked cold prior to starting the engine after an overhaul. The valve clearances should then be checked with the engine hot once it has been warmed-up to normal operating temperature

Camshaft and followers

Drive	Chain
Number of bearings	5

Endfloat:
Standard	0.070 to 0.143 mm
Service limit	0.20 mm

Camshaft lobe height:
Inlet	39.380 to 39.570 mm

Exhaust:
1.4 litre engines	39.380 to 39.570 mm
1.6 litre engines	39.880 to 40.070 mm

Bearing journal outer diameter:
No 1 bearing	27.935 to 27.955 mm
Nos 2 to 5 bearings	23.935 to 23.955 mm

Camshaft cylinder head bearing journal internal diameter:
No 1 bearing	28.000 to 28.021 mm
Nos 2 to 5 bearings	24.000 to 24.021 mm

Camshaft journal-to-bearing clearance:
Standard	0.045 to 0.086 mm
Service limit	0.15 mm

Camshaft run-out:
Standard	Less than 0.02 mm
Service limit	0.1 mm
Camshaft follower outer diameter	29.960 to 29.975 mm
Camshaft follower cylinder head bore internal diameter	30.000 to 30.021 mm
Camshaft follower-to-cylinder head bore clearance	0.025 to 0.061 mm

Lubrication system

Oil pump type	Gear-type, driven off crankshaft right-hand end

Minimum oil pressure at normal operating temperature:
At idle	0.59 bar
At 6000 rpm	4.22 bars

Oil pump clearances:
Outer gear-to-cover clearance	0.11 to 0.20 mm
Outer gear-to-crescent clearance	0.21 to 0.32 mm
Inner gear-to-crescent clearance	0.21 to 0.32 mm
Outer gear endfloat	0.05 to 0.11 mm
Inner gear endfloat	0.05 to 0.09 mm
Inner gear flange-to-cover bearing surface clearance	0.045 to 0.091 mm

Torque wrench settings

	Nm	lbf ft
Big-end bearing cap nuts:		
Stage 1	15	11
Stage 2 (if an angle tightening gauge is available)	Angle-tighten by 35° to 40°	
Stage 2 (if an angle-tightening gauge is not available)	25	18
Camshaft bearing cap bolts (see text):		
Bolts 1 to 14	11	8
Bolt 15 (1.6 litre engines only)	8	6
Camshaft sprocket retaining bolts	115	85
Centre member bolts	50	37
Crankshaft oil seal housing bolts	8	6
Crankshaft pulley bolt	142	105

Torque wrench settings (continued)

	Nm	lbf ft
Cylinder head bolts:		
Main bolts:		
Stage 1	29	21
Stage 2	59	44
Stage 3	Fully slacken all the bolts	
Stage 4	29	21
Stage 5 (if an angle tightening gauge is available)	Angle-tighten by 50° to 55°	
Stage 5 (if an angle-tightening gauge is not available)	59 ± 5	44 ± 4
6.0 mm bolts	8	6
Cylinder head camshaft sprocket access cover nuts/bolts	5	4
Cylinder head cover bolts	4	3
Driveplate (automatic transmission)	98	72
Engine-to-transmission fixing bolts:		
Bolts less than 30.0 mm in length	19	14
Bolts 30.0 mm in length and longer	35	26
Flywheel (manual transmission)	88	65
Front engine/transmission mounting bolts	50	37
Left-hand engine/transmission mounting:		
Through-bolt	49	36
Mounting-to-transmission bolts	49	36
Lower timing chain front guide bolts	16	12
Lower timing chain rear guide pivot bolt	16	12
Lower timing chain tensioner bolts	9	7
Main bearing cap bolts	50	37
Oil pump:		
Cover retaining screws	5	4
Cover retaining bolt	8	6
Regulator valve bolt	50	37
Rear engine/transmission mounting:		
Mounting-to-centre member bolts	50	37
Through-bolt	69	51
Mounting bracket retaining bolts	80	59
Right-hand engine/transmission mounting bolts	50	37
Sump drain plug	35	26
Sump nuts and bolts	8	6
Timing chain idler sprocket centre bolt	50	37
Upper timing chain tensioner bolts	8	6

1 General information

Using this Chapter

Chapter 2 is divided into two Parts; A and B. Repair operations that can be carried out with the engine in the car are described in Part A. Part B covers the removal of the engine/transmission as a unit, and describes the engine dismantling and overhaul procedures.

Note that, while it may be possible physically to overhaul items such as the piston/connecting rod assemblies while the engine is in the car, such tasks are not normally carried out as separate operations. Usually, several additional procedures (not to mention the cleaning of components and of oilways) have to be carried out. For this reason, all such tasks are classed as major overhaul procedures, and are described in Part B of this Chapter.

In Part A the assumption is made that the engine is installed in the car, with all ancillaries connected. If the engine has been removed for overhaul, the preliminary dismantling information which precedes each operation may be ignored.

Engine description

The 1.4 litre (1392cc) and 1.6 litre (1597cc) engines are from the GA series of Nissan engines. They are of the sixteen-valve, in-line four-cylinder, double overhead camshaft (DOHC) type, mounted transversely at the front of the car with the transmission attached to the left-hand end.

The crankshaft runs in five main bearings. Thrustwashers are fitted to No 3 main bearing (upper half) to control crankshaft endfloat.

The connecting rods rotate on horizontally-split bearing shells at their big-ends. The pistons are attached to the connecting rods by gudgeon pins, which are a sliding fit in the connecting rod small-end eyes and retained in the pistons by circlips. The aluminium-alloy pistons are fitted with three piston rings – two compression rings and an oil control ring.

The cylinder block is made of cast iron and the cylinder bores are an integral part of the block. On this type of engine the cylinder bores are sometimes referred to as having dry liners.

The inlet and exhaust valves are each closed by coil springs, and operate in guides pressed into the cylinder head; the valve seat inserts are also pressed into the cylinder head, and can be renewed separately if worn.

The camshaft is driven by a timing chain, and operates the sixteen valves via bucket-type followers. The followers are situated directly below the camshafts. Valve clearances are adjusted by shims. The camshafts rotate directly in the cylinder head.

Lubrication is by means of an oil pump, which is driven off the right-hand end of the crankshaft. It draws oil through a strainer located in the sump, and then forces it through an externally-mounted filter into galleries in the cylinder block/crankcase. From there, the oil is distributed to the crankshaft (main bearings) and camshaft. The big-end bearings are supplied with oil via internal drillings in the crankshaft, while the camshaft bearings also receive a pressurised supply. The camshaft lobes and valves are lubricated by splash, as are all other engine components.

Repairs with engine in car

The following work can be carried out with the engine in the car:

a) Compression pressure – testing.
b) Cylinder head cover – removal and refitting.
c) Timing chain cover – removal and refitting.
d) Timing chains – removal, inspection and refitting.
e) Timing chain tensioners, guides and sprockets – removal, inspection and refitting.
f) Camshaft and followers – removal, inspection and refitting.
g) Valve clearances – adjustment.
h) Cylinder head – removal and refitting.
i) Cylinder head and pistons – decarbonising.
j) Sump – removal and refitting.
k) Oil pump – removal, inspection and refitting.
l) Crankshaft oil seals – renewal.
m) Engine/transmission mountings – inspection and renewal.
n) Flywheel/driveplate – removal, inspection and refitting.

2 Compression test – description and interpretation

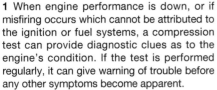

1 When engine performance is down, or if misfiring occurs which cannot be attributed to the ignition or fuel systems, a compression test can provide diagnostic clues as to the engine's condition. If the test is performed regularly, it can give warning of trouble before any other symptoms become apparent.

2 The engine must be fully warmed-up to normal operating temperature, the battery must be fully-charged, and the aid of an assistant will also be required.

3 Depressurise the fuel system by removing the fuel pump fuse (typically fuse 17) from the passenger compartment fusebox – the fuses can be identified from the label inside the fusebox cover, or from the wiring diagrams at the end of this manual. With the fuse removed, start the engine, and allow it to run until it stalls. Try to start the engine at least twice more, to ensure that all residual pressure has been relieved.

4 Disable the ignition system by disconnecting the wiring connector(s) from the distributor body. On early models there is a large connector on the top of the distributor and a smaller connector on the side of the unit. On later models, only the single large connector is used.

5 Remove the spark plugs as described in Chapter 1.

6 Fit a compression tester to the No 1 cylinder spark plug hole – the type of tester which screws into the plug thread is to be preferred.

7 Have the assistant hold the throttle wide open, and crank the engine on the starter motor; after one or two revolutions, the compression pressure should build-up to a maximum figure, and then stabilise. Record the highest reading obtained.

8 Repeat the test on the remaining cylinders, recording the pressure in each.

9 All cylinders should produce very similar pressures; any difference greater than that specified indicates the existence of a fault. Note that the compression should build-up quickly in a healthy engine; low compression on the first stroke, followed by gradually increasing pressure on successive strokes, indicates worn piston rings. A low compression reading on the first stroke, which does not build-up during successive strokes, indicates leaking valves or a blown head gasket (a cracked head could also be the cause). Deposits on the undersides of the valve heads can also cause low compression.

10 If the pressure in any cylinder is reduced to the specified minimum or less, carry out the following test to isolate the cause. Introduce a teaspoonful of clean oil into that cylinder through its spark plug hole and repeat the test.

11 If the addition of oil temporarily improves the compression pressure, this indicates that bore or piston wear is responsible for the pressure loss. No improvement suggests that leaking or burnt valves, or a blown head gasket, may be to blame.

12 A low reading from two adjacent cylinders is almost certainly due to the head gasket having blown between them; the presence of coolant in the engine oil will confirm this.

13 If one cylinder is about 20 percent lower than the others and the engine has a slightly rough idle, a worn camshaft lobe could be the cause.

14 If the compression reading is unusually high, the combustion chambers are probably coated with carbon deposits. If this is the case, the cylinder head should be removed and decarbonised.

15 On completion of the test, refit the spark plugs and fuel pump fuse, then reconnect the distributor wiring connector(s).

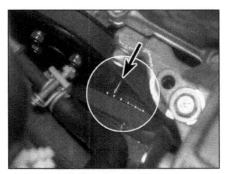

3.6 Crankshaft pulley TDC notch aligned with timing chain cover pointer (arrowed)

3 Top dead centre (TDC) for No 1 piston – locating

1 Disconnect the battery negative terminal (refer to Disconnecting the battery in the Reference Chapter), then remove all the spark plugs as described in Chapter 1.

2 Trace No 1 spark plug (HT) lead from the plug back to the distributor cap, and use chalk or similar to mark the distributor body or engine casting nearest to the cap's No 1 terminal. Undo the distributor cap retaining screws, remove the cap and recover the seal.

3 Apply the handbrake and ensure that the transmission is in neutral, then jack up the front of the car and support it on axle stands (see Jacking and vehicle support). Remove the right-hand roadwheel.

4 From underneath the front of the car, undo the retaining screws and remove the wheel arch liner from underneath the wing to gain access to the crankshaft pulley. If necessary, also undo the retaining screws and remove the engine undershield to improve access.

5 The timing marks are in the form of notches on the crankshaft pulley rim which align with a pointer on the timing chain cover. The notches are spaced at intervals of 5°, and go from 20° before top dead centre (BTDC) to 5° after top dead centre (ATDC). The TDC mark is highlighted with yellow paint to aid identification.

6 Using a spanner (or socket and extension bar) applied to the crankshaft pulley bolt, rotate the crankshaft clockwise until the TDC notch on the crankshaft pulley rim is aligned with the pointer on the timing chain cover **(see illustration)**.

7 With the crankshaft in this position, Nos 1 and 4 cylinders are now at TDC, one of them on the compression stroke. If the distributor rotor arm is pointing at (the previously-marked) No 1 terminal, then No 1 cylinder is correctly positioned; if the rotor arm is pointing at No 4 terminal, rotate the crankshaft one full turn (360°) clockwise until the arm points at the marked terminal. No 1 cylinder will then be at TDC on the compression stroke.

4 Cylinder head cover – removal and refitting

Removal

1 Disconnect the battery negative terminal (refer to Disconnecting the battery in the Reference Chapter), then disconnect the spark plug (HT) leads from the plugs, and free them from their retaining clips on the top of the cover.

2 Release the retaining clips and disconnect the breather hoses from the rear of the cover.

3 Working in the **reverse** of the tightening

sequence **(see illustration 4.11a)**, slacken and remove the cylinder head cover retaining screws and washers.

4 Lift off the cylinder head cover, and recover the rubber seal from the outer edge of the cover, and the circular seal from each of the cover spark plug holes.

5 Inspect the cover seals for signs of damage and deterioration, and renew as necessary.

Refitting

6 Carefully clean the cylinder head and cover mating surfaces, and remove all traces of oil.

7 Apply a bead of sealant to the circular cut-outs on the right-hand end of the cylinder head **(see illustration)**.

8 Fit the rubber seal to the cylinder head cover groove, ensuring that it is correctly located along its entire length, and install the four spark plug hole seals, ensuring that they are the correct way round **(see illustrations)**. If necessary, the seals can be held in position using a smear of suitable sealant.

9 Apply a bead of sealant to the cover seal, approximately 10.0 mm either side of the left-hand exhaust camshaft bearing cap circular cut-out edges **(see illustration)**.

10 Carefully refit the cylinder head cover to the engine, taking great care not to displace any of the rubber seals.

11 Make sure the cover is correctly seated, then install the retaining screws and washers. Working in sequence, tighten all the cover screws to the specified torque **(see illustrations)**.

12 Reconnect the breather hoses to the cylinder head cover, and secure them in position with the retaining clips.

13 Connect the HT leads to the correct spark plugs, and clip the leads back into the retaining clips. Reconnect the battery negative terminal.

5 Crankshaft pulley – removal and refitting

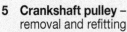

Removal

1 Remove the auxiliary drivebelt(s) as described in Chapter 1.

2 If necessary, position No 1 cylinder at TDC on its compression stroke as described in Section 3.

3 To prevent crankshaft rotation while the pulley bolt is unscrewed on manual transmission models, select top gear and have an assistant apply the brakes firmly. On automatic transmission models or if the engine has been removed from the car, lock the flywheel **(see illustration 15.2)**.

4 Unscrew the pulley bolt, along with its washer, and remove the pulley from the crankshaft **(see illustration)**. If the pulley Woodruff key is a loose fit, remove it and store it with the pulley for safe-keeping.

4.7 Apply a bead of sealant to the cylinder head cut-outs

4.8b ... and fit the spark plug hole seals

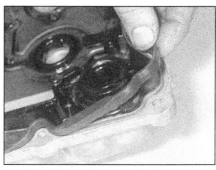

4.8a Fit the rubber seal to the cover groove ...

4.9 Apply a bead of sealant to the seal on 10.0 mm either side of the exhaust camshaft bearing cap cut-out

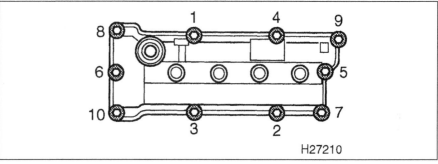

H27210

4.11a Cylinder head cover bolt tightening sequence

Refitting

5 Refit the Woodruff key (where removed).

6 Align the crankshaft pulley groove with the key, then slide the sprocket onto the crankshaft, and refit the retaining bolt and washer.

7 Lock the crankshaft by the method used on removal, and tighten the pulley retaining bolt to the specified torque.

8 Refit the auxiliary drivebelt(s) and adjust them as described in Chapter 1.

4.11b Working in the specified sequence, tighten the cover screws to the specified torque

5.4 Unscrew the pulley bolt and washer, and withdraw the crankshaft pulley

6.4 Removing the power steering pump mounting bracket

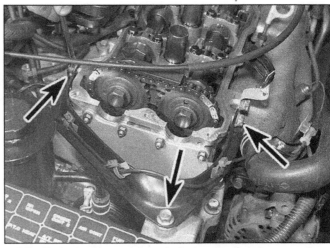

6.5a Undo the three retaining bolts (arrowed) and remove the bracket from the engine mounting

6 Timing chain cover – removal and refitting

Note: *If the timing chain cover is to be removed without disturbing the cylinder head, there is a slight risk of oil leakage from the chain cover-to-cylinder head joint after refitting. Bearing in mind this information, it is up to the individual owner to decide whether or not it is worth renewing the head gasket when the chain cover is removed.*

Removal

1 Remove the coolant pump as described in Chapter 3.
2 Remove the crankshaft pulley as described in Section 5.
3 Remove the sump and oil pump pick-up/strainer as described in Section 12.
4 Undo the two bolts securing the power steering pump mounting bracket to the front of the timing chain cover, then undo the nut/bolt securing the bracket to the pump and remove the bracket **(see illustration)**.
5 Slacken and remove the three retaining

bolts, and remove the mounting bracket from the top of the right-hand engine/transmission mounting; where necessary, free the wiring from its retaining clips on the bracket. Unscrew the retaining nuts and bolts, and remove the camshaft sprocket access cover from the right-hand end of the cylinder head **(see illustrations)**.
6 Unscrew the two retaining bolts, and remove the lower timing chain tensioner from the rear of the chain cover. Recover the tensioner gasket, noting which way up its is fitted. **Note:** *Do not rotate the engine whilst the chain tensioner is removed.*
7 Slacken and remove the four bolts securing the right-hand end of the cylinder head to the top of the timing chain cover.
8 Place a jack with interposed block of wood beneath the engine to take the weight of the engine. Alternatively, attach a hoist or support bar to the engine lifting eyes, and take the weight of the engine.
9 Unscrew the through-bolt from the right-hand engine/transmission mounting, then undo the three retaining bolts and remove the mounting assembly from the engine compartment **(see illustrations)**. Recover the rubbers fitted to the body mounting bracket, if they are loose.

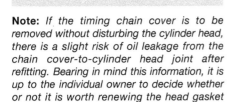

6.5b Undo the retaining nuts and bolts . . .

6.5c . . . and remove the camshaft sprocket access cover

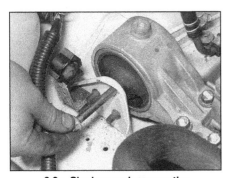

6.9a Slacken and remove the through-bolt . . .

6.9b . . . then undo the three retaining bolts (arrowed) . . .

6.9c . . . and remove the right-hand engine/transmission mounting

6.10 Undo the retaining bolts and remove the right-hand engine/transmission mounting bracket from the engine

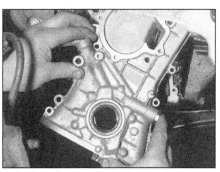

6.12 Removing the timing chain cover

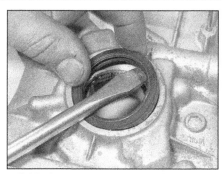

6.14a Lever out the crankshaft oil seal with a large screwdriver . . .

10 Undo the bracket retaining bolts and the bolt securing the alternator adjuster strap to the bracket, and remove the right-hand engine/transmission mounting bracket from the engine **(see illustration)**.
11 Slacken and remove the timing chain cover retaining bolts. Note the correct fitted location of each bolt, as the bolts are of different lengths.
12 Slide the timing chain cover off the end of the crankshaft, and manoeuvre it out of the engine compartment **(see illustration)**. Recover the special sealing collars from the oil galleries and discard them; new ones must be used on refitting. If the cover locating dowels are a loose fit, remove them and store them with the cover for safe-keeping.
13 Slide the oil pump drive spacer off the end of the crankshaft.

Refitting

14 Prior to refitting the cover, it is recommended that the crankshaft oil seal should be renewed. Carefully lever the old seal out of the cover using a large flat-bladed screwdriver. Fit the new seal to the cover, making sure its sealing lip is facing inwards. Drive the seal into position until it seats on its locating shoulder, using a suitable tubular drift, such as a socket, which bears only on the hard outer edge of the seal **(see illustrations)**.
15 Ensure that the cover and crankcase mating surfaces are clean and dry.
16 Fit the new special sealing collars to the cylinder block oil galleries **(see illustration)**.
17 Apply a thin coat of suitable sealant to the timing cover crankcase mating surface, not forgetting to apply sealant to the area around the coolant pump passage in the centre of the cover **(see illustration)**. If the cylinder head is in position, also apply sealant to the upper face of the cover.
18 Refit the cover locating dowels to the cylinder block (where removed).
19 Align the oil pump drive spacer groove with the key, then slide the spacer onto the crankshaft **(see illustration)**.
20 Offer up the cover, and position the oil pump inner rotor so that it will engage with the drive spacer as the cover is refitted **(see illustration)**. Slide the cover over the end of

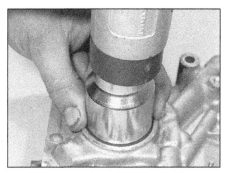

6.14b . . . and tap the new seal into position with a suitable socket

the crankshaft, taking great care not to damage the oil seal lip, and seat it on its locating dowels.
21 Refit the cover retaining bolts in their original locations and tighten them securely,

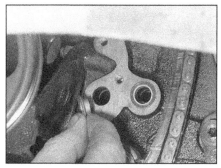

6.16 Fit the new special sealing collars to the cylinder block oil galleries

working in several stages.
22 Fit the four 6.0 mm bolts securing the cylinder head to the chain cover, and tighten them to the specified torque setting **(see illustration)**.

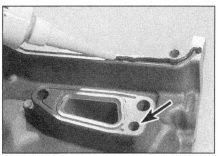

6.17 Apply sealant to the timing cover mating surface, not forgetting the area around the coolant pump passage (arrowed)

6.19 Align the oil pump drive spacer with the key, and slide it onto the crankshaft (cover locating dowel locations arrowed)

6.20 Engage the oil pump with the drive spacer, and slide the timing cover into position

6.22 Refit the bolts securing the cylinder head to the top of the timing cover, and tighten them to the specified torque

6.23a Fit the gasket, ensuring that its cut-out is correctly aligned with the tensioner oil hole (arrowed) . . .

6.23b . . . then refit the tensioner to the engine

23 Fit a new gasket to the lower chain tensioner, making sure it is fitted the correct way up so its cut-out is aligned with the tensioner oil hole. Install the tensioner and gasket, and tighten its retaining bolts to the specified torque setting **(see illustrations)**.
24 Refit right-hand engine/transmission mounting bracket to the engine, and securely tighten its retaining bolts. Fit the rubbers to the body mounting bracket (where removed), ensuring that their pins are correctly seated in the bracket holes, and fit the mounting, tightening its retaining bolts to the specified torque setting.
25 Align the right-hand mounting with its body bracket, then insert the through-bolt and tighten its nut to the specified torque setting. Remove the jack from underneath the engine.
26 Ensure that the sprocket access cover and head mating surfaces are clean and dry. Where the cover was originally fitted with a

gasket, fit a new gasket to the cylinder head; where the cover was originally fitted using sealant, apply a continuous bead of suitable sealant to the cover mating surface. Install the cover and tighten its retaining nuts and bolts to the specified torque setting.
27 Install the bracket to the top of the right-hand engine/transmission mounting, and tighten its bolts to the specified torque setting.
28 Refit the power steering pump to the engine, and securely tighten its mounting bracket bolts.
29 Install the oil pump pick-up/strainer and sump as described in Section 12.
30 Refit the crankshaft pulley as described in Section 5.
31 Refit the coolant pump as described in Chapter 3.
32 Replenish the engine oil and coolant as described in Chapter 1.

7 Timing chains – removal, inspection and refitting

Removal

1 Position No 1 cylinder at TDC on its compression stroke, as described in Section 3.
2 Remove the cylinder head cover as described in Section 4.
3 Remove the timing chain cover as described in Section 6.
4 Retract the upper chain tensioner, and hold it in position by inserting a small-diameter rod in front of the tensioner pad. Undo the two retaining bolts, and remove the tensioner from the end of the cylinder head.
5 Slacken the camshaft sprocket retaining bolts, whilst retaining the camshafts with a large open-ended spanner fitted to the flats on the right-hand end of each shaft. Remove each bolt along with its washer **(see illustrations)**.
6 Disengage each sprocket from the end of its respective camshaft, and manoeuvre them out from the cylinder head. If the sprocket locating pins are a loose fit in the camshaft ends, remove them and store them with the sprockets.
7 Disengage the upper timing chain from the idler sprocket, and manoeuvre it out of the cylinder head **(see illustration)**.
8 Unscrew the pivot bolt, and remove the lower chain rear guide from the crankcase **(see illustration)**.
9 Slacken and remove the idler sprocket centre bolt and washer, then lower the sprocket out of position and remove it from the bottom of the cylinder head **(see illustration)**. Recover the idler shaft from the rear of the sprocket.
10 Disengage the lower timing chain from the crankshaft sprocket, and remove it from the engine.
11 Slide the crankshaft sprocket off the end of the crankshaft. Remove the Woodruff key (if loose) from the crankshaft, and store it with the sprocket for safe-keeping. **Note:** *Do not rotate the crankshaft or camshafts whilst the timing chains are removed.*

7.5a Slacken the sprocket retaining bolt whilst holding the camshaft with an open-ended spanner . . .

7.5b . . . then remove the bolt along with its washer

7.7 Manoeuvre the upper timing chain out through the head aperture

7.8 Unscrew the pivot bolt, and remove the lower chain rear guide from the cylinder block

7.9 Unscrew the centre bolt and washer, and remove the idler sprocket from the engine

Inspection

12 Examine the teeth on the camshaft, idler and crankshaft sprockets for any sign of wear or damage such as chipped, hooked or missing teeth. If there is any sign of wear or damage on either sprocket, *all* sprockets and *both* timing chains should be renewed as a set.

13 Inspect the links of the timing chains for signs of wear or damage on the rollers. The extent of wear can be judged by checking the amount by which the chain can be bent sideways; a new chain will have very little sideways movement. If there is an excessive amount of side play in either timing chain, it must be renewed.

14 Note that it is a sensible precaution to renew the timing chains, regardless of their apparent condition, if the engine has covered a high mileage, or if it has been noted that the chain(s) have sounded noisy when the engine running. Although not strictly necessary, it is always worth renewing the chains and sprockets as a matched set, since it is false economy to run a new chain on worn sprockets and *vice versa*. If there is any doubt about the condition of the timing chains and sprockets, seek the advice of a Nissan dealer service department, who will be able to advise you as to the best course of action.

15 Examine the chain guides for signs of wear or damage to their chain contact faces, renewing any which are badly marked **(see illustration)**.

16 Check the upper chain tensioner pad for signs of wear, and check that the plunger is free to slide freely in the tensioner body. The condition of the tensioner spring can only be judged in comparison to a new component. Renew the tensioner if its pad is worn or there is any doubt about the condition of its tensioning spring.

Refitting

17 Check the crankshaft is still positioned at

7.15 Examine the chain guides for signs of wear

7.18 . . . and slide on the crankshaft sprocket

TDC (the keyway will be in the 12 o'clock position, seen from the right-hand end of the engine) and refit the Woodruff key to the crankshaft groove **(see illustration)**.

18 Ensure that the crankshaft sprocket is positioned the correct way around, with its timing mark facing away from the crankcase, then align its groove with the key, and slide the sprocket onto the crankshaft **(see illustration)**.

19 Apply a smear of clean engine oil to the idler sprocket shaft, and fit the shaft to the rear of the sprocket so that its flange is

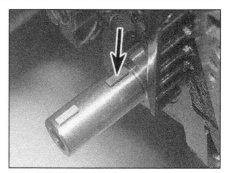

7.17 Ensure that the crankshaft is correctly positioned, then fit the Woodruff key (arrowed) . . .

7.19a Fit the shaft to the rear of the idler sprocket . . .

positioned between the sprocket and crankcase. Engage the idler sprocket with the lower timing chain, aligning its outer timing mark with one of the chain's coloured (silver) links **(see illustrations)**.

20 Manoeuvre the chain and idler sprocket assembly into position, engaging the chain with the crankshaft sprocket so that its second coloured (silver) link is aligned with the timing mark on the crankshaft sprocket; the timing mark is in the form of a small cut-out on the sprocket hub **(see illustration)**.

7.19b . . . then engage the sprocket with the timing chain, aligning its timing mark with one of the chain's coloured links (arrowed)

7.20 Engage the chain with the crankshaft sprocket so that its coloured link is correctly aligned with the sprocket timing mark (arrowed)

7.21 Check that the timing marks are correctly positioned then fit the idler sprocket bolt and washer, and tighten it to the specified torque

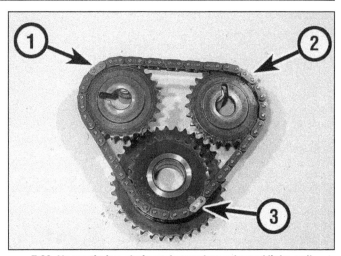

7.22 Upper timing chain and sprocket coloured links and timing marks

1   *Inlet camshaft sprocket link and mark*
2   *Exhaust camshaft sprocket link and mark*
3   *Idler sprocket link and mark*

7.23 Manoeuvre the upper chain into position, and engage it with the idler sprocket so that the relevant coloured link is correctly aligned with its timing mark (arrowed)

21 Check that the idler sprocket and crankshaft sprocket marks are still correctly aligned with the lower chain silver links, then install the idler sprocket centre bolt and washer, and tighten it to the specified torque setting **(see illustration)**.
22 The upper timing chain has three coloured links, one for each of the sprocket timing marks. Note, however, that the links are not spaced at regular intervals. On early models, all three links are silver; the first (inlet camshaft sprocket) link and second (exhaust camshaft sprocket) link are 16 rollers apart, the second and third (idler sprocket) link are also 16 rollers apart, but the gap between the third silver link and the first link is 22 rollers **(see illustration)**. On later models, the gaps between the links are the same, but the idler

gear link is easily identified since it is gold in colour (camshaft sprocket links remain silver).
23 Bearing in mind paragraph 22, lower the upper chain into position, making sure its coloured links are facing outwards. Engage the chain with the idler sprocket, aligning its appropriate link with the sprocket inner timing mark **(see illustration)**.
24 Manoeuvre both the inlet and exhaust camshaft sprockets into position, ensuring that their timing marks are facing outwards. **Note:** *Both sprockets are identical.* Engage them with the chain, aligning the inlet sprocket timing mark with the first chain silver link and the exhaust timing mark with the second. Check that all the timing marks are correctly aligned with the upper chain links **(see illustrations)**.

7.24a Engage the inlet camshaft sprocket with the chain, aligning its timing mark (arrowed) with the relevant coloured link . . .

7.24b . . . then install the exhaust camshaft sprocket and align its timing mark (arrowed) with the last coloured link

7.25 Locate the sprockets on the camshafts, and check that all the timing marks and coloured links are correctly aligned (arrowed)

7.26 Refit the sprocket retaining bolts and washers, and tighten them to the specified torque setting

25 Locate the sprockets on the camshafts, aligning their cut-outs with the locating pins. Check that the chain silver links are correctly aligned with each sprocket's timing mark. If not, disengage the sprocket(s) from the chain, and make the necessary adjustments **(see illustration)**.

26 With the timing marks correctly positioned, install the camshaft sprocket retaining bolts and washers, and tighten them both to the specified torque **(see illustration)**.

27 Fit the lower chain rear guide, and tighten its pivot bolt to the specified torque.

28 Retract the tensioner pad and hold it in position with a suitable rod. Fit the tensioner to the cylinder head, and tighten its retaining bolts to the specified torque. Withdraw the rod, and check that the tensioner pad is forced against the chain.

29 Refit the cylinder head cover as described in Section 4.

30 Refit the timing chain cover as described in Section 6.

---

8 Timing chain tensioners, guides and sprockets – removal, inspection and refitting

Removal

Lower timing chain tensioner

1 Firmly apply the handbrake, then jack up the front of the vehicle and support it securely on axle stands (see *Jacking and vehicle support*).

2 From underneath the vehicle, unscrew the two retaining bolts and remove the lower timing chain tensioner from the rear of the chain cover. Recover the tensioner gasket, noting which way up it is fitted. **Note:** *Do not rotate the engine whilst the chain tensioner is removed.*

Upper timing chain tensioner

3 Slacken and remove the three retaining bolts, and remove the mounting bracket from

the top of the right-hand engine/transmission mounting.

4 Unscrew the retaining nuts and bolts, and remove the camshaft sprocket access cover from the right-hand end of the cylinder head. Recover the cover gasket (where fitted).

5 Retract the chain tensioner, and hold it in position by inserting a small diameter rod in front of the tensioner pad.

6 Undo the two retaining bolts and remove the tensioner from the end of the cylinder head. **Note:** *Do not rotate the engine whilst the chain tensioner is removed.*

Lower timing chain front guide

7 Remove the timing chains as described in Section 7.

8 Slacken and remove the retaining bolts, and remove the lower chain front guide from the side of the crankcase.

Lower timing chain rear guide

9 Remove the timing chain cover as described in Section 6.

10 Unscrew the pivot bolt, and remove the lower chain rear guide from the crankcase (see illustration).

Camshaft sprockets

11 Position No 1 cylinder at TDC on its compression stroke as described in Section 3.

12 Remove the cylinder head cover as described in Section 4.

13 Remove the camshaft sprocket access cover as described in paragraphs 3 and 4 of this Section.

14 Remove both camshaft sprockets and the upper timing chain as described in paragraphs 4 to 6 of Section 7.

Idler sprocket and crankshaft sprocket

15 Remove the timing chains and sprockets as described in Section 7.

Inspection

16 Refer to Section 7.

Refitting

Lower timing chain tensioner

17 Ensure that the tensioner and cover mating surfaces are clean and dry.

18 Fit a new tensioner gasket to the cover, making sure it is fitted the correct way up so its cut-out is aligned with the tensioner oil hole **(see illustration 6.23a)**.

19 Install the chain tensioner, and tighten its retaining bolts to the specified torque setting.

Upper timing chain tensioner

20 Retract the tensioner pad, and hold it in position with a suitable rod. Fit the tensioner to the head, and tighten its retaining bolts to the specified torque. Withdraw the rod, and check that the tensioner pad is forced against the chain.

21 Ensure that the sprocket access cover and head mating surfaces are clean and dry. Where the cover was originally fitted with a gasket, fit a new gasket to the cylinder head; where the cover was originally fitted using sealant, apply a continuous bead of suitable sealant to the cover mating surface. Install the cover, and tighten its retaining nuts and bolts to the specified torque setting.

22 Install the bracket to the top of the right-hand engine/transmission mounting, and tighten its bolts to the specified torque setting.

8.10 Remove the pivot bolt, and withdraw the lower chain rear guide from the cylinder block

9.5 Checking a valve clearance

Lower timing chain front guide

23 Fit the guide to the crankcase, and tighten its retaining bolts to the specified torque.
24 Refit the timing chains as described in Section 7.

Lower timing chain rear guide

25 Fit the guide to the crankcase, and tighten its pivot bolt to the specified torque.
26 Refit the timing chain cover as described in Section 6.

Camshaft sprockets

27 Refit the camshaft sprockets and upper timing chain as described in paragraphs 22 to 26 of Section 7.
28 Refit the upper chain tensioner as described in paragraphs 20 to 22.
29 Refit the cylinder head cover as described in Section 4.

Idler sprocket and crankshaft sprocket

30 Refit the timing chains and sprockets as described in Section 7.

---

9 Valve clearances – checking and adjustment

Note: *The valve clearances must always be checked with the engine 'hot'. Although Nissan quote valve clearances for a 'hot' and 'cold' engine, the valve clearances should only be checked 'cold', prior to starting the engine after an overhaul. The valve clearances should then be checked again once the engine has been warmed-up to normal operating temperature.*
Note: *This is not a routine operation. It should only be necessary at high mileage, after overhaul, or when investigating noise or power loss which may be attributable to the valvegear.*

1 The importance of having the valve clearances correctly adjusted cannot be overstressed, as they vitally affect the performance of the engine. The clearances are checked as follows.
2 Draw the outline of the engine on a piece of paper, numbering the cylinders 1 to 4, with No 1 cylinder at the timing chain end of the engine. Show the position of each valve, together with the specified valve clearance.

Above each valve, draw two lines for noting the actual clearance and the amount of adjustment required. **Note:** *Nissan quote two sets of clearance tolerances; one for* **checking** *the clearance, and a second one to use when* **adjusting/setting** *the clearance. The checking tolerance is exceptionally large and, although not strictly necessary, it is desirable to have all valve clearances within the adjusting/setting tolerances.*
3 Warm the engine up to normal operating temperature, then switch off. Remove the cylinder head cover as described in Section 4.
4 Position No 1 cylinder at TDC on its compression stroke, as described in Section 3.
5 Using feeler gauges, measure the clearance between the base of the cam and the follower of the following valves, recording each clearance on the paper **(see illustration)**.
  No 1 cylinder inlet and exhaust valves
  No 2 cylinder inlet valves
  No 3 cylinder exhaust valves
6 Rotate the crankshaft through one complete turn (360º) clockwise until the TDC notch on the crankshaft pulley is realigned with the pointer. No 4 cylinder is now at TDC on its compression stroke.
7 Check the clearances of the following valves, and record them on the paper.
  No 2 cylinder exhaust valves
  No 3 cylinder inlet valves
  No 4 cylinder inlet and exhaust valves
8 Calculate the difference between each measured clearance and the desired value, and record it on the piece of paper. Where a valve clearance differs from the specified value, then the shim for that valve must be substituted with a thinner or thicker shim accordingly.
9 To remove the shim, the follower has to be pressed down against valve spring pressure just far enough to allow the shim to be slid out. To do this, make sure that the cam lobe of the valve on which the shim is to be removed is pointing away from the follower, then rotate the follower so that its notch is at a right-angle to the camshaft centre-line.
10 Using a suitable C-spanner or stout screwdriver, carefully lever down between the camshaft and the edge of the follower until the follower is depressed sufficiently to allow the shim to be slid out of position. With the shim removed, slowly release the follower. If difficulty is experienced in removing the shims, it will be necessary to remove the camshaft(s) as described in Section 10.
11 The shim size is stamped on the bottom face of the shim (eg, 224 indicates the shim is 2.24 mm thick), but it is advisable to use a micrometer to measure the true thickness of any shim removed, as it may have been reduced by wear. **Note:** *Shims are available in thicknesses between 2.00 mm and 2.98 mm, in steps of 0.02 mm.* The size of shim required is calculated as follows.
12 If the measured clearance is less than specified, subtract the measured clearance from the specified clearance, and subtract the result from the thickness of the existing shim.

For example:
Sample calculation –
inlet valve clearance too small
Clearance measured = 0.26 mm
Desired clearance = 0.36 mm
$\qquad$ *(0.32 to 0.40 mm)*
Difference = 0.10 mm
Shim thickness fitted = 2.50 mm
Shim thickness required =
$\qquad$ *2.50 – 0.10 = 2.40 mm*
13 If the measured clearance is greater than specified, subtract the specified clearance from the measured clearance, and add the result to the thickness of the existing shim.
For example:
Sample calculation –
exhaust valve clearance too big
Clearance measured = 0.51 mm
Desired clearance = 0.41 mm
$\qquad$ *(0.37 to 0.45 mm)*
Difference = 0.10 mm
Shim thickness fitted = 2.76 mm
Shim thickness required =
$\qquad$ *2.76 + 0.10 = 2.86 mm*
14 Depress the follower, then slide the required size of shim into position, so that it is fitted with its marked face facing downwards. Ensure that the shim is correctly seated, then repeat the procedure (as required) for the remaining valve(s) which require adjustment.
15 Once all valves have been adjusted, rotate the crankshaft through at least four complete turns in the correct direction of rotation, to settle all disturbed shims in position, then recheck the clearances as described above.
16 With all valve clearances correctly adjusted, refit the cylinder head cover as described in Section 4, and refit all components removed to gain access to the crankshaft pulley.

---

10 Camshafts and followers – removal, inspection and refitting

Removal

1 Remove the distributor as described in Chapter 5B.
2 Remove the camshaft sprockets as described in Section 8.
3 The camshaft right- and left-hand end bearing caps are noticeably different to the others, however, the centre bearing caps are all similar. The caps should have identification markings stamped into their top surface; the exhaust camshaft caps being marked E2 to E5 and the inlet camshaft caps being marked I2 to I5; the No 2 caps are fitted nearest the timing chain end of the engine **(see illustration)**. The caps should also have an arrow stamped on them which should point to the timing chain end of the engine. If the caps are not marked, suitable identification marks should be made prior to removal. Using white paint or a suitable marker pen, mark each cap in some way to indicate its correct fitted orientation

and position. This will avoid the possibility of installing the caps in the wrong positions and/or the wrong way around on refitting.

4 Working in the **reverse** of the tightening sequence **(see illustration 10.23)**, evenly and progressively slacken the fourteen camshaft bearing cap retaining bolts by one turn at a time, to relieve the pressure of the valve springs on the bearing caps gradually and evenly. Note that on 1.6 litre engines, there is an additional bolt (bolt 15) securing the right-hand end bearing cap. Once the valve spring pressure has been relieved, the bolts can be fully unscrewed and removed. Remove the end bearing caps first, then remove the exhaust camshaft caps followed by the inlet camshaft caps.

5 Lift the camshafts out of the cylinder head.

6 Obtain sixteen small, clean plastic containers, and number them 1 to 16. Alternatively, divide a larger container into sixteen compartments. Using a rubber sucker, withdraw each shim and follower in turn, and place it in its respective container. Do not interchange the cam followers, or the rate of wear will be increased.

Inspection

7 Inspect the cam bearing surfaces of the head and the bearing caps. Look for score marks and deep scratches. Check the camshaft lobes for heat discoloration (blue appearance), score marks, chipped areas or flat spots.

8 Camshaft run-out can be checked by supporting each end of the camshaft on V-blocks, and measuring any run-out at the centre of the shaft using a dial gauge. If the run-out exceeds the specified limit, a new camshaft will be required.

9 Measure the height of each lobe with a micrometer, and compare the results to the figures given in the Specifications. If damage is noted or wear is excessive, new camshaft(s) must be fitted.

10 The camshaft bearing oil clearance should now be checked. There are two possible ways of checking this; the first method is by direct measurement (see paragraphs 11 and 16) and the second by the use of a product called Plastigauge (see paragraphs 12 to 16).

11 If the direct measurement method is to be used, fit the bearing caps to the head, using the identification markings or the marks made on removal to ensure that they are correctly positioned. Tighten the retaining bolts to the specified torque in sequence **(see illustration 10.23)**. Measure the diameter of each bearing cap journal, and compare the measurements obtained with the results given in the Specifications at the start of this Chapter. If any journal is worn beyond the service limit, the cylinder head must be renewed. The camshaft bearing oil clearance can then be calculated by subtracting the camshaft bearing journal diameter from the bearing cap journal diameter.

12 If the Plastigauge method is to be used,

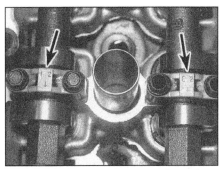

10.3 Inlet and exhaust camshaft bearing cap markings (arrowed)

clean the camshafts, the bearing surfaces in the cylinder head and the bearing caps with a clean, lint-free cloth, then lay the camshafts in place in the cylinder head.

13 Cut strips of Plastigauge, and lay one piece on each bearing journal, parallel with the camshaft centreline. Ensuring that the camshafts are not rotated at all, refit both camshafts as described in paragraphs 19 to 23, ignoring the remark about applying sealant to the bearing cap.

14 Now unscrew the bolts as described in paragraph 4 and carefully lift off the bearing caps, again making sure the camshafts are not rotated.

15 To determine the oil clearance, compare the crushed Plastigauge (at its widest point) on each journal to the scale printed on the Plastigauge container.

16 Compare the results with the figures given in the Specifications. If the oil clearance is greater than specified, measure the diameter of the cam bearing journal with a micrometer. If the journal diameter is less than the specified limit, renew the camshaft and recheck the clearance. If the clearance is still too great, renew the cylinder head and bearing caps.

17 Check the cam follower and cylinder head bearing surfaces for signs of wear or damage.

If the necessary measuring equipment is available, the amount of wear can be assessed by direct measurement. Compare the measurement of each follower and its cylinder head bore with the measurements given in the Specifications at the start of this Chapter. Renew worn components as necessary.

Refitting

18 Liberally oil the cylinder head cam follower bores and the followers. Carefully refit the followers to the cylinder head, ensuring that each follower is refitted to its original bore. Some care will be required to enter the followers squarely into their bores. Ensure that all the shims are correctly seated in the top of each follower, then liberally oil the camshaft bearing and lobe contact surfaces.

19 Refit the camshafts to their correct locations in the cylinder head. The exhaust camshaft is easily distinguished by the distributor drive slot on its left-hand end. The shafts should also have identification markings – the inlet camshaft is marked I and the exhaust camshaft E.

20 Check that the crankshaft pulley TDC notch is still aligned with the pointer on the timing chain cover. Position each camshaft so that its No 1 cylinder lobes are pointing away from their valves. With the shafts in this position, the sprocket locating pin in the inlet camshaft's right-hand end will be in the 9 o'clock position when viewed from the right-hand end of the engine, while that of the exhaust camshaft will be in the 12 o'clock position.

21 Ensure that the bearing cap and head mating surfaces are completely clean, unmarked and free from oil. Apply a smear of suitable sealant to the exhaust camshaft's left-hand end bearing cap mating surface.

22 Refit the bearing caps, using the identification markings or the marks made on removal to ensure that each is installed the correct way round and in its original location.

23 Working in sequence **(see illustration)**, evenly and progressively tighten the camshaft

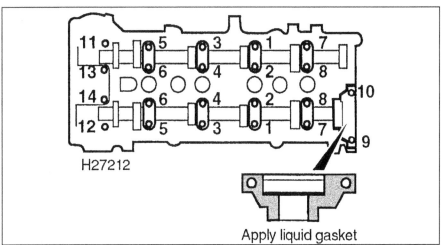

H27212

Apply liquid gasket

10.23 Camshaft bearing cap bolt tightening sequence. Apply sealant to the shaded area of the left-hand exhaust camshaft bearing (inset) cap prior to refitting

bearing cap bolts by one turn at a time until the caps touch the cylinder head. Then go round again and tighten all the bolts to the specified torque setting. Work only as described, to impose the pressure of the valve springs gradually and evenly on the bearing caps. Note that on 1.6 litre engines, the torque setting for the additional bolt (bolt 15) securing the right-hand end bearing cap is different from the rest.

24 Refit the camshaft sprockets as described in Section 8. **Note**: *If the cylinder head/camshafts have been overhauled, check the valve clearances 'cold' prior to refitting the cylinder head cover (see Section 9).*

25 Refit the distributor as described in Chapter 5B.

26 Check the valve clearances as described in Section 9.

11 Cylinder head –
removed and refitting

 To aid refitting, make notes on the locations of all relevant brackets and the routing of hoses and cables before removal.

Removal

1 Depressurise the fuel system as described in Chapter 4A.

2 Disconnect the battery negative terminal (refer to *Disconnecting the battery* in the Reference Chapter).

3 Remove the timing chains as described in Section 7.

4 Remove the camshafts as described in Section 10.

5 Carry out the following operations as described in Chapter 4A.

a) *Remove the air cleaner assembly.*
b) *Disconnect the exhaust system front pipe from the manifold.*
c) *Disconnect the fuel feed and return hoses from the fuel rail (plug all openings, to prevent loss of fuel and entry of dirt into the fuel system).*
d) *Disconnect the accelerator cable.*
e) *Disconnect the relevant electrical connectors from the throttle housing, inlet manifold and associated components.*
f) *Disconnect the vacuum servo unit hose, coolant hose(s) and all the other relevant/breather hoses from the manifold and associated valves.*
g) *Remove the inlet manifold support brackets.*
h) *Disconnect the exhaust gas sensor wiring connector.*

6 Slacken the retaining clip(s) and disconnect the coolant hose(s) from the cylinder head.

7 Slacken and remove the 6.0 mm bolt from the front, left-hand corner of the cylinder head.

8 Working in the **reverse** of the tightening sequence **(see illustration 11.22)**, progressively slacken the ten main cylinder head bolts by half a turn at a time, until all bolts can be unscrewed by hand.

9 Lift out the cylinder head bolts and recover the washers, noting which way around they are fitted.

10 Lift the cylinder head away with the aid of an assistant, as it is a heavy assembly. Remove the gasket from the top of the block, noting the two locating dowels and the oil jet fitted to the top of the cylinder block. If they are a loose fit in the block, remove the locating dowels and oil jet, noting which way round they are fitted, and store them with the head for safe-keeping.

11 If the cylinder head is to be dismantled for overhaul, then refer to Part B of this Chapter.

Preparation for refitting

12 Check the condition of the cylinder head bolts, and particularly their threads, whenever they are removed. Wash the bolts and wipe dry, then check each for any sign of visible wear or damage, renewing any bolt if necessary. Although Nissan do not specify that the bolts must be renewed, it is strongly recommended that the bolts should be renewed as a complete set whenever they are disturbed.

13 The mating faces of the cylinder head and cylinder block/crankcase must be perfectly clean before refitting the head. Use a hard plastic or wood scraper to remove all traces of gasket and carbon; also clean the piston crowns. Take particular care, as the surfaces are damaged easily. Also, make sure that the carbon is not allowed to enter the oil and water passages – this is particularly important for the lubrication system, as carbon could block the oil supply to any of the engine's components. Using adhesive tape and paper, seal the water, oil and bolt holes in the cylinder block/crankcase. To prevent carbon entering the gap between the pistons and bores, smear a little grease in the gap. After cleaning each piston, use a small brush to remove all traces of grease and carbon from the gap, then wipe away the remainder with a clean rag. Clean all the pistons in the same way.

14 Check the mating surfaces of the cylinder block/crankcase and the cylinder head for nicks, deep scratches and other damage. If slight, they may be removed carefully with a file, but if excessive, machining may be the only alternative to renewal.

15 If warpage of the cylinder head gasket surface is suspected, use a straight-edge to check it for distortion. Refer to Part B of this Chapter if necessary.

Refitting

16 Wipe clean the mating surfaces of the cylinder head and cylinder block/crankcase. Check that the two locating dowels are in position at each end of the cylinder block/crankcase surface, and refit the oil jet to the centre of the block.

17 Fit a new gasket to the cylinder block/crankcase surface, aligning it with the oil jet and locating dowels.

18 With the aid of an assistant, carefully refit the cylinder head assembly to the block, aligning it with the locating dowels.

19 Apply a smear of clean oil to the threads, and to the underside of the heads, of the ten main cylinder head bolts.

20 Fit the washer to each head bolt, making sure it is fitted with its tapered edge uppermost.

21 Carefully enter each bolt into its relevant hole (*do not drop them in*) and screw in, by hand only, until finger-tight.

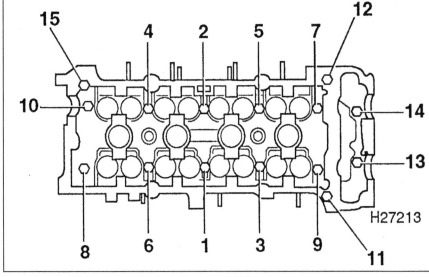

H27213

11.22 Cylinder head bolt tightening sequence

22 Working progressively and in sequence, tighten the **ten main** cylinder head bolts (Nos 1 to 10) to their Stage 1 torque setting, using a torque wrench and suitable socket **(see illustration)**. **Note:** *Bolts 11 to 15 in the sequence (the 6.0 mm bolts) should not be tightened now, but only after the ten main bolts have been tightened to Stage 5.*

23 Once the ten main bolts have been tightened to their Stage 1 setting, go around again in the specified sequence and tighten them to the specified Stage 2 torque setting.

24 Leave the bolts a minute then, working in the **reverse** of the specified sequence, progressively slacken the head bolts by half a turn at a time, until all bolts can be unscrewed by hand.

25 Tighten the ten main bolts again by hand, then go around again in the specified sequence and tighten these ten bolts to the specified Stage 4 torque setting.

26 Finally, go around again in the specified sequence and tighten the ten main head bolts either through the specified Stage 5 angle setting **or**, if an angle-measuring gauge is not available, to the specified Stage 5 torque setting.

27 With the main cylinder head bolts correctly tightened, fit the five 6.0 mm bolts (Nos 11 to 15 in the tightening sequence) and tighten them in sequence to their specified torque setting.

28 Reconnect the coolant hose to the cylinder head and securely tightening its retaining clip.

29 Working as described in Chapter 4A, carry out the following operations:

a) *Refit all disturbed wiring, hoses and control cable(s) to the inlet manifold and fuel system components.*

b) *Reconnect and adjust the accelerator cable.*

c) *Reconnect the exhaust system front pipe to the manifold, and reconnect the exhaust gas sensor wiring connector.*

d) *Refit the inlet manifold support brackets*

e) *Refit the air cleaner assembly and inlet duct.*

30 Refit the camshafts to the cylinder head as described in Section 10.

31 Fit the timing chains and sprockets as

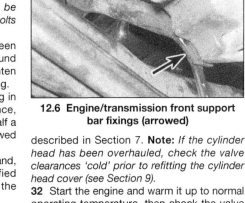

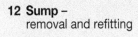

12.6 Engine/transmission front support bar fixings (arrowed)

described in Section 7. **Note:** *If the cylinder head has been overhauled, check the valve clearances 'cold' prior to refitting the cylinder head cover (see Section 9).*

32 Start the engine and warm it up to normal operating temperature, then check the valve clearances as described in Section 9.

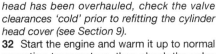

12 Sump –
removal and refitting

Removal

1 Disconnect the battery negative terminal (refer to *Disconnecting the battery* in the Reference Chapter).

2 Firmly apply the handbrake, then jack up the front of the vehicle and support it securely on axle stands (see *Jacking and vehicle support*).

3 To improve access, slacken and remove the retaining screws and remove the plastic undershield(s) from beneath the engine.

4 Drain the engine oil, then clean and refit the engine oil drain plug, tightening it to the specified torque. If the engine is nearing its service interval when the oil and filter are due for renewal, it is recommended that the filter is also removed, and a new one fitted. After reassembly, the engine can then be refilled with fresh oil. Refer to Chapter 1 for further information.

5 Remove the exhaust system front pipe as described in Chapter 4A.

12.8 Removing the sump from the engine

6 Undo the retaining bolts, and remove the small support bars linking the transmission to the cylinder block, from the front and rear of the block **(see illustration)**.

7 Progressively slacken and remove all the sump retaining nuts and bolts.

8 Break the joint by striking the sump with the palm of your hand, then lower the sump and withdraw it from underneath the vehicle **(see illustration)**.

9 While the sump is removed, undo the two retaining bolts and remove the oil pump pick-up/strainer and O-ring from the base of the timing cover **(see illustration)**.

10 Wash the strainer in a suitable solvent, and check it for signs of clogging or splitting. Renew it if the pick-up/strainer is damaged in any way.

Refitting

11 Clean all traces of sealant from the mating surfaces of the cylinder block/crankcase and sump, then use a clean rag to wipe out the sump and the engine's interior.

12 Fit a new O-ring to the recess in the top of the pick-up/strainer, and fit the strainer to the base of the timing cover **(see illustration)**. Securely tighten the strainer bolts.

13 Ensure that the sump and cylinder block/crankcase mating surfaces are clean and dry. Apply a continuous bead of suitable sealant to the mating surface of the sump. Apply the sealant to the groove in the centre of the mating surface between the holes, and around the inner edge of each bolt hole **(see illustration)**.

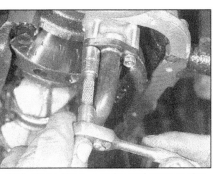

12.9 Undo the two bolts and remove the oil pump pick-up/strainer from the base of the timing chain cover

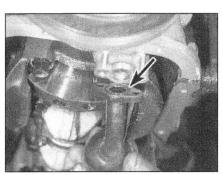

12.12 Fit a new O-ring (arrowed) to the oil pump pick-up/strainer recess

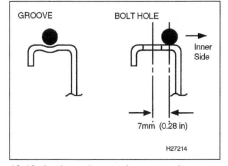

GROOVE   BOLT HOLE

Inner Side

7mm (0.28 in)

H27214

12.13 Apply sealant to the groove between the sump mounting bolt holes, and to the inner side of each hole

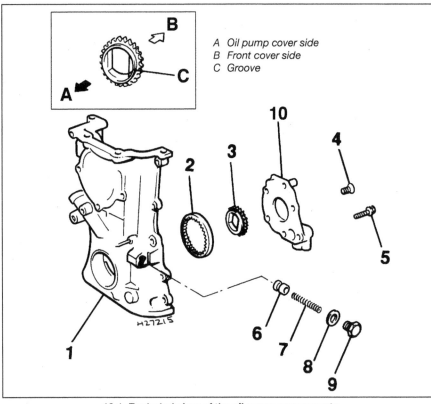

A Oil pump cover side
B Front cover side
C Groove

13.1 **Exploded view of the oil pump components**

1	Timing chain cover	5	Pump cover bolt	8	Sealing washer
2	Outer gear	6	Oil pressure regulator	9	Oil pressure regulator
3	Inner gear		valve piston		valve bolt
4	Pump cover screw	7	Spring	10	Pump cover

14 Offer up the sump, locating it on its retaining studs, and refit its retaining nuts and bolts. Tighten the nuts and bolts evenly and progressively to the specified torque.
15 Refit the support bars, tightening their retaining bolts securely.
16 Refit the exhaust front pipe as described in Chapter 4A.
17 Refit the undershields (if removed) and securely tighten their retaining screws.
18 Replenish the engine oil as described in Chapter 1.

13 Oil pump – removal, inspection and refitting

Removal

1 The oil pump is an integral part of the timing chain cover **(see illustration)**. Remove the cover as described in Section 6. **Note:** *If necessary, the pressure regulator valve can be dismantled without removing the timing cover from the engine.*

Inspection

2 Unscrew the retaining screws and bolt, and remove the pump cover from the rear of the timing chain cover **(see illustrations)**.
3 Remove both the oil pump gears from the cover **(see illustrations)**.
4 Unscrew the oil pressure regulator valve bolt from the front edge of the cover, and recover its sealing washer. Withdraw the spring and the valve piston, noting which way around the piston is fitted **(see illustrations)**.

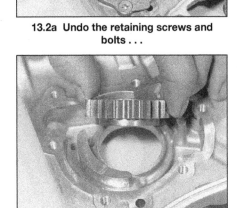

13.2a **Undo the retaining screws and bolts . . .**

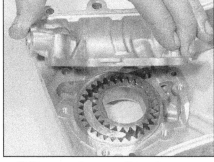

13.2b **. . . and remove the pump cover from the rear of the timing chain cover**

13.3a **. . . then lift out the pump outer gear . . .**

13.3b **. . . and inner gear**

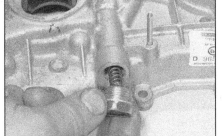

13.4a **Unscrew the pressure regulator valve bolt and sealing washer . . .**

13.4b **. . . and withdraw the spring and piston from the cover**

13.6a Measuring outer gear-to-cover clearance

13.6b Measuring outer gear-to-crescent clearance

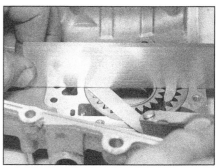

13.7 Checking gear endfloat with a straight-edge and feeler gauge

5 Inspect the pump gears, regulator valve piston and the cover for obvious signs of wear or damage.

6 Fit the gears to the cover and, using feeler blades of the appropriate thickness, measure the clearance between the outer gear and cover, and between the tips of inner and outer gear teeth and the crescent which is positioned between them **(see illustrations)**.

7 Using feeler gauge blades and a straight-edge placed across the top of the cover and the gears, measure the inner and outer gear endfloat **(see illustration)**.

8 If access to the necessary measuring equipment can be gained, measure the diameter of the inner gear and cover bearing surfaces. Subtract the gear outer diameter from the cover inner diameter, and calculate the gear-to-housing clearance.

9 If any measurement is outside the specified limits, or the gears, valve or cover are damaged, the complete timing chain cover assembly should be renewed.

10 If the pump is found to be worn, also check the oil pressure relief valve. To gain access to the relief valve, unscrew the oil filter from the front of the cylinder block. Inspect the valve ball for signs of wear or damage. Depress the valve ball, and check that it moves smoothly and easily, and returns quickly under spring pressure. If not the valve must be renewed. Pull the valve out from the cylinder block, noting which way round it is fitted. Press the new one into position using a suitable tubular spacer. Fit a new oil filter to the engine (see Chapter 1).

11 Lubricate the gears with clean engine oil, and refit them to the pump body. Ensure that the inner gear is fitted with its flange towards the cover.

12 Ensure that the mating surfaces are clean and dry, and refit the pump cover. Fit the cover retaining bolt and screws, and tighten them to the specified torque settings.

13 Fit the pressure regulator valve piston, ensuring it is the correct way around, and install the spring. Fit a new sealing washer to the valve bolt, and tighten the bolt to the specified torque setting.

Refitting

14 Refit the timing chain cover as described in Section 6.

14 Crankshaft oil seals – renewal

Timing chain cover oil seal

1 Remove the crankshaft pulley as described in Section 5.

2 Carefully lever the oil seal out of position, using a large flat-bladed screwdriver, taking care not to damage the oil pump gears or timing cover.

3 Clean the seal housing, and polish off any burrs or raised edges which may have caused the seal to fail in the first place.

4 Lubricate the lips of the new seal with a smear of grease and offer up the seal, ensuring its sealing lip is facing inwards. Carefully ease the seal into position, taking care not to damage its sealing lip. Drive the seal into position until it seats on its locating shoulder, using a suitable tubular drift, such as a socket, which bears only on the hard outer edge of the seal. Take care not to damage the seal lips during fitting. Note that the seal lips should face inwards.

5 Wash off any traces of oil, then refit the crankshaft pulley as described in Section 5.

Flywheel/driveplate oil seal

6 Remove the flywheel or driveplate, as applicable, as described in Section 15.

7 Taking care not to mark either the crankshaft or any part of the cylinder block/crankcase, lever the seal evenly out of its housing using a large flat-bladed screwdriver.

8 Clean the seal housing, and polish off any burrs or raised edges which may have caused the seal to fail in the first place.

9 Lubricate with grease the lips of the new seal and the crankshaft shoulder, then offer up the seal to the cylinder block/crankcase.

10 Ease the sealing lip of the seal over the crankshaft shoulder by hand only, and press the seal evenly into its housing until its outer flange seats evenly on the housing lip. If necessary, a soft-faced mallet can be used to tap the seal gently into place.

11 Wash off any traces of oil, then refit the flywheel/driveplate as described in Section 15.

15 Flywheel/driveplate – removal, inspection and refitting

Removal

1 Remove the transmission as described in Chapter 7A or 7B, as applicable then, on manual transmission models, remove the clutch assembly as described in Chapter 6.

2 Prevent the flywheel/driveplate from turning by locking the ring gear teeth **(see illustration)**. Alternatively, bolt a strap between the flywheel/driveplate and the cylinder block.

3 Slacken and remove the retaining bolts, and remove the flywheel/driveplate from the end of the crankshaft. Do not drop it, as it is very heavy.

4 If necessary, remove the cover plate from the cylinder block, noting which way round it is fitted. If the cover plate dowels are a loose fit in the block, remove them and store them with the plate for safe-keeping.

Inspection

5 On manual transmission models, if the flywheel's clutch mating surface is deeply scored, cracked or otherwise damaged, the flywheel must be renewed. However, it may be possible to have it surface-ground; seek the advice of a Nissan dealer or engine reconditioning specialist.

6 If the ring gear is badly worn or has missing teeth, it must be renewed. This job is best left

15.2 Use the fabricated tool shown to lock the flywheel ring gear and prevent crankshaft rotation

to a Nissan dealer or engine reconditioning specialist. The temperature to which the new ring gear must be heated for installation is critical and, if not done accurately, the hardness of the teeth will be destroyed.

Refitting

7 Install the locating dowels (where removed) and refit the cover plate to the cylinder block.

8 Clean the mating surfaces of the flywheel/driveplate and crankshaft.

9 Offer up the flywheel/driveplate, and refit the retaining bolts.

10 Lock the ring gear using the method employed on dismantling, and tighten the retaining bolts to the specified torque.

11 On manual transmission models, refit the clutch as described in Chapter 6. Remove the locking tool, and refit the transmission as described in Chapter 7A or 7B.

16 Engine/transmission mountings – inspection and renewal

Inspection

1 If improved access is required, firmly apply the handbrake, then jack up the front of the vehicle and support it securely on axle stands (see *Jacking and vehicle support*).

2 Check the mounting rubber to see if it is cracked, hardened or separated from the metal at any point; renew the mounting if any such damage or deterioration is evident **(see illustration)**.

3 Check that all the mounting's fasteners are securely tightened; use a torque wrench to check if possible.

4 Using a large screwdriver or a crowbar, check for wear in the mounting by carefully levering against it to check for free play. Where this is not possible, enlist the aid of an assistant to move the engine/transmission back-and-forth, or from side-to-side, while you watch the mounting. While some free play is to be expected even from new components, excessive wear should be obvious. If excessive free play is found, check first that the fasteners are correctly secured, then renew any worn components as described below.

Renewal

Right-hand mounting

5 Disconnect the battery negative terminal (refer to *Disconnecting the battery* in the Reference Chapter).

6 Place a jack beneath the engine, with a block of wood on the jack head. Raise the jack until it is supporting the weight of the engine.

7 Slacken and remove the three retaining bolts, and remove the mounting bracket from the top of the right-hand engine/transmission mounting.

8 Unscrew the nut and through-bolt, then undo the three retaining bolts and remove the right-hand mounting assembly from the engine compartment. Recover the rubbers which are fitted to each side of the body mounting bracket.

9 If necessary, undo the retaining bolts and remove the right-hand engine/transmission mounting bracket from the engine.

10 Check carefully for signs of wear or damage on all components, and renew them where necessary.

11 On refitting, fit the mounting bracket (where removed) to the engine, and securely tighten its retaining bolts.

12 Fit the rubbers to the body mounting bracket, ensuring that their pins are correctly seated in the bracket holes. Fit the mounting to the top of the engine bracket, tightening its retaining bolts to the specified torque setting.

13 Align the right-hand mounting with the body bracket, then insert the through-bolt and tighten its nut to the specified torque setting. Remove the jack from underneath the engine.

14 Refit the bracket to the top of the right-hand engine/transmission mounting, and tighten its bolts to the specified torque setting

15 Reconnect the battery negative terminal.

Left-hand mounting

16 Remove the battery as described in Chapter 5A.

17 Place a jack and block of wood beneath the transmission, and raise the jack to take the weight of the transmission.

18 Slacken and remove the through-bolt, then undo the three bolts and remove the left-hand mounting from the transmission. Recover the rubbers from each side of the mounting bracket and, if necessary, unbolt the mounting bracket from the vehicle body.

19 Check carefully for signs of wear or damage on all components, and renew them where necessary.

20 On refitting, fit the mounting bracket (where removed) and securely tighten its retaining bolts.

21 Refit the rubbers to the mounting bracket, ensuring that their pins are correctly seated in the bracket holes, and manoeuvre the mounting into position. Fit the bolts securing the mounting to the transmission, and tighten them to the specified torque setting.

22 Align the left-hand mounting with its bracket, then insert the through-bolt and tighten its nut to the specified torque setting.

23 Remove the jack from underneath the engine, and refit the battery.

Front mounting

24 If not already done, firmly apply the handbrake, then jack up the front of the vehicle and support it securely on axle stands (see *Jacking and vehicle support*).

25 Disconnect the battery negative terminal (refer to *Disconnecting the battery* in the Reference Chapter), then undo the retaining screws and, if necessary, remove the engine undershields to improve access.

26 Using a suitable marker pen, mark the outline of the front engine/transmission through-bolt on the mounting bracket to use as a guide on refitting.

27 Place a jack beneath the engine, with a block of wood on the jack head. Raise the jack until it is supporting the weight of the engine.

28 Slacken and remove the nut and through-bolt securing the mounting to the mounting bracket, then slacken and remove the bolt securing the lower end of the mounting to the centre member. Withdraw the mounting from its location.

29 Undo the retaining bolts, and remove the mounting bracket from the front of the engine.

30 Check carefully for signs of wear or damage on all components, and renew them where necessary.

31 On refitting, fit the mounting bracket to the engine, and tighten its retaining bolts to the specified torque.

32 Locate the mounting in position, fit the lower retaining bolt and lightly tighten it.

33 Refit the through-bolt to the mounting, and lightly tighten its nut. Position the engine/transmission so that the front mounting through-bolt is correctly aligned with the mark made prior to removal, then tighten the through-bolt to the specified torque.

34 Tighten the lower retaining bolt to the specified torque.

35 Remove the jack from underneath the transmission.

36 Refit the undershields, tighten their fasteners securely, then lower the vehicle to the ground and reconnect the battery.

Rear mounting

37 If not already done, firmly apply the handbrake, then jack up the front of the vehicle and support it securely on axle stands (see *Jacking and vehicle support*).

38 Disconnect the battery negative terminal (refer to *Disconnecting the battery* in the Reference Chapter).

39 Slacken and remove the through-bolt from the rear engine/transmission mounting.

40 Undo the bolts securing the mounting bracket in position, and manoeuvre it away from the engine/transmission.

41 Undo the two retaining bolts, and remove the mounting assembly from the centre member.

42 Check carefully for signs of wear or damage on all components, and renew them where necessary.

43 On reassembly, fit the rear mounting to the centre member, and tighten its retaining bolts to the specified torque.

44 Refit the mounting bracket and tighten its retaining bolts to the specified torque.

45 Align the rear mounting with its bracket, then insert the through-bolt and tighten its nut to the specified torque setting.

46 Lower the vehicle to the ground and reconnect the battery.

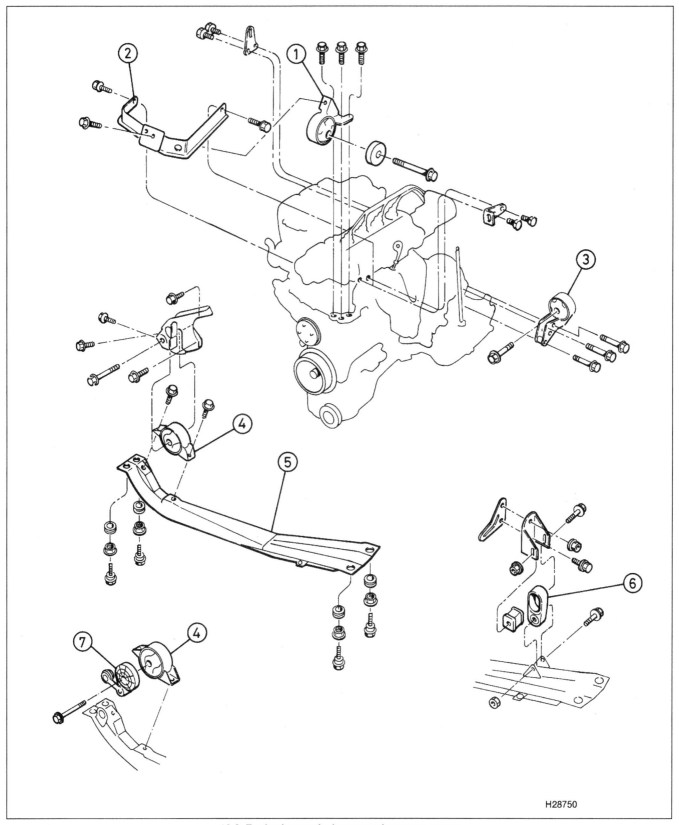

16.2 Engine/transmission mounting components

1 Right-hand mounting
2 Right-hand mounting bracket
3 Left-hand mounting
4 Rear mounting
5 Centre member
6 Front mounting
7 Roll damper (where fitted)

H28750

Chapter 2 Part B:
Engine removal and overhaul procedures

Contents

Degrees of difficulty

Easy, suitable for novice with little experience	Fairly easy, suitable for beginner with some experience	Fairly difficult, suitable for competent DIY mechanic	Difficult, suitable for experienced DIY mechanic	Very difficult, suitable for expert DIY or professional

Specifications

Cylinder head

Maximum gasket face distortion . . . . 0.1 mm
Cylinder head height . . . . 117.8 to 118.0 mm

Valves

Valve head diameter:
  Inlet:
    1.4 litre engine . . . . 28.9 to 29.1 mm
    1.6 litre engine . . . . 29.9 to 30.1 mm
  Exhaust . . . . 23.9 to 24.1 mm
Valve stem diameter:
  Inlet . . . . 5.465 to 5.480 mm
  Exhaust . . . . 5.445 to 5.460 mm
Overall length:
  Inlet . . . . 92.00 to 92.50 mm
  Exhaust . . . . 92.37 to 92.87 mm
Valve stem-to-guide clearance:
  Inlet . . . . 0.020 to 0.050 mm
  Exhaust . . . . 0.040 to 0.070 mm
Valve spring free length . . . . 41.19 mm
Valve spring out-of-square limit . . . . 1.80 mm

Cylinder block

Cylinder bore diameter:
  1.4 litre engine:
    Nominal size . . . . 73.0 mm
    Oversizes available . . . . 0.5 mm and 1.0 mm
  1.6 litre engine:
    Nominal size . . . . 76.0 mm
    Oversizes available . . . . 0.5 mm and 1.0 mm

Pistons

Piston diameter:
1.4 litre engine (measured 9.5 mm up from the base of skirt):
    Nominal size .... 73.575 to 74.105 mm
    0.5 mm oversize piston .... 74.075 to 74.105 mm
    1.0 mm oversize piston .... 74.575 to 74.605 mm
1.6 litre engine (measured 9.5 mm up from the base of skirt):
    Nominal size .... 75.975 to 76.005 mm
    0.5 mm oversize piston .... 76.475 to 76.505 mm
    1.0 mm oversize piston .... 76.975 to 77.005 mm
Piston-to-bore clearance .... 0.015 to 0.035 mm

Piston rings

Ring-to-groove clearance:
Standard:
    Top compression ring .... 0.040 to 0.085 mm
    Second compression ring .... 0.030 to 0.070 mm
Service limit .... 0.2 mm
End gaps:
Standard:
    Top compression ring .... 0.20 to 0.40 mm
    Second compression ring .... 0.35 to 0.55 mm
    Oil control ring .... 0.25 to 1.00 mm
Service limit:
    Top compression ring .... 0.49 mm
    Second compression ring .... 0.64 mm
    Oil control ring .... 1.09 mm

Crankshaft

Endfloat:
    Standard .... 0.06 to 0.18 mm
    Service limit .... 0.3 mm
Main bearing running clearance:
    Standard .... 0.018 to 0.042 mm
    Service limit .... 0.1 mm
Big-end bearing running clearance:
    Standard .... 0.010 to 0.035 mm
    Service limit .... 0.1 mm

Torque wrench settings

Refer to Chapter 2A Specifications.

1 General information

Included in this Part of Chapter 2 are details of removing the engine/transmission from the vehicle, and general overhaul procedures for the cylinder head, cylinder block/crankcase and all other engine internal components.

The information given ranges from advice concerning preparation for an overhaul and the purchase of new parts, to detailed step-by-step procedures covering removal, inspection, renovation and refitting of engine internal components.

After Section 5, all instructions are based on the assumption that the engine has been removed from the vehicle. For information concerning in-car engine repair, as well as the removal and refitting of those external components necessary for full overhaul, refer to Part A of this Chapter and to Section 5. Ignore any preliminary dismantling operations described in Part A that are no longer relevant once the engine has been removed.

Apart from torque wrench settings, which are given at the beginning of Part A, all specifications relating to engine overhaul are at the beginning of this Part of Chapter 2.

Engine overhaul

It is not always easy to determine when, or if, an engine should be completely overhauled, as a number of factors must be considered.

High mileage is not necessarily an indication that an overhaul is needed, while low mileage does not preclude the need for an overhaul. Frequency of servicing is probably the most important consideration. An engine which has had regular and frequent oil and filter changes, as well as other required maintenance, should give many thousands of miles of reliable service. Conversely, a neglected engine may require an overhaul very early in its life.

Excessive oil consumption is an indication that piston rings, valve seals and/or valve guides are in need of attention. Make sure that oil leaks are not responsible before deciding that the rings and/or guides are worn. Perform a compression test, as described in Part A of this Chapter, to determine the likely cause of the problem.

Check the oil pressure with a gauge fitted in place of the oil pressure switch, and compare it with that specified in Part A of this Chapter. If it is extremely low, the main and big-end bearings, and/or the oil pump, are probably worn out.

Loss of power, rough running, knocking or metallic engine noises, excessive valve gear noise, and high fuel consumption may also point to the need for an overhaul, especially if they are all present at the same time. If a complete service does not remedy the situation, major mechanical work is the only solution.

An engine overhaul involves restoring all internal parts to the specification of a new engine. During an overhaul, the cylinder bores are rebored (where necessary) and the pistons and piston rings are renewed. New main and big-end bearings are generally fitted; if necessary, the crankshaft may be reground, to restore the journals. The valves are also serviced as well, since they are usually in less-than-perfect condition at this point. The end result should be an as-new engine that will give many trouble-free miles.

Note: *Critical cooling system components such as the hoses, thermostat and coolant pump should be renewed when an engine is overhauled. The radiator should be checked carefully, to ensure that it is not clogged or leaking. Also, it is a good idea to renew the oil pump whenever the engine is overhauled.*

Before beginning the engine overhaul, read through the entire procedure, to familiarise yourself with the scope and requirements of the job. Check on the availability of parts, and make sure that any necessary special tools and equipment are obtained in advance. Most work can be done with typical hand tools, although a number of precision measuring tools are required for inspecting parts to determine if they must be renewed.

The services provided by an engineering machine shop or engine reconditioning specialist will almost certainly be required, particularly if major repairs such as crankshaft regrinding or cylinder reboring are necessary. Apart from carrying out machining operations, these establishments will normally handle the inspection of parts, offer advice concerning reconditioning or renewal and supply new components such as pistons, piston rings and bearing shells. It is recommended that the establishment used is a member of the Federation of Engine Re-Manufacturers, or a similar society.

Always wait until the engine has been completely dismantled, and until all components (especially the cylinder block and the crankshaft) have been inspected, before deciding what service and repair operations must be performed by an automotive engineering works. The condition of these components will be the major factor to consider when determining whether to overhaul the original engine, or to buy a reconditioned unit. Do not, therefore, purchase parts or have overhaul work done on other components until they have been thoroughly inspected.

As a final note, to ensure maximum life and minimum trouble from a reconditioned engine, everything must be assembled with care, in a spotlessly-clean environment.

2 Engine removal – methods and precautions

If you have decided that the engine must be removed for overhaul or major repair work, several preliminary steps should be taken.

Locating a suitable place to work is extremely important. Adequate work space, along with storage space for the vehicle, will be needed. If a workshop or garage is not available, at the very least, a flat, level, clean work surface is required.

Cleaning the engine compartment and

engine/transmission before beginning the removal procedure will help keep tools clean and organised.

An engine hoist will also be necessary. Make sure the equipment is rated in excess of the combined weight of the engine and transmission. Safety is of primary importance, considering the potential hazards involved in removing the engine/transmission from the vehicle.

The help of an assistant is essential. Apart from the safety aspects involved, there are many instances when one person cannot simultaneously perform all of the operations required during engine/transmission removal.

Plan the operation ahead of time. Before starting work, arrange for the hire of, or obtain, all of the tools and equipment you will need. Some of the equipment necessary to perform engine/transmission removal and installation safely (in addition to an engine hoist) is as follows: a heavy-duty trolley jack, complete sets of spanners and sockets as described at the rear of this manual, wooden blocks, and plenty of rags and cleaning solvent for mopping-up spilled oil, coolant and fuel. If the hoist must be hired, make sure that you arrange for it in advance, and perform all of the operations possible without it beforehand. This will save you money and time.

Plan for the vehicle to be out of use for quite a while. An engineering machine shop or engine reconditioning specialist will be required to perform some of the work which cannot be accomplished without special equipment. These places often have a busy schedule, so it would be a good idea to consult them before removing the engine, in order to accurately estimate the amount of time required to rebuild or repair components that may need work.

During the engine/transmission removal procedure, it is advisable to make notes of the locations of all brackets, cable ties, earthing points, etc, as well as how the wiring harnesses, hoses and electrical connections are attached and routed around the engine and engine compartment. An effective way of doing this is to take a series of photographs of the various components before they are disconnected or removed. A simple inexpensive disposable camera is ideal for this and the resulting photographs will prove invaluable when the engine is refitted.

Always be extremely careful when removing and refitting the engine/transmission. Serious injury can result from careless actions. Plan ahead and take your time, and a job of this nature, although major, can be accomplished successfully.

The engine and transmission assembly is removed downwards from the engine compartment on all models described in this manual.

3 Engine and manual transmission – removal, separation, reconnection and refitting

Note: *The engine can be removed from the car only as a complete unit with the transmission; the two are then separated for overhaul. The engine/transmission unit is lowered out of position, and withdrawn from under the vehicle. To allow adequate clearance there should be at least 75 cm between the front bumper and the ground when the vehicle is raised and supported.*

Removal

1 Depressurise the fuel system as described in Chapter 4A.

2 Disconnect the battery negative terminal (refer to *Disconnecting the battery* in the Reference Chapter).

3 Firmly apply the handbrake, then jack up the front of the vehicle and support it securely on axle stands, bearing in mind the note at the start of this Section (see *Jacking and vehicle support*). Remove both front roadwheels.

4 Undo all the retaining screws, and remove the undershields from underneath and around the engine.

5 Remove the bonnet as described in Chapter 11.

6 Drain the cooling system (see Chapter 1), saving the coolant if it is fit for re-use.

7 Drain the transmission oil as described in Chapter 1. Refit the drain and filler plugs, and tighten them to their specified torque settings.

8 If the engine is to be dismantled, working as described in Chapter 1, drain the oil and if required remove the oil filter. Clean and refit the drain plug, tightening it to the specified torque.

9 Remove the radiator, complete with hoses and cooling fans, as described in Chapter 3.

10 Working as described in Chapter 5A, remove the alternator and disconnect the wiring from the starter motor.

11 Remove the power steering pump as described in Chapter 10.

12 Carry out the following operations as described in Chapter 4A.

a) *Remove the air cleaner assembly.*

b) *Disconnect the fuel feed and return hoses from the fuel rail (plug all openings, to prevent loss of fuel and entry of dirt into the fuel system).*

c) *Disconnect the accelerator cable.*

d) *Disconnect the relevant electrical connectors from the throttle housing, inlet manifold and associated components. Free the wiring from the manifold, and position it clear of the cylinder head so that it does not hinder removal.*

e) *Disconnect the vacuum servo unit hose, coolant hose(s), and all the other relevant/breather hoses from the manifold and associated valves.*

f) *Remove the inlet manifold support bracket(s).*

g) *Remove the exhaust front pipe.*

3.20a Prior to removal, mark the position of the front engine/ transmission mounting through-bolt (arrowed) on its bracket

3.20b Slacken and remove the rear engine/transmission mounting through-bolt (arrowed)

13 Slacken the retaining clips, and disconnect the heater hoses and all other relevant cooling system hoses from the engine, noting each hose's correct fitted location.

14 On models with air conditioning, unbolt the compressor and position it clear of the engine. Support the weight of the compressor by tying it to the vehicle body, to prevent any excess strain being placed on the compressor lines whilst the engine is removed. **Do not** disconnect the refrigerant lines from the compressor (see the warnings given in Chapter 3).

15 Referring to Chapter 5B, disconnect the wiring connectors from the distributor body. Free the wiring loom from any relevant retaining clips, so that it is free from the cylinder head and will not hinder the removal procedure. Also undo the retaining bolts, and disconnect all the relevant earth leads from the head and inlet manifold.

16 Working as described in Chapter 8, remove the driveshafts.

17 Disconnect the clutch cable from the transmission as described in Chapter 6.

18 Working as described in Chapter 7A, disconnect the gearchange linkage link rods and the wiring connector(s) from the transmission.

19 Manoeuvre the engine hoist into position, and attach it to the cylinder head using suitable lifting brackets. Raise the hoist until it is supporting the weight of the engine.

20 Using a suitable marker pen, mark the outline of the front engine/transmission through-bolt on the mounting bracket to use as a guide on refitting. Slacken and remove the nut and withdraw the through-bolt from the mounting. Slacken and remove the nut and through-bolt from the rear engine/ transmission mounting **(see illustrations)**.

21 Slacken and remove the four bolts and washers securing the centre member to the

vehicle body, and lower the assembly away from the engine **(see illustrations)**.

22 Undo the three retaining bolts and remove the mounting bracket from the top of the right-hand engine/transmission mounting.

23 Unscrew the nut and through-bolt from the right-hand engine/transmission mounting, then undo the retaining bolts and remove the mounting assembly from the engine compartment. Recover the rubbers which are fitted to each side of the body mounting bracket, if they are loose.

24 Slacken and remove the through-bolt from the left-hand engine/transmission mounting. Undo the three bolts securing the mounting to the transmission, and manoeuvre the mounting out of position. Recover the rubbers from each side of the mounting bracket, if they are loose.

25 Make a final check that any components which would prevent the removal of the engine/transmission from the car have been

3.21a Undo the four bolts and washers . . .

3.21b . . . and remove the centre member from underneath the engine/transmission

3.26 Carefully lower the engine/transmission downwards and out of position

removed or disconnected. Ensure that components such as the gearchange link rods are secured so that they cannot be damaged on removal.

26 If available, a low trolley should be placed under the engine/transmission assembly, to facilitate its easy removal from under the vehicle. Lower the engine/transmission assembly, making sure that nothing is trapped or damaged. Note that it may be necessary to tilt the assembly slightly to clear the body panels **(see illustration)**. Great care must be taken to ensure that no components are trapped and damaged during the removal procedure.

27 Withdraw the assembly from under the vehicle.

Separation

28 Unscrew the retaining bolts, and remove the starter motor from the transmission.

29 Ensure that both engine and transmission are adequately supported, then slacken and remove the bolts securing the transmission housing to the engine. Note the correct fitted positions of each bolt (and, where fitted, the relevant brackets) as they are removed, to use as a reference on refitting.

30 Carefully withdraw the transmission from the engine, ensuring that the weight of the transmission is not allowed to hang on the input shaft while it is engaged with the clutch friction disc.

31 If they are loose, remove the locating dowels from the engine or transmission, and keep them in a safe place.

Reconnection

32 Apply a smear of high-melting-point grease to the splines of the transmission input shaft. Do not apply too much, otherwise there is a possibility of the grease contaminating the clutch friction plate.

33 Ensure that the locating dowels are correctly positioned in the engine or transmission, and that the release bearing is correctly engaged with the fork.

34 Carefully offer the transmission to the engine, until the locating dowels are engaged. Ensure that the weight of the transmission is not allowed to hang on the input shaft as it is engaged with the clutch friction plate.

35 Refit the transmission housing-to-engine

bolts, ensuring that all the necessary brackets are correctly positioned, and tighten them to the specified torque setting.

36 Refit the starter motor and tighten the retaining bolts.

Refitting

37 Position the engine/transmission assembly under the vehicle, then reconnect the hoist and lifting tackle to the engine lifting brackets.

38 Lift the assembly up into the engine compartment, making sure that it clears the surrounding components.

39 Refit the rubbers to the left-hand engine/transmission mounting bracket, ensuring that their pins are correctly seated in the bracket holes. Manoeuvre the mounting into position, then fit the bolts securing it to the transmission and tighten them to the specified torque setting. Insert the through-bolt and nut, tightening it by hand only at this stage.

40 Fit the rubbers to the right-hand body mounting bracket, ensuring that their pins are correctly seated in the bracket holes. Refit the mounting to the top of its bracket, and tighten its retaining bolts to the specified torque setting. Insert the through-bolt and nut, tightening it by hand only at this stage.

41 Ensure that all the mounting rubbers are in position. Manoeuvre the centre member into place, aligning it with the engine mountings, and refit its mounting bolts and washers. Tighten the centre member mounting bolts to the specified torque.

42 Refit the through-bolt and nut to the rear engine/transmission mounting, tightening it by hand only.

43 Refit the through-bolt and nut to the front engine/transmission mounting. Position the engine/transmission so that the front mounting through-bolt is correctly aligned with the mark made prior to removal, then tighten to the specified torque setting.

44 Rock the engine/transmission to settle it in position, then tighten all the engine/transmission mounting through-bolts to their specified torque settings.

45 Refit the mounting bracket to the top of the right-hand mounting, and tighten its retaining bolts to the specified torque.

46 The remainder of the refitting procedure is a direct reversal of the removal sequence, noting the following points:

a) Ensuring that the wiring harness is correctly routed and retained by all the relevant retaining clips, and all connectors are correctly and securely reconnected.

b) Prior to refitting the driveshafts to the transmission, renew the driveshaft oil seals as described in Chapter 7A.

c) Ensure that all coolant hoses are correctly reconnected and securely retained by their retaining clips.

d) Adjust the accelerator cable as described in Chapter 4A.

e) Connect and adjust the clutch cable as described in Chapter 6.

f) Refill the engine and transmission unit with correct quantity and type of lubricant, as described in the relevant Sections of Chapter 1.

g) Refill the cooling system as described in Chapter 1.

h) On completion, start the engine and check for leaks.

4 Engine and automatic transmission – removal, separation, reconnection and refitting

Note: *The engine can be removed from the car only as a complete unit with the transmission; the two are then separated for overhaul. The engine/transmission is lowered out of position, and withdrawn from under the front of the vehicle. To allow adequate clearance, there should be at least 75 cm between the front bumper and the ground when the vehicle is raised and supported.*

Removal

1 Carry out the operations described in paragraphs 1 to 16 of Section 3.

2 Carry out the following operations as described in Chapter 7B.

a) Disconnect the selector cable from the transmission.

b) Disconnect the kickdown cable from the throttle body/housing.

c) Disconnect the transmission wiring connectors.

d) Disconnect the fluid cooler hoses from the transmission.

3 Manoeuvre the engine hoist into position, and attach it to the cylinder head using suitable lifting brackets bolted. Raise the hoist until it is supporting the weight of the engine.

4 Slacken and remove the nut and through-bolt from the rear engine/transmission mounting.

5 Slacken and remove the four bolts and washers securing the centre member to the vehicle body, and lower the assembly away from the engine. Recover the stopper ring which is fitted between the rear engine/transmission mounting and its bracket.

6 Unscrew the nut and through-bolt from the right-hand engine/transmission mounting, then undo the retaining bolts and remove the mounting assembly from the engine compartment. Recover the rubbers which are fitted to each side of the body mounting bracket.

7 Slacken and remove the through-bolt from the left-hand engine/transmission mounting. Undo the three bolts securing the mounting to the transmission, and manoeuvre the mounting out of position. Recover the rubbers from each side of the mounting bracket.

8 Make a final check that any components which would prevent the removal of the engine/transmission from the car have been removed or disconnected.

9 If available, a low trolley should be placed under the engine/transmission assembly to facilitate its removal from under the vehicle. Lower the engine/transmission assembly, making sure that nothing is trapped or damaged. Note that it may be necessary to tilt the assembly slightly to clear the body panels. Great care must be taken to ensure that no components are trapped and damaged during the removal procedure.

10 Withdraw the assembly from under the vehicle.

Separation

11 Unscrew the retaining bolts, and remove the starter motor from the transmission.

12 Undo the retaining bolts, and remove the cover plate from the sump flange to gain access to the torque converter retaining bolts. Slacken and remove the visible bolt then, using a socket and extension bar to rotate the crankshaft pulley, undo the remaining bolts securing the torque converter to the driveplate as they become accessible. There are four bolts in total.

13 To ensure that the torque converter does not fall out as the transmission is removed, secure it in position using a length of metal strip bolted to one of the starter motor bolt holes.

14 Ensure that both engine and transmission are adequately supported, then slacken and remove the bolts securing the transmission housing to the engine. Note the correct fitted positions of each bolt (and, where fitted, the relevant brackets) as they are removed, to use as a reference on refitting.

15 Carefully withdraw the transmission from the engine. If they are loose, remove the locating dowels from the engine or transmission, and keep them in a safe place.

Reconnection

16 Prior to joining, ensure that the torque converter is correctly engaged with the transmission. This can be checked by measuring the distance from the converter mounting bolt holes to the transmission mating surface; if the converter is correctly seated, this distance will be at least 21.1 mm.

17 Ensure that the locating dowels are correctly positioned in the engine or transmission.

18 Carefully offer the transmission to the engine, and engage it on the locating dowels. Refit the transmission housing-to-engine bolts, ensuring that all the necessary brackets are correctly positioned, and tighten them to the specified torque settings.

19 Remove the torque converter retaining strap (where fitted) installed prior to removal. Align the torque converter holes with the those in the driveplate, and install the retaining bolts.

20 Tighten the torque converter retaining bolts to the specified torque setting, then refit the cover plate to the sump and securely tighten its retaining bolts.

21 Refit the starter motor and tighten the retaining bolts.

Refitting

22 Position the engine/transmission assembly under the vehicle, then reconnect the hoist and lifting tackle to the engine lifting brackets.

23 Lift the assembly up into the engine compartment, making sure that it clears the surrounding components.

24 Refit the rubbers to the left-hand engine/transmission mounting bracket, ensuring that their pins are correctly seated in the bracket holes. Manoeuvre the mounting into position, then fit the bolts securing it to the transmission and tighten them to the specified torque setting. Insert the through-bolt and nut, tightening it by hand only at this stage.

25 Fit the rubbers to the right-hand body mounting bracket, ensuring that their pins are correctly seated in the bracket holes. Refit the mounting to the top of its bracket, and tighten its retaining bolts to the specified torque setting. Insert the through-bolt and nut, tightening it by hand only at this stage.

26 Ensure that all the mounting rubbers are in position. Manoeuvre the centre member into place, aligning it with the engine mounting, and refit its mounting bolts and washers. Tighten the mounting bolts to the specified torque.

27 Refit the through-bolt and nut to the rear engine/transmission mounting, tightening it by hand only.

28 Rock the engine/transmission to settle it in position, then tighten all the engine/transmission mounting through-bolts to their specified torque settings.

29 The remainder of the refitting procedure is a direct reversal of the removal sequence, noting the following points:

a) *Ensuring that the wiring harness is correctly routed and retained by all the relevant retaining clips, and all connectors are correctly and securely reconnected.*

b) *Prior to refitting the driveshafts to the transmission, renew the driveshaft oil seals as described in Chapter 7B.*

c) *Ensure that all coolant hoses are correctly reconnected and securely retained by their retaining clips.*

d) *Adjust the accelerator cable as described in the relevant Part of Chapter 4.*

e) *Connect and adjust the selector and kickdown cables as described in Chapter 7B.*

f) *Refill the engine and transmission with correct quantity and type of lubricant, as described in the relevant Sections of Chapter 1.*

g) *Refill the cooling system as described in Chapter 1.*

h) *On completion, start the engine and check for leaks.*

5 Engine overhaul –
dismantling sequence

It is preferable to dismantle and work on the engine with it mounted on a portable engine stand. These stands can often be hired from a tool hire shop. Before the engine is mounted on a stand, the flywheel/driveplate should be removed, so that the stand bolts can be tightened into the end of the cylinder block/crankcase.

If a stand is not available, it is possible to dismantle the engine with it blocked up on a sturdy workbench, or on the floor. Be extra-careful not to tip or drop the engine when working without a stand.

If a reconditioned engine is to be obtained, or if the original engine is to be overhauled, the external components in the following list must be removed first. These components can then be transferred to the reconditioned engine, or refitted to the existing engine after overhaul **(see illustration overleaf)**.

a) *Alternator, power steering pump and/or air conditioning compressor mounting brackets (as applicable).*

b) *Distributor, HT leads and spark plugs (Chapters 1 and 5B).*

c) *Coolant pump and thermostat/coolant outlet housing(s) (Chapter 3).*

d) *Fuel system components (Chapter 4A).*

e) *All electrical switches and sensors, and the engine wiring harness.*

f) *Inlet and exhaust manifolds (Chapter 4A).*

g) *Engine mountings (Part A of this Chapter).*

h) *Flywheel/driveplate (Part A of this Chapter).*

Note: *When removing the external components from the engine, pay close attention to details that may be helpful or important during refitting. Note the fitted position of gaskets, seals, spacers, pins, washers, bolts, and other small items.*

If a 'short' engine is to be obtained (cylinder block, crankshaft, pistons and connecting rods all assembled), then the cylinder head, sump, oil pump, and timing chains will have to be removed also.

If a complete overhaul of the existing engine is being undertaken, the engine can be dismantled, in the order given below, referring to Part A of this Chapter unless otherwise stated.

a) *Inlet and exhaust manifolds (Chapter 4A).*

b) *Sump.*

c) *Timing chain(s) and sprockets.*

d) *Cylinder head.*

e) *Flywheel/driveplate.*

f) *Piston/connecting rod assemblies (Section 9 of this Chapter).*

g) *Crankshaft (Section 10 of this Chapter).*

6 Cylinder head –
dismantling

Note: *New and reconditioned cylinder heads are available from the manufacturer, and from engine reconditioning specialists. Some specialist tools are required for dismantling*

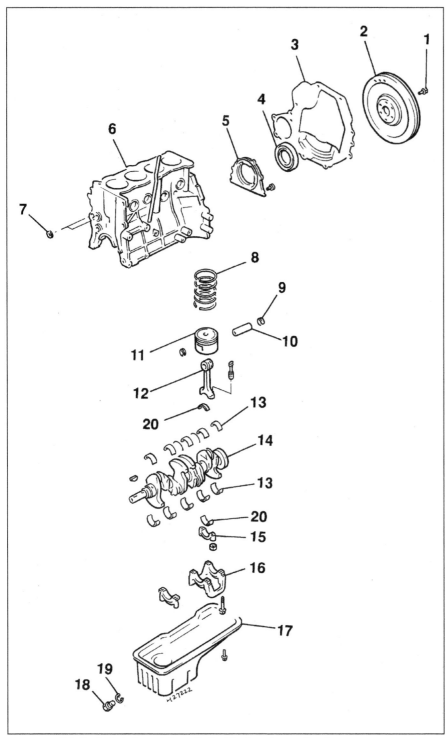

5.3 Exploded view of the cylinder block and associated components

1 Flywheel/driveplate retaining bolt
2 Flywheel/driveplate
3 Cover plate
4 Oil seal
5 Oil seal housing
6 Cylinder block
7 Locating dowels and O-rings

8 Piston rings
9 Circlips
10 Gudgeon pin
11 Piston
12 Connecting rod
13 Main bearing inserts and thrustwashers
14 Crankshaft

15 Connecting rod big-end cap
16 Main bearing caps
17 Sump
18 Drain plug
19 Sealing washer
20 Big-end bearing inserts

and inspection, and new components may not be readily available. It may therefore be more practical and economical to obtain a reconditioned head, rather than overhaul the original head.

1 Remove the cylinder head as described in Part A of this Chapter.

2 If not already done, remove the camshaft followers and shims as described in Part A of this Chapter.

3 Remove the inlet and exhaust manifolds with reference to Chapter 4A.

4 Using a valve spring compressor, compress each valve spring in turn until the split collets can be removed. Release the compressor, and lift off the spring retainer, spring and spring seat. Using a pair of pliers, carefully extract the valve stem seal from the top of the guide **(see illustration)**.

5 If, when the valve spring compressor is screwed down, the spring retainer refuses to free and expose the split collets, gently tap the top of the tool, directly over the retainer, with a light hammer. This will free the retainer.

6 Withdraw the valve through the combustion chamber.

7 It is essential that each valve is stored together with its collets, retainer, spring, and spring seat. The valves should also be kept in their correct sequence, unless they are so badly worn that they are to be renewed. If they are going to be kept and used again, place each valve assembly in a labelled polythene bag or similar small container **(see illustration)**. Note that No 1 valve is nearest to the timing chain end of the engine.

6.4 Pull the valve stem oil seal off the guide using a pair of pliers

6.7 Place each valve and its associated components in a labelled polythene bag

7 Cylinder head and valves – cleaning and inspection

Cleaning

1 Scrape away all traces of old gasket material from the cylinder head.
2 Scrape away the carbon from the combustion chambers and ports, then wash the cylinder head thoroughly with paraffin or a suitable solvent.
3 Scrape off any heavy carbon deposits that may have formed on the valves, then use a power-operated wire brush to remove deposits from the valve heads and stems.

Inspection

Cylinder head

4 Inspect the head very carefully for cracks, evidence of coolant leakage, and other damage. If significant defects are found, a new cylinder head should be obtained.
5 Use a straight-edge and feeler blade to check for distortion of the cylinder head gasket surface **(see illustration)**. If the head is distorted beyond the limit given in the Specifications, seek the advice of an engine reconditioning specialist as to whether machining is possible.
6 Examine the valve seats in each of the combustion chambers. If they are severely pitted, cracked, or burned, they will need to be renewed or recut by an engine reconditioning specialist. If they are only slightly pitted, this can be removed by grinding-in the valve heads and seats with fine valve-grinding compound, as described below.
7 Check the valve guides for wear by inserting the relevant valve, and checking for side-to-side motion of the valve. A very small amount of movement is acceptable. If the movement seems excessive, remove the valve. Measure the valve stem diameter (see below), and renew the valve if it is worn. If the valve stem is not worn, the wear must be in the valve guide, and the guide must be renewed. The renewal of valve guides should be carried out by an engine reconditioning specialist, who will have the necessary tools available.
8 If renewing the valve guides, the valve seats are to be recut or reground only *after* the guides have been fitted.

Valves

9 Examine the head of each valve for pitting, burning, cracks, and general wear. Check the valve stem for scoring and wear ridges. Rotate the valve, and check for any obvious indication that it is bent. Look for pits and excessive wear on the tip of each valve stem. Renew any valve that shows any such signs of wear or damage.
10 If the valve appears satisfactory at this stage, measure the valve stem diameter at

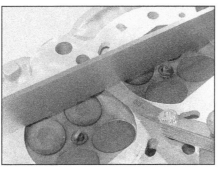

7.5 Using a straight-edge and feeler blade to measure cylinder head gasket face distortion

several points using a micrometer **(see illustration)**. Any significant difference in the readings obtained indicates wear of the valve stem. Should any of these conditions be apparent, the valve(s) must be renewed.
11 If the valves are in satisfactory condition, they should be ground (lapped) into their respective seats, to ensure a smooth, gas-tight seal. If the seat is only lightly pitted, or if it has been recut, fine grinding compound *only* should be used to produce the required finish. Coarse valve-grinding compound should *not* be used, unless a seat is badly burned or deeply pitted. If this is the case, the cylinder head and valves should be inspected by a specialist, to decide whether seat recutting, or even the renewal of the valve or seat insert (where possible) is required.
12 Valve grinding is carried out as follows, with the head supported upside-down on blocks.
13 Smear a trace of (the appropriate grade of) valve-grinding compound on the seat face, and press a suction grinding tool onto the valve head. With a semi-rotary action, grind the valve head to its seat, lifting the valve occasionally to redistribute the grinding compound **(see illustration)**. A light spring placed under the valve head will greatly ease this operation.
14 If coarse grinding compound is being used, work only until a dull, matt even surface is produced on both the valve seat and the valve, then wipe off the used compound, and repeat the process with fine

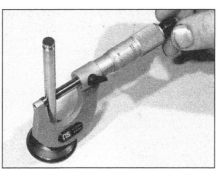

7.10 Measuring a valve stem diameter

compound. When a smooth unbroken ring of light grey matt finish is produced on both the valve and seat, the grinding operation is complete. *Do not* grind-in the valves any further than absolutely necessary, or the seat will be prematurely sunk into the cylinder head.
15 When all the valves have been ground-in, carefully wash off *all* traces of grinding compound using paraffin or a suitable solvent, before reassembling the cylinder head.

Valve components

16 Examine the valve springs for signs of damage and discoloration. The specified Nissan procedure for checking the condition of valve springs involves measuring the force necessary to compress each spring to a specified height. This is not possible without the use of the Nissan special test equipment, and therefore spring checking must be entrusted to a Nissan dealer. A rough idea of the condition of the spring can be gained by measuring the spring free length, and comparing it with the dimension given in the Specifications.
17 Stand each spring on a flat surface, and position a square alongside the edge of the spring. Measure the gap between the upper edge of the spring and the square, and compare it to the out-of-square limit given in the Specifications.
18 If any of the springs are damaged, distorted or have lost their tension, obtain a complete new set of springs. It is normal to renew the valve springs as a matter of course if a major overhaul is being carried out.
19 Renew the valve stem oil seals regardless of their apparent condition.

8 Cylinder head – reassembly

1 Refit the spring seat then, working on the first valve, dip the new valve stem seal in fresh engine oil. Place it on the valve guide and use a suitable socket or metal tube to press the seal firmly onto the guide **(see illustrations)**.
2 Lubricate the stems of the valves, and

7.13 Grinding-in a valve

8.1a Fit the spring seat . . .

8.1b . . . and press a new oil seal onto the valve guide using a suitable socket

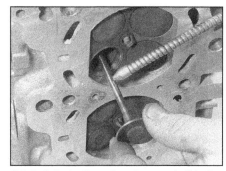

8.2 Lubricate the valve stem, and slide the valve into its respective guide

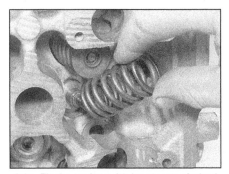

8.3a Fit the spring with its closer-pitched coils at the bottom . . .

8.3b . . . and fit the spring retainer

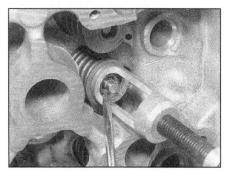

8.4 Install the collets, noting the use of grease to help keep the collets in position

insert the valves into their original locations (see illustration). If new valves are being fitted, insert them into the locations to which they have been ground.

3 Locate the valve spring on top of its seat, ensuring that the spring is fitted with its closer-pitched coils at the bottom, then refit the spring retainer (see illustrations).

4 Compress the valve spring, and locate the split collets in the recess in the valve stem (see illustration). Release the compressor, then repeat the procedure on the remaining valves.

 HAYNES HiNT *Use a little dab of grease to hold the collets in position on the valve stem while the spring compressor is released.*

9.3 Connecting rods should be stamped with their relevant cylinder number (arrowed)

5 With all the valves installed, place the cylinder head on blocks on the bench and, using a hammer and interposed block of wood, tap the end of each valve stem to settle the components.

6 The cylinder head and associated components may now be refitted as described in Part A of this Chapter, and in Chapter 4A.

9 Piston/connecting rod assembly – removal

1 Remove the sump, timing chain(s) and cylinder head as described in Part A of this Chapter.

2 If there is a pronounced wear ridge at the top of any bore, it may be necessary to remove it with a scraper or ridge reamer, to avoid piston damage during removal. Such a ridge indicates excessive wear of the cylinder bore.

3 Each connecting rod and bearing cap should be stamped with its respective cylinder number, No 1 cylinder being at the timing chain end of the engine (see illustration). If no markings are visible, using quick-drying paint or similar, mark each connecting rod and big-end bearing cap with its respective cylinder number on the flat machined surface provided.

4 Turn the crankshaft to bring pistons 1 and 4 to BDC (bottom dead centre).

5 Unscrew the nuts from No 1 piston big-end bearing cap. Take off the cap, and recover the

bottom half bearing shell. If the bearing shells are to be re-used, tape the cap and the shell together.

6 Using a hammer handle, push the piston up through the bore, and remove it from the top of the cylinder block. Recover the bearing shell, and tape it to the connecting rod for safe-keeping.

7 Loosely refit the big-end cap to the connecting rod, and secure with the nuts – this will help to keep the components in their correct order.

8 Remove No 4 piston assembly in the same way.

9 Turn the crankshaft through 180° to bring pistons 2 and 3 to BDC (bottom dead centre), and remove them in the same way.

10 Crankshaft – removal

1 Remove the sump, timing chain(s) and flywheel as described in Part A of this Chapter.

2 Remove the pistons and connecting rods, as described in Section 10. If no work is to be done on the pistons and connecting rods, there is no need to remove the cylinder head, or to push the pistons out of the cylinder bores. The pistons should just be pushed far enough up the bores that they are positioned clear of the crankshaft journals.

3 Check the crankshaft endfloat as described in Section 13, then proceed as follows.

4 Undo the retaining bolts, and remove the oil seal housing from the left-hand (flywheel) end of the cylinder block. If the locating dowels are a loose fit, remove them and store them with the housing for safe-keeping.

5 The main bearing caps should be numbered 1 to 5 from the timing chain end of the engine **(see illustration)**. If not, using quick-drying paint or similar, mark each cap so as to indicate its correct fitted orientation and position.

6 Working in the **reverse** of the tightening sequence **(see illustration 17.10)**, slacken the main bearing cap retaining bolts by a turn at a time. Once all bolts are loose, unscrew and remove them from the cylinder block.

7 Withdraw the bearing caps, and recover the lower main bearing shells. Tape each shell to its respective cap for safe-keeping.

8 Carefully lift out the crankshaft, taking care not to displace the upper main bearing shells.

9 Recover the upper bearing shells from the cylinder block, and tape them to their respective caps for safe-keeping. Remove the thrustwasher halves from the side of No 3 main bearing, and store them with the bearing cap.

11 Cylinder block/crankcase – cleaning and inspection

Cleaning

1 Remove all external components and electrical switches/sensors from the block.

2 Scrape all traces of sealant from the cylinder block/crankcase, taking care not to damage the gasket sealing surfaces.

3 Remove all oil gallery plugs (where fitted). The plugs are usually very tight – they may have to be drilled out, and the holes re-tapped. Use new plugs when the engine is reassembled.

4 If any of the castings are extremely dirty, all should be steam-cleaned, or cleaned with a suitable degreasing agent.

5 After cleaning, clean all oil holes and oil galleries one more time. Flush all internal passages with warm water until the water runs clear. Dry thoroughly, and apply a light film of oil to the cylinder bores to prevent rusting. If possible, use compressed air to speed up the drying process, and to blow out all the oil holes and galleries.

⚠️ *Warning: Wear eye protection when using compressed air.*

6 If the castings are not very dirty, you can do an adequate cleaning job with very hot, soapy water and a stiff brush. Take plenty of time, and do a thorough job. Regardless of the cleaning method used, be sure to clean all oil holes and galleries very thoroughly, and to dry all components well. Protect the cylinder bores as described above, to prevent rusting.

7 All threaded holes must be clean, to ensure

10.5 Main bearing cap identification numbers (arrowed)

accurate torque readings during reassembly. To clean the threads, run the correct-size tap into each of the holes to remove rust, corrosion, thread sealant or sludge, and to restore damaged threads **(see illustration)**. If possible, use compressed air to clear the holes of debris produced by this operation.

> **HAYNES HINT** *A good alternative to compressed air is to inject an aerosol water-dispersant lubricant into each hole, using the long tube usually supplied.*
> ⚠️ *Warning: Wear eye protection when cleaning out these holes in this way.*

8 Apply suitable sealant to the new oil gallery plugs, and insert them into the holes in the block. Tighten them securely.

9 If the engine is not going to be reassembled right away, cover it with a large plastic bag to keep it clean; protect all mating surfaces and the cylinder bores as described above, to prevent rusting.

Inspection

10 Visually check the casting for cracks and corrosion. Look for stripped threads in the threaded holes. If there has been any history of internal water leakage, it may be worthwhile having an engine reconditioning specialist check the cylinder block/crankcase with special equipment. If defects are found, have them repaired if possible, or obtain a new block.

11 Check each cylinder bore for scuffing and

11.7 Cleaning a cylinder block threaded hole using a suitable tap

scoring. Check for signs of a wear ridge at the top of the cylinder, indicating that the bore is excessively worn.

12 Accurate measuring of the cylinder bores requires specialised equipment and experience. We recommend having the bores measured by an automotive engineering workshop, who will also be able to supply appropriate pistons should a rebore be necessary.

13 If the cylinder bores and pistons are in reasonably good condition, and not worn to the specified limits, and if the piston-to-bore clearances can be maintained properly, then it will only be necessary to renew the piston rings. If this is the case, the bores should be honed, to allow the new rings to bed in correctly and provide the best possible seal. An engine reconditioning specialist will carry out this work at moderate cost.

12 Piston/connecting rod assembly – inspection

1 Before the inspection process can begin, the piston/connecting rod assemblies must be cleaned, and the original piston rings removed from the pistons. **Note:** *Always use new piston rings when the engine is reassembled.*

2 Carefully expand the old rings over the top of the pistons. The use of two or three old feeler blades will be helpful in preventing the rings dropping into empty grooves **(see illustration)**. Be careful not to scratch the piston with the ends of the ring. The rings are brittle, and will snap if they are spread too far. They're also very sharp – protect your hands and fingers.

3 Scrape away all traces of carbon from the top of the piston. A hand-held wire brush (or a piece of fine emery cloth) can be used, once the majority of the deposits have been scraped away.

4 Remove the carbon from the ring grooves in the piston, using an old ring. Break the ring in half to do this (be careful not to cut your fingers – piston rings are sharp). Be careful to remove only the carbon deposits – do not remove any metal, and do not nick or scratch the sides of the ring grooves.

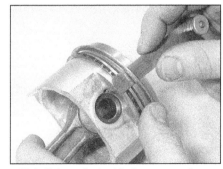

12.2 Using a feeler blade to ease piston ring removal

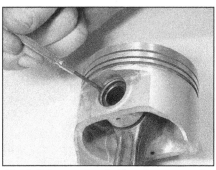

12.14 Prise out the circlips, then press the gudgeon pin out of the piston

5 Once the deposits have been removed, clean the piston/connecting rod assembly with paraffin or a suitable solvent, and dry thoroughly. Make sure that the oil return holes in the ring grooves are clear.

6 If the pistons and cylinder bores are not damaged or worn excessively, and if the cylinder block does not need to be rebored, the original pistons can be refitted. Normal piston wear shows up as even vertical wear on the piston thrust surfaces, and slight looseness of the top ring in its groove. New piston rings, however, should always be used when the engine is reassembled.

7 Carefully inspect each piston for cracks

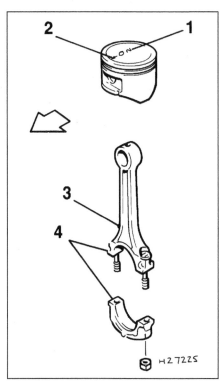

12.16 Correct connecting rod/piston fitting orientation

1 Piston grade number
2 Front marking (arrowed)
3 Connecting rod oil hole
4 Connecting rod cylinder number marking (maybe on opposite side of rod)

Gudgeon pin removal will be considerably eased if the piston is warmed (to approximately 60 to 70°C) first. Warm the piston by submerging it in a pan of hot water, then remove the assembly and press the gudgeon pin out, taking great care not to burn your hands.

around the skirt, around the gudgeon pin holes, and at the piston ring 'lands' (between the ring grooves).

8 Look for scoring and scuffing on the piston skirt, holes in the piston crown, and burned areas at the edge of the crown. If the skirt is scored or scuffed, the engine may have been suffering from overheating, and/or abnormal combustion which caused excessively high operating temperatures. The cooling and lubrication systems should be checked thoroughly. Scorch marks on the sides of the pistons show that blow-by has occurred. A hole in the piston crown, or burned areas at the edge of the piston crown, indicates that abnormal combustion (pre-ignition, knocking, or detonation) has been occurring. If any of the above problems exist, the causes must be investigated and corrected, or the damage will occur again.

9 Corrosion of the piston, in the form of pitting, indicates that coolant has been leaking into the combustion chamber and/or the crankcase. Again, the cause must be corrected, or the problem may persist in the rebuilt engine.

10 Measure the piston ring-to-groove clearance by placing a new piston ring in each ring groove and measuring the clearance with a feeler blade. Check the clearance at three or

12.17 Ensure that the piston and connecting rod are correctly mated, then press the gudgeon pin into position . . .

four places around each groove. If the measured clearance is greater than specified, new pistons will be required.

11 Accurate measurement of the pistons requires specialised equipment and experience. We recommend having the piston measured by an automotive engineering workshop, who will also be able to supply appropriate pistons should a rebore be necessary.

12 Check the fit of the gudgeon pin by twisting the piston and connecting rod in opposite directions. Any noticeable play indicates excessive wear of the gudgeon pin, piston, or connecting rod small-end bearing.

13 If necessary, the pistons and connecting rods can be separated and reassembled as follows.

14 Using a small screwdriver, prise out the circlips, and push out the gudgeon pin **(see illustration and Haynes Hint)**. If necessary, support the piston, and tap the pin out using a suitable hammer and punch, taking great care not to mark the piston/connecting rod bores. Identify the piston, gudgeon pin and rod to ensure correct reassembly. Discard the circlips – new ones *must* be used on refitting.

15 Examine each connecting rod carefully for signs of damage, such as cracks around the big-end and small-end bearings. Check that the rod is not bent or distorted. Damage is highly unlikely, unless the engine has been seized or badly overheated. Detailed checking of the connecting rod assembly and any remedial action necessary can only be carried out by an engine reconditioning specialist with the necessary equipment.

16 To refit the pistons, position the piston so that the front marking on the piston crown (either in the form of an arrow or a dot) is positioned correctly in relation to the oil hole in the connecting rod shaft **(see illustration)**. With the piston and rod correctly mated, the piston front marking will face towards the timing chain end of the engine, and the connecting rod oil hole will face the rear of the cylinder block.

17 Apply a smear of clean engine oil to the gudgeon pin. Slide it into the piston and through the connecting rod small-end **(see illustration)**. **Note:** *Gudgeon pin installation will be greatly eased if the piston is first warmed.* If necessary, tap the pin into position

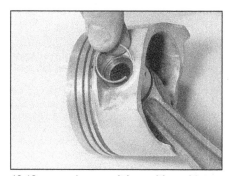

12.18 . . . and secure it in position with two new circlips

using a hammer and suitable punch, whilst ensuring that the piston is securely supported.
18 Check that the piston pivots freely on the rod, then secure the gudgeon pin in position with two new circlips **(see illustration)**. Ensure that each circlip is correctly located in its groove in the piston.

13 Crankshaft – inspection

Checking endfloat

1 If the crankshaft endfloat is to be checked, this must be done when the crankshaft is still installed in the cylinder block/crankcase, but is free to move (see Section 10).
2 Check the endfloat using a dial gauge in contact with the end of the crankshaft. Push the crankshaft fully one way, and then zero the gauge. Push the crankshaft fully the other way, and check the endfloat **(see illustration)**. The result can be compared with the specified amount, and will give an indication as to whether new thrustwashers are required.
3 If a dial gauge is not available, feeler blades can be used. First push the crankshaft fully towards the flywheel/driveplate end of the engine, then use feeler blades to measure the gap between the No 4 crankpin web and No 3 main bearing thrustwasher.

Inspection

4 Clean the crankshaft using paraffin or a suitable solvent, and dry it, preferably with compressed air if available. Be sure to clean the oil holes with a pipe cleaner or similar probe, to ensure that they are not obstructed.

 Warning: Wear eye protection when using compressed air.

5 Check the main and big-end bearing journals for uneven wear, scoring, pitting and cracking.
6 Big-end bearing wear is accompanied by distinct metallic knocking when the engine is running (particularly noticeable when the engine is pulling from low speed) and some loss of oil pressure.
7 Main bearing wear is accompanied by severe engine vibration and rumble – getting progressively worse as engine speed increases – and again by loss of oil pressure.
8 Check the bearing journal for roughness by running a finger lightly over the bearing surface. Any roughness (which will be accompanied by obvious bearing wear) indicates that the crankshaft requires regrinding (where possible) or renewal.
9 Accurate measurement of the crankshaft requires specialised equipment and experience. We recommend having the crankshaft measured by an automotive engineering workshop, who will also be able to supply appropriate journal bearings should a regrind be necessary.

10 If the crankshaft has been reground, check for burrs around the crankshaft oil holes (the holes are usually chamfered, so burrs should not be a problem unless regrinding has been carried out carelessly). Remove any burrs with a fine file or scraper, and thoroughly clean the oil holes as described previously.
11 Nissan produce undersize bearing shells for both the main bearings and big-end bearings. There are two undersizes of main bearing shells (0.25 and 0.50 mm) and three undersizes of big-end bearing shells (0.08, 0.12 and 0.25 mm). If the crankshaft journals have not already been reground, it may be possible to have the crankshaft reconditioned, and to fit the undersize shells. If the crankshaft has worn beyond the specified limits, it will have to be renewed. Consult your Nissan dealer or engine reconditioning specialist for further information on parts availability.

14 Main and big-end bearings – inspection

1 Even though the main and big-end bearings should be renewed during the engine overhaul, the old bearings should be retained for close examination, as they may reveal valuable information about the condition of the engine. The bearing shells are graded by thickness, the grade of each shell being indicated by the colour code marked on it.
2 Bearing failure can occur due to lack of lubrication, the presence of dirt or other foreign particles, overloading the engine, or corrosion. Regardless of the cause of bearing failure, the cause must be corrected (where applicable) before the engine is reassembled, to prevent it from happening again **(see illustration)**.
3 When examining the bearing shells, remove them from the cylinder block/crankcase, the main bearing caps, the connecting rods and the connecting rod big-end bearing caps. Lay them out on a clean surface in the same general position as their location in the engine. This will enable you to match any bearing problems with the corresponding crankshaft journal. Do not touch any shell's bearing surface with your fingers while checking it, or the delicate surface may be scratched.
4 Dirt and other foreign matter gets into the engine in a variety of ways. It may be left in the engine during assembly, or it may pass through filters or the crankcase ventilation system. It may get into the oil, and from there into the bearings. Metal chips from machining operations and normal engine wear are often present. Abrasives are sometimes left in engine components after reconditioning, especially when parts are not thoroughly cleaned using the proper cleaning methods.

13.2 Using a dial gauge to measure crankshaft endfloat

Whatever the source, these foreign objects often end up embedded in the soft bearing material, and are easily recognised. Large particles will not embed in the bearing, and will score or gouge the bearing and journal. The best prevention for this cause of bearing failure is to clean all parts thoroughly, and keep everything spotlessly-clean during engine assembly. Frequent and regular engine oil and filter changes are also recommended.
5 Lack of lubrication (or lubrication breakdown) has a number of interrelated causes. Excessive heat (which thins the oil), overloading (which squeezes the oil from the bearing face) and oil leakage (from excessive bearing clearances, worn oil pump or high engine speeds) all contribute to lubrication breakdown. Blocked oil passages, which usually are the result of misaligned oil holes in a bearing shell, will also oil-starve a bearing, and destroy it. When lack of lubrication is the cause of bearing failure, the bearing material is wiped or extruded from the steel backing of the bearing. Temperatures may increase to the point where the steel backing turns blue from overheating.

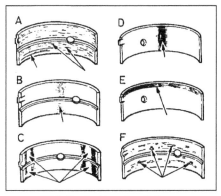

14.2 Typical bearing failures

A *Scratched by dirt; dirt embedded in bearing material*
B *Lack of oil; overlay wiped out*
C *Improper seating; bright (polished) sections*
D *Tapered journal; overlay gone from entire surface*
E *Radius ride*
F *Fatigue failure; craters or pockets*

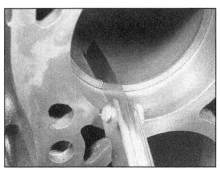

16.4 Measuring a piston ring end gap

16.9a Fit the oil control ring expander . . .

16.9b . . . then install the side rails as described in text

6 Driving habits can have a definite effect on bearing life. Full-throttle, low-speed operation (labouring the engine) puts very high loads on bearings, tending to squeeze out the oil film. These loads cause the bearings to flex, which produces fine cracks in the bearing face (fatigue failure). Eventually, the bearing material will loosen in pieces, and tear away from the steel backing.

7 Short-distance driving leads to corrosion of bearings, because insufficient engine heat is produced to drive off the condensed water and corrosive gases. These products collect in the engine oil, forming acid and sludge. As the oil is carried to the engine bearings, the acid attacks and corrodes the bearing material.

8 Incorrect bearing installation during engine assembly will lead to bearing failure as well. Tight-fitting bearings leave insufficient bearing running clearance, and will result in oil starvation. Dirt or foreign particles trapped behind a bearing shell result in high spots on the bearing, which lead to failure.

9 *Do not* touch any shell's bearing surface with your fingers during reassembly; there is a risk of scratching the delicate surface, or of depositing particles of dirt on it.

10 As mentioned at the beginning of this Section, the bearing shells should be renewed as a matter of course during engine overhaul; to do otherwise is false economy. Refer to Sections 17 and 18 for details of bearing shell selection.

15 Engine overhaul –
reassembly sequence

Before reassembly begins, ensure that all new parts have been obtained, and that all necessary tools are available. Read through the entire procedure, to familiarise yourself with the work involved, and to ensure that all items necessary for reassembly of the engine are at hand. In addition to all normal tools and materials, thread-locking compound will be needed. A suitable tube of liquid sealant will also be required for the joint faces that are fitted without gaskets; it is recommended that Nissan's Genuine Liquid Gasket (available from your Nissan dealer) is used.

In order to save time and avoid problems, engine reassembly can be carried out in the following order:
a) *Crankshaft (Section 17).*
b) *Piston/connecting rod assemblies (Section 18).*
c) *Cylinder head (See Part A of this Chapter).*
d) *Timing chain(s) and cover (See Part A of this Chapter).*
e) *Sump (See Part A of this Chapter).*
f) *Flywheel/driveplate (See Part A of this Chapter).*
g) *Engine external components.*

At this stage, all engine components should be absolutely clean and dry, with all faults repaired. The components should be laid out (or in individual containers) on a completely clean work surface.

16 Piston rings –
refitting

1 Before fitting new piston rings, the ring end gaps must be checked as follows.

2 Lay out the piston/connecting rod assemblies and the new piston ring sets, so that the ring sets will be matched with the same piston and cylinder during the end gap measurement and subsequent engine reassembly.

3 Insert the top ring into the first cylinder, and push it down the bore using the top of the piston. This will ensure that the ring remains

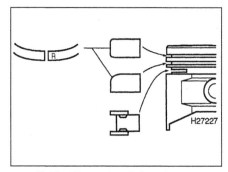

16.10a Piston ring fitting diagram

square with the cylinder walls. Push the ring down into the bore until the piston skirt is level with the block mating surface, then withdraw the piston.

4 Measure the end gap using feeler gauges, and compare the measurements with the figures given in the Specifications **(see illustration)**.

5 If the gap is too small (unlikely if reputable parts are used), it must be enlarged, or the ring ends may contact each other during engine operation, causing serious damage. Ideally, new piston rings providing the correct end gap should be fitted. As a last resort, the end gap can be increased by filing the ring ends very carefully with a fine file. Mount the file in a vice with soft jaws, slip the ring over the file with the ends contacting the file face, and slowly move the ring to remove material from the ends. Take care, as piston rings are sharp, and are easily broken.

6 With new piston rings, it is unlikely that the end gap will be too large. If the gaps are too large, check that you have the correct rings for the engine and for the particular cylinder bore size.

7 Repeat the checking procedure for each ring in the first cylinder, and then for the rings in the remaining cylinders. Remember to keep rings, pistons and cylinders matched up.

8 Once the ring end gaps have been checked and if necessary corrected, the rings can be fitted to the pistons. **Note:** *Always follow any instructions supplied with the new piston ring sets – different manufacturers may specify different procedures. Do not mix up the top and second compression rings, as they have different cross-sections.*

9 The oil control ring (lowest on the piston) is installed first. It is composed of three separate components. Slip the expander into the groove, then install the upper side rail into the groove between the expander and the ring land, then install the lower side rail in the same manner **(see illustrations)**.

10 Install the second ring next. **Note:** *The second ring and top ring are different, and can be identified by their cross-sections.* Making sure the ring is the correct way up, fit the ring into the middle groove on the piston, taking care not to expand the ring any more than is necessary **(see illustrations)**.

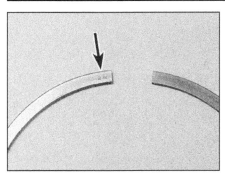

16.10b Where necessary, ensure that the rings are installed with their identification marking (arrowed) uppermost

11 Install the top ring in the same way, making sure the ring is the correct way up. Where the ring is symmetrical, fit it with its identification marking facing upwards.

12 With all the rings in position on the piston, space the ring end gaps correctly **(see illustration)**.

13 Repeat the above procedure for the remaining pistons and rings.

17 Crankshaft – bearing selection and refitting

Bearing selection

1 Main bearings for the engines described in this Chapter are available in standard sizes and a range of undersizes to suit reground crankshafts. Refer to your Nissan dealer or automotive engineering workshop for details.

Refitting

2 Clean the backs of the bearing shells, and the bearing locations in both the cylinder block and the main bearing caps.

3 Press the bearing shells into their locations, ensuring that the tab on each shell engages in the notch in the cylinder block/crankcase. Take care not to touch any shell's bearing surface with your fingers. Note that all the upper bearing shells are grooved, and have oil holes in them; the lower shells are plain **(see illustrations)**.

17.5 ... then fit the thrustwasher halves, making sure that their grooved faces are facing away from the crankcase

OIL RING
UPPER RAIL

TOP RING

OIL RING
EXPANDER

FRONT

OIL RING
LOWER RAIL

2ND RING

H27229

16.12 Position the piston ring end gaps

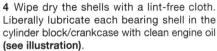

17.3b ... and fit the lower plain shells to the caps, also aligning their tabs with the cap cut-outs (arrowed)

4 Wipe dry the shells with a lint-free cloth. Liberally lubricate each bearing shell in the cylinder block/crankcase with clean engine oil **(see illustration)**.

5 Using a little grease, stick the upper thrustwashers to each side of the No 3 main bearing upper location; ensure that the oilway grooves on each thrustwasher face outwards (away from the cylinder block) **(see illustration)**.

6 Lower the crankshaft into position, and check the crankshaft endfloat as described in Section 13.

7 Thoroughly degrease the mating surfaces of the cylinder block and the main bearing caps.

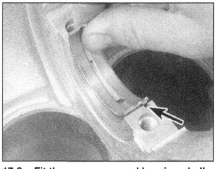

17.3a Fit the upper grooved bearing shells, aligning their tabs with the crankcase cut-outs (arrowed) ...

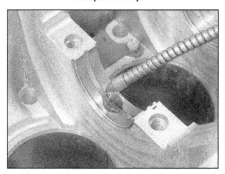

17.4 Lubricate the main bearing shells with clean engine oil ...

8 Lubricate the lower bearing shells in the main bearing caps with clean engine oil. Make sure that the locating lugs on the shells engage with the corresponding recesses in the caps.

9 Fit the main bearing caps, using the identification marks to ensure that they are installed in the correct locations and are fitted the correct way round. Insert the retaining bolts, tightening them by hand only.

10 Working in the sequence shown, tighten the bearing cap retaining bolts to approximately half the specified torque setting **(see illustration)**. Then go around in the same sequence and tighten the bolts to the full

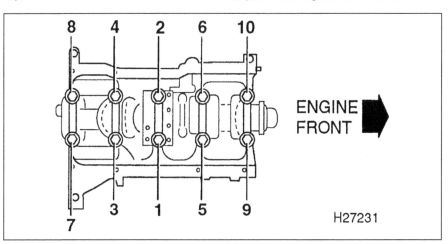

ENGINE FRONT

H27231

17.10 Main bearing cap bolt tightening sequence

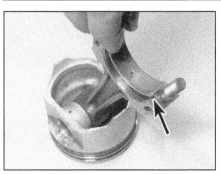

18.3 Fit each bearing shell to its connecting rod, aligning its tab with the rod cut-out (arrowed)

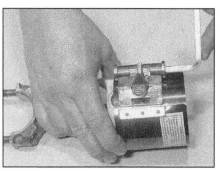

18.7 Ensure that the piston rings end gaps are correctly spaced, then clamp them in position with piston ring compressor

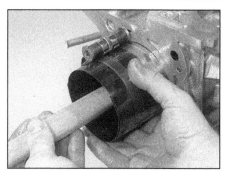

18.8 Insert the piston/connecting rod assembly into its respective cylinder, and gently tap it into position

specified torque setting. Check that the crankshaft rotates freely before proceeding any further.

11 Fit the piston/connecting rod assemblies as described in Section 18.

12 Ensure that the mating surfaces of the oil seal housing and cylinder block are clean and dry. Note the correct fitted depth of the oil seal then, using a large flat-bladed screwdriver, lever the seal out of the housing.

13 Fit the new crankshaft seal to the housing, making sure that its sealing lip is facing inwards. Tap the seal squarely into the housing until it is positioned at the same depth as the original was noted prior to removal.

14 Apply a bead of suitable sealant to the oil seal housing mating surface, and make sure that the locating dowels are in position. Slide the housing over the end of the crankshaft, and into position on the cylinder block. Tighten the housing retaining bolts to the specified torque setting.

15 Refit the flywheel, timing chains and sump as described in Part A of this Chapter.

18 Piston/connecting rod assembly – bearing selection and refitting

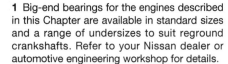

Bearing selection

1 Big-end bearings for the engines described in this Chapter are available in standard sizes and a range of undersizes to suit reground crankshafts. Refer to your Nissan dealer or automotive engineering workshop for details.

Refitting

2 Clean the backs of the bearing shells, and the bearing locations in both the connecting rod and bearing cap.

3 Press the bearing shells into their locations, ensuring that the tab on each shell engages in the recess in the connecting rod and cap **(see illustration)**. Take care not to touch any

shell's bearing surface with your fingers, and ensure that the shells are correctly installed so that the upper shell oil hole is correctly aligned with connecting rod oil hole.

4 Note that the following procedure assumes that the crankshaft and main bearing caps are in place (see Section 17).

5 Wipe dry the shells and connecting rods with a lint-free cloth.

6 Lubricate the cylinder bores, the pistons, and piston rings, then lay out each piston/connecting rod assembly in its respective position.

7 Start with assembly No 1. Make sure that the piston rings are still spaced as described in Section 16, then clamp them in position with a piston ring compressor **(see illustration)**.

8 Insert the piston/connecting rod assembly into the top of cylinder No 1. Ensure that the piston front marking (in the form of either an arrow or a dot) on the piston crown is on the timing chain side of the bore. Using a block of wood or hammer handle against the piston crown, tap the assembly into the cylinder until the piston crown is flush with the top of the cylinder **(see illustration)**.

9 Ensure that the bearing shell is still correctly installed. Liberally lubricate the crankpin and both bearing shells. Taking care not to mark the cylinder bores, tap the piston/connecting rod assembly down the bore and onto the crankpin. Refit the big-end bearing cap, tightening its retaining nuts finger-tight at first. Note that the faces with the identification marks must match (which means that the bearing shell locating tabs abut each other).

10 Tighten the bearing cap retaining nuts to their Stage 1 torque setting, using a torque wrench and suitable socket. Then tighten them either through the specified Stage 2 angle setting or, if an angle-measuring gauge is not available, to the specified Stage 2 torque setting.

11 Rotate the crankshaft. Check that it turns freely; some stiffness is to be expected if new

components have been fitted, but there should be no signs of binding or tight spots.

12 Refit the remaining three piston/connecting rod assemblies in the same way.

13 Refit the cylinder head, timing chain(s) and sump as described in Part A of this Chapter.

19 Engine – initial start-up after overhaul

1 With the engine refitted in the vehicle, double-check the engine oil and coolant levels. Make a final check that everything has been reconnected, and that there are no tools or rags left in the engine compartment.

2 Start the engine, noting that this may take a little longer than usual, due to the fuel system components having been disturbed. Make sure that the oil pressure warning light goes out then allow the engine to idle.

3 While the engine is idling, check for fuel, water and oil leaks. Don't be alarmed if there are some odd smells and smoke from parts getting hot and burning off oil deposits.

4 Assuming all is well, keep the engine idling until hot water is felt circulating through the top hose, then switch off the engine.

5 Check the ignition timing and the idle speed settings (see *Weekly checks*), then switch the engine off.

6 After a few minutes, recheck the oil and coolant levels as described in Chapter 1, and top-up as necessary.

7 If they were tightened as described, there is no need to retighten the cylinder head bolts once the engine has first run after reassembly.

8 If new pistons, rings or crankshaft bearings have been fitted, the engine must be treated as new, and run-in for the first 500 miles (800 km). *Do not* operate the engine at full-throttle, or allow it to labour at low engine speeds in any gear. It is recommended that the oil and filter be changed at the end of this period.

Chapter 3
Cooling, heating and ventilation systems

Contents

Degrees of difficulty

Easy, suitable for novice with little experience	Fairly easy, suitable for beginner with some experience	Fairly difficult, suitable for competent DIY mechanic	Difficult, suitable for experienced DIY mechanic	Very difficult, suitable for expert DIY or professional

Specifications

General
Radiator cap opening pressure .... 0.78 to 0.98 bars

Thermostat
Opening temperature:
    Starts to open .... 76.5°C
    Fully open .... 90°C
Maximum valve lift (approximate) .... 8.0 mm

Electric cooling fan
Cut-in temperature .... 82 to 88°C

Engine coolant temperature sensor
Resistance:
    At 20°C .... 2.1 to 2.9 kilohms
    At 90°C .... 0.24 to 0.26 kilohms
    At 110°C .... 0.14 to 0.15 kilohms

Torque wrench settings	Nm	lbf ft
Coolant pump pulley securing bolts	7	5
Coolant pump securing bolts	7	5
Thermostat cover securing bolts	7	5
Thermostat housing securing bolts	7	5

1 General information and precautions

General information

The cooling system is of pressurised type, comprising a coolant pump driven by the auxiliary drivebelt from the crankshaft pulley, crossflow radiator, coolant expansion tank, electric cooling fan(s), thermostat, heater matrix, and all associated hoses and switches.

The system functions as follows. The coolant pump pumps cold coolant around the cylinder block and head passages, and through the inlet manifold, heater and throttle housing to the thermostat housing.

When the engine is cold, the coolant is returned from the thermostat housing to the coolant pump. When the coolant reaches a predetermined temperature, the thermostat opens, and the coolant passes through the top hose to the radiator. As the coolant circulates through the radiator, it is cooled by the inrush of air when the car is in forward motion. The airflow is supplemented by the action of the electric cooling fan(s) when necessary. Upon reaching the bottom of the radiator, the coolant has now cooled, and the cycle is repeated.

When the engine is at normal operating temperature, the coolant expands, and some of it is released through the valve in the radiator pressure cap into the expansion tank. Coolant collects in the tank, and is returned to the radiator when the system cools.

A single or twin electric cooling fan arrangement is used according to model and equipment fitted. The fan assembly is mounted behind the radiator and controlled by the engine management electronic control unit in conjunction with the engine coolant temperature sensor.

Precautions

 Warning: Do not attempt to remove the radiator pressure cap, or to disturb any part of the cooling system, while the engine is hot, as there is a high risk of scalding. If the radiator pressure cap must be removed before the engine and radiator have fully cooled (even though this is not recommended), the pressure in the cooling system must first be relieved. Cover the cap with a thick layer of cloth, to avoid scalding, and slowly unscrew the pressure cap until a hissing sound is heard. When the hissing has stopped, indicating that the pressure has reduced, slowly unscrew the pressure cap until it can be removed; if more hissing sounds are heard, wait until they have stopped before unscrewing the cap completely. At all times, keep your face well away from the pressure cap opening, and protect your hands.

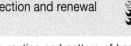 **Warning: Do not allow antifreeze to come into contact with your skin, or with the painted surfaces of the vehicle. Rinse off spills immediately, with plenty of water. Never leave antifreeze lying around in an open container, or in a puddle in the driveway or on the garage floor. Children and pets are attracted by its sweet smell, but antifreeze can be fatal if ingested.**

Warning: If the engine is hot, the electric cooling fan may start rotating even if the engine is not running. Be careful to keep your hands, hair, and any loose clothing well clear when working in the engine compartment.

Warning: Refer to Section 10 for precautions to be observed when working on models equipped with air conditioning.

2 Cooling system hoses – disconnection and renewal

1 The number, routing and pattern of hoses will vary according to model, but the same basic procedure applies. Before commencing work, make sure that the new hoses are to hand, along with new hose clips if needed. It is good practice to renew the hose clips at the same time as the hoses.

2 Drain the cooling system, as described in Chapter 1, saving the coolant if it is fit for re-use. Squirt a little penetrating oil onto the hose clips if they are corroded.

3 Release the hose clips from the hose concerned. Three types of clip are used; worm-drive, spring and 'sardine-can'. The worm-drive clip is released by turning its screw anti-clockwise. The spring clip is released by squeezing its tags together with pliers, at the same time working the clip away from the hose stub. The 'sardine-can' clip is not re-usable, and is best cut off with snips or side-cutters.

4 Unclip any wires, cables or other hoses which may be attached to the hose being removed. Make notes for reference when reassembling if necessary.

5 Release the hose from its stubs with a twisting motion. Be careful not to damage the stubs on delicate components such as the radiator. If the hose is stuck fast, the best course is often to cut it off using a sharp knife, but again be careful not to damage the stubs.

6 Before fitting the new hose, smear the stubs with washing-up liquid or a suitable rubber lubricant to aid fitting. Do not use oil or grease, which may attack the rubber.

7 Fit the hose clips over the ends of the hose, then fit the hose over its stubs. Work the hose into position. When satisfied, locate and tighten the hose clips.

8 Refill the cooling system as described in Chapter 1. Run the engine, and check that there are no leaks.

9 Recheck the tightness of the hose clips on any new hoses after a few hundred miles.

10 Top-up the coolant level if necessary.

3 Radiator – removal, inspection and refitting

Note: *If leakage is the reason for removing the radiator, bear in mind that minor leaks can often be cured using a radiator sealant with the radiator in situ.*

Removal

1 Disconnect the battery negative terminal (refer to *Disconnecting the battery* in the Reference Chapter).

2 Drain the cooling system as described in Chapter 1.

3 Disconnect the remaining coolant hose(s) from the radiator (on models with automatic transmission, the fluid cooler hoses are connected to the bottom of the radiator) **(see illustration)**.

4 On models with air conditioning, unbolt the refrigerant pipe support bracket from the radiator upper left-hand mounting bracket.

5 Disconnect the fan motor wiring connector(s) and release the wiring harness from the cable ties on the fan shroud.

6 Working at the top of the radiator, unscrew the nuts securing the radiator mounting brackets to the upper body panel **(see illustration)**.

3.3 On models with automatic transmission, disconnect the fluid cooler hoses (arrowed) from the radiator

3.6 Unscrew the radiator top mounting bracket securing nuts (arrowed)

7 Carefully tilt the radiator back towards the engine, then lift the radiator from the engine compartment **(see illustration)**. Recover the lower mounting rubbers if they are loose.

Inspection

8 If the radiator has been removed due to suspected blockage, reverse-flush it as described in Chapter 1. Clean dirt and debris from the radiator fins, using an air line (in which case, wear eye protection) or a soft brush. Be careful, as the fins are sharp, and easily damaged.

9 If necessary, a radiator specialist can perform a 'flow test' on the radiator, to establish whether an internal blockage exists.

10 A leaking radiator must be referred to a specialist for permanent repair. Do not attempt to weld or solder a leaking radiator, as damage to the plastic components may result.

11 Inspect the condition of the radiator mounting rubbers, and renew them if necessary.

Refitting

12 Refitting is a reversal of removal, bearing in mind the following points:
  a) *Ensure that the radiator lower lugs engage correctly with the lower mounting rubbers.*
  b) *On completion, refill the cooling system as described in Chapter 1.*

4 Thermostat – removal, testing and refitting

Note: *Suitable sealant (liquid gasket) will be required when refitting the thermostat housing.*

Removal

1 The thermostat is located in a housing bolted to the side of the coolant pump, at the timing chain end of the engine.

2 Disconnect the battery negative terminal (refer to *Disconnecting the battery* in the Reference Chapter).

3 Drain the cooling system as described in Chapter 1.

4 Disconnect the coolant hose from the thermostat cover **(see illustration)**.

5 Unscrew the securing bolts, and remove the thermostat cover from the housing. If the cover is stuck to the housing, tap it gently, or rock it back and forth to free it – **do not** lever between the mating faces.

6 Lift the thermostat from the housing, noting that the bleed valve is located at the top.

Testing

7 A rough test of the thermostat may be made by suspending it with a piece of string in a container full of water. Heat the water to bring it to the boil – the thermostat must open by the time the water boils. If not, renew it.

8 If a thermometer is available, the precise opening temperature of the thermostat may be determined; compare with the figures given in the Specifications. The opening temperature is also marked on the thermostat.

9 A thermostat which fails to close as the water cools must also be renewed.

Refitting

10 Commence refitting by thoroughly cleaning the mating faces of the cover and the housing.

11 Refit the thermostat to the housing, noting that it should fit with the bleed valve uppermost **(see illustration)**.

12 Apply a continuous bead of sealant (liquid gasket) to the housing mating face of the thermostat cover, taking care not to apply excess sealant, which may enter the cooling system.

13 Fit the cover to the thermostat housing, then refit the securing bolts, and tighten to the specified torque.

14 Reconnect the coolant hose to the thermostat cover.

15 Refill the cooling system as described in Chapter 1.

16 Reconnect the battery negative terminal.

5 Electric cooling fan(s) – testing, removal and refitting

Testing

1 Battery voltage for operation of the cooling fan(s) is supplied via a relay which is energised by the ignition switch through a fuse. The circuit is completed by the engine management electronic control unit applying an earth to activate the relay. A single cooling fan is used on manual transmission models without air conditioning. On all other models, two cooling fans are fitted.

2 If the fan does not appear to work, run the engine until normal operating temperature is reached, then allow it to idle. The fan should cut in within a few minutes (before the temperature gauge needle enters the red section). If the fan does not operate, switch the ignition off and remove the cooling fan

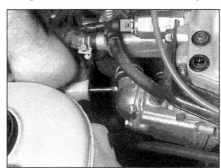

4.4 Disconnecting the coolant hose from the thermostat housing

3.7 Carefully lift the radiator from the engine compartment – arrows indicate lower mountings

relay from the relay box on the right-hand side of the engine compartment.

3 Switch the ignition on, and using a voltmeter, check for battery voltage between a vehicle earth and relay terminals 1 and 3 in the relay box. If battery voltage is present, test the fan motor as described in paragraph 5. If the fan motor operates with voltage applied directly to it, then either the relay, the wiring harness between the relay box and the fan motor, or the engine management ECU are at fault.

4 If no voltage is present at the relay box terminals, check the condition of the cooling fan fuse (typically fuse 25 or 16) in the passenger compartment fusebox. If the fuse is satisfactory the fault is likely to be in the wiring harness or harness connectors between the fusebox and relay box, or between the relay box and the battery.

5 The fan motor itself can be tested by disconnecting it from the wiring loom, and connecting a 12 volt supply directly to it. The motor should operate – if not, the motor, or the motor wiring, is faulty.

Removal

6 Disconnect the battery negative terminal (refer to *Disconnecting the battery* in the Reference Chapter).

7 Disconnect the motor wiring connector(s) **(see illustration)**.

8 Unscrew the two securing bolts from the top of the shroud, then lift out the assembly to release the lower clips **(see illustrations)**.

4.11 Thermostat bleed valve (arrowed) must be uppermost

5.7 Disconnecting a cooling fan motor wiring connector

5.8a Remove the two securing bolts . . .

5.8b . . . and lift out the cooling fan assembly

Refitting

9 Refitting is a reversal of removal.

6 Cooling system electrical switches – testing, removal and refitting

Coolant temperature sensor

Testing

1 The sensor is located in the inlet manifold, at the timing chain end of the cylinder head **(see illustration)**.

2 The unit contains a thermistor – an electronic component whose electrical resistance decreases at a predetermined rate as its temperature rises.

3 The engine management ECU supplies the sensor with a set voltage and then, by measuring the current flowing in the sensor circuit, it determines the engine temperature. This information is then used, in conjunction with other inputs, to control the engine management system and associated components.

4 If the sensor circuit should fail to provide plausible information, the ECU back-up facility will override the sensor signal. In this event, the ECU assumes a predetermined setting which will allow the engine management system to operate, albeit at reduced efficiency. When this occurs, the engine warning light on the instrument panel will come on, and the advice of a Nissan dealer should be sought. The sensor itself can

be tested by removing it, and checking the resistances at various temperatures using an ohmmeter (heat the sensor in a container of water, and monitor the temperature with a thermometer). The resistance values are given in the Specifications. *Do not* attempt to test the circuit with the sensor fitted to the engine, and the wiring connector fitted, as there is a high risk of damaging the ECU.

5 Refer to Chapter 4A for further details of the engine management system.

Removal

6 Disconnect the battery negative terminal (refer to *Disconnecting the battery* in the Reference Chapter).

7 Partially drain the cooling system to just below the level of the sensor (see Chapter 1). Alternatively, have ready a suitable bung to plug the aperture in the housing when the sensor is removed.

8 Disconnect the wiring connector from the sensor.

9 Carefully unscrew the sensor and recover the sealing ring. If the system has not been drained, plug the sensor aperture to prevent further coolant loss.

Refitting

10 Check the condition of the sealing ring and renew it if necessary.

11 Refitting is a reversal of removal, but refill (or top-up) the cooling system as described in Chapter 1 and *Weekly checks*.

12 On completion, start the engine and run it until it reaches normal operating temperature. Continue to run the engine until the cooling fan cuts in and out correctly.

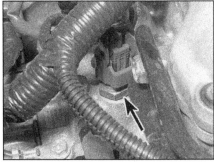

6.1 Engine coolant temperature sensor (arrowed)

6.13 Coolant temperature gauge sender (arrowed)

Temperature gauge sender

Testing

13 The sender is located in the rear of the cylinder head, at the transmission end of the engine **(see illustration)**.

14 The temperature gauge is supplied with battery voltage from the instrument panel main feed (via the ignition switch, ignition relay and a fuse). The gauge earth is controlled by the sender. The sender contains a thermistor – an electronic component whose electrical resistance decreases at a predetermined rate as its temperature rises. When the coolant is cold, the sender resistance is high, current flow through the gauge is reduced, and the gauge needle points towards the blue (cold) end of the scale. As the coolant temperature rises and the sender resistance falls, current flow increases, and the gauge needle moves towards the upper end of the scale.

15 If the gauge needle remains at the 'cold' end of the scale when the engine is hot, disconnect the sender wiring connector, and earth the wire to the cylinder head. If the needle then deflects when the ignition is switched on, the sender unit is proved faulty, and should be renewed. If the needle still does not move, remove the instrument panel (see Chapter 12) and check the continuity of the wire between the sender unit and the gauge, and the feed to the gauge unit. If continuity is shown, and the fault still exists, then the gauge is faulty, and the gauge should be renewed.

16 If the gauge needle remains at the 'hot' end of the scale when the engine is cold, disconnect the sender wiring connector. If the needle then returns to the 'cold' end of the scale when the ignition is switched on, the sender unit is proved faulty, and should be renewed. If the needle still does not move, check the remainder of the circuit as described previously.

Removal and refitting

17 The procedure is similar to that described previously in this Section for the engine coolant temperature sensor. On certain models, access to the switch is poor, and other components may need to be removed (or hoses, wiring, etc, moved to one side) before the sender unit can be reached.

Air conditioning temperature sensor

Testing

18 The sensor is integral with the air conditioning temperature control unit, located on the side of the heater/air conditioning unit behind the facia. Testing should be entrusted to a Nissan dealer.

Removal and refitting

19 Removal and refitting should be entrusted to a Nissan dealer suitably equipped to test the unit.

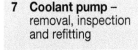

7 Coolant pump – removal, inspection and refitting

Note: *Suitable sealant (liquid gasket) will be required when refitting the coolant pump.*

Removal

1 Disconnect the battery negative terminal (refer to *Disconnecting the battery* in the Reference Chapter).
2 Drain the cooling system as described in Chapter 1.
3 Remove the auxiliary drivebelt(s) as described in Chapter 1.
4 Unscrew the securing bolts, and remove the coolant pump pulley **(see illustration)**. It will be necessary to counterhold the pulley in order to unscrew the bolts, and this is most easily achieved by wrapping an old drivebelt tightly around the pulley to act in a similar manner to a strap wrench.
5 Unbolt the thermostat housing from the stub on the coolant pump, and move the housing to one side **(see illustration)**. If the housing is stuck, tap it gently with a soft-faced mallet – **do not** lever between the mating faces. If necessary to improve access, disconnect the coolant hose, and remove the housing completely.
6 Unscrew the coolant pump securing bolts, noting the different bolt lengths and their locations, and withdraw the pump. If the pump is stuck, tap it gently using a soft-faced mallet – **do not** lever between the pump and cylinder block mating faces.

Inspection

7 Check the pump body and impeller for signs of excessive corrosion. Turn the impeller, and check for stiffness due to corrosion, or roughness due to excessive end play.
8 No spare parts are available for the pump, and if faulty, worn or corroded, a new pump should be fitted.

Refitting

9 Commence refitting by thoroughly cleaning all traces of sealant from the mating faces of the pump and cylinder block, and from the thermostat housing.

7.4 Removing the coolant pump pulley

10 Apply a continuous bead of sealant (liquid gasket) to the cylinder block mating face of the pump, taking care not to apply excessive sealant, which may enter the pump itself. Similarly, apply a bead of sealant to the thermostat housing mating face of the pump **(see illustrations)**.
11 Place the pump in position in the cylinder block, refit the bolts to their correct locations and tighten to the specified torque.
12 Refit the pump pulley, then refit the securing bolts and tighten to the specified torque. Counterhold the pulley using an old drivebelt as during removal.
13 Apply a continuous bead of sealant (liquid gasket) to the coolant pump mating face of the thermostat housing, again taking care not to apply excessive sealant.
14 Place the thermostat housing in position on the pump stub, then refit the securing bolts, and tighten to the specified torque.
15 Refit and tension the auxiliary drivebelt(s) as described in Chapter 1.
16 Refill the cooling system as described in Chapter 1.
17 Reconnect the battery negative terminal.

8 Heater/ventilation system – general information

1 The heater/ventilation system consists of a four-speed blower motor (housed behind the facia), face level vents in the centre and at each end of the facia, and air ducts to the front footwells.

7.10a Apply sealant to the coolant pump's cylinder block mating face . . .

7.5 Unbolt the thermostat housing (arrowed) from the coolant pump

2 The control unit is located in the facia, and the controls operate flap valves to deflect and mix the air flowing through the various parts of the heating/ventilation system. The flap valves are contained in the air distribution housing, which acts as a central distribution unit, passing air to the various ducts and vents.
3 Cold air enters the system through the grille at the rear of the engine compartment. If required, the airflow is boosted by the blower fan, and then flows through the various ducts, according to the settings of the controls. Stale air is expelled through ducts at the rear of the vehicle. If warm air is required, the cold air is passed over the heater matrix, which is heated by the engine coolant.
4 On models fitted with air conditioning, a recirculation switch enables the outside air supply to be closed off, while the air inside the vehicle is recirculated. This can be useful to prevent unpleasant odours entering from outside the vehicle, but should only be used briefly, as the recirculated air inside the vehicle will soon become stale.

9 Heater/ventilation components – removal and refitting

Heater/ventilation control unit

Removal

1 Disconnect the battery negative terminal (refer to *Disconnecting the battery* in the Reference Chapter).

7.10b . . . and thermostat housing mating face

9.4 Heater/ventilation control unit securing screws (arrowed)

9.5 Withdraw the control unit, and disconnect the wiring plugs

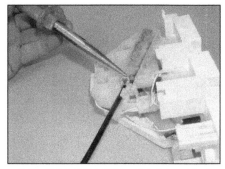

9.7a Extract the relevant outer control cable securing clip . . .

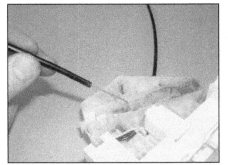

9.7b . . . and disconnect the inner cable from the switch lever

9.10 Lift up the sill trim and remove the lower side trim panel

9.12a Pull out the plastic hinge pin each side . . .

2 Remove the facia centre switch/ventilation nozzle housing as described in Chapter 11, Section 29.

3 Working under the facia at the heater unit, mark the outer control cable locations with respect to the cable clips, then release the clips and disconnect the cables from the heater unit.

4 Undo the four screws securing the control unit to the facia **(see illustration)**.

5 Withdraw the control unit from the facia, and disconnect the wiring plugs from the rear of the unit **(see illustration)**.

6 Remove the control unit noting the routing of the two control cables.

7 With the unit removed, the control cables can be disconnected, if required. To do this, extract the relevant outer cable securing clip, and disconnect the inner cable from the switch lever **(see illustrations)**.

Refitting

8 Refitting is a reversal of removal, ensuring that the control cables are refitted in the positions marked prior to removal. If adjustment is required, release the outer cable clips and move the heater controls to either their maximum or minimum positions. Move the operating levers on the heater unit to the corresponding positions, then refit the outer cable clips.

Heater blower motor

Removal

9 Disconnect the battery negative terminal (refer to *Disconnecting the battery* in the Reference Chapter).

10 Extract the plastic clip securing the lower side trim panel to the front door pillar below the glovebox. Lift up the sill trim and remove

the lower side trim panel by pulling it outward to release the metal retaining clip **(see illustration)**.

11 Undo the screw, now exposed, securing the lower edge of the facia to the body.

12 Working under the glovebox, pull out the plastic hinge pin each side and remove the glovebox from the facia **(see illustrations)**.

13 Disconnect the cooling air hose from the bottom of the blower motor casing.

14 Undo the three screws securing the blower motor to the housing. Carefully pull the lower corner of the facia outward slightly, then withdraw the motor assembly from the housing **(see illustrations)**. Disconnect the wiring connector and remove the blower motor.

Refitting

15 Refitting is a reversal of removal.

9.12b . . . and remove the glovebox from the facia

9.14a Undo the three blower motor securing screws . . .

9.14b . . . pull the lower corner of the facia outward slightly, then withdraw the motor assembly from the housing

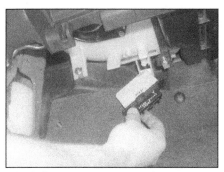

9.19 Remove the two securing screws, and withdraw the resistor from the blower motor housing

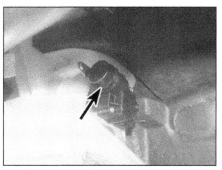

9.21 Heater air flap motor location (arrowed) at the top of the blower motor housing

9.27 Undo the screws and remove the glovebox aperture reinforcement plate from the facia

Heater blower motor resistor

Removal

16 The resistor is located at the bottom of the blower motor housing, behind the blower motor.

17 Disconnect the battery negative terminal (refer to *Disconnecting the battery* in the Reference Chapter).

18 Working under the facia, behind the glovebox, disconnect the wiring plug from the resistor.

19 Remove the two securing screws, and withdraw the resistor from the motor housing **(see illustration)**.

Refitting

20 Refitting is a reversal of removal.

Heater air flap motor

Removal

21 The air flap motor is located at the top of the blower motor housing **(see illustration)**.

22 Working under the glovebox, pull out the plastic hinge pin each side and remove the glovebox from the facia.

23 Disconnect the motor wiring connector, then unscrew the retaining bolts and withdraw the motor.

Refitting

24 Refitting is a reversal of removal, but make sure that the air flap motor lever is fitted to the slit portion of the air flap actuating link.

Pollen filter

Removal

25 The pollen filter is located in the air intake housing between the blower motor and the heater assembly.

26 Working under the glovebox, pull out the plastic hinge pin each side and remove the glovebox from the facia.

27 Undo the four screws and remove the glovebox aperture reinforcement plate from the facia **(see illustration)**.

28 If the filter has not been previously removed, it will be necessary to cut out a portion of the glovebox aperture in the facia to provide sufficient clearance for removal. To do this, cut out the lower centre of the

glovebox aperture, along the moulded cutting lines, using a hacksaw **(see illustration)**.

29 Remove the metal retaining clip and slide out the lower portion of the two-piece pollen filter. With the lower portion removed, the upper portion can now be withdrawn.

Refitting

30 Slide the upper portion of the filter into the housing and move it up to its fitted position. Hold the it in place and slide in the lower portion. Secure the lower portion with the retaining clip.

31 Refit the glovebox aperture reinforcement plate and secure with the four screws.

32 The cut-out portion of the glovebox aperture can be re-attached to the reinforcement plate using self-tapping screws through the holes provided in the plate.

33 Locate the glovebox in position and refit the two plastic hinge pins.

Heater matrix

34 Removal and refitting of the heater matrix can only be carried out with the complete heater assembly removed from the car. This entails removal of the centre console, complete facia assembly, facia support braces and cross-tube, electronic control units and all associated wiring harnesses. This is an extremely complex operation which is considered beyond the scope of this manual. Should it be necessary to remove and refit the matrix or any of the heater casing components, this work should be entrusted to a Nissan dealer.

9.28 Using a hacksaw to cut out the lower centre of the glovebox aperture

10 Air conditioning system – general information and precautions

General information

An air conditioning system is available on the majority of Almera models covered by this manual. It enables the temperature of incoming air to be lowered, and also dehumidifies the air, which makes for rapid demisting and increased comfort.

The cooling side of the system works in the same way as a domestic refrigerator. Refrigerant gas is drawn into a belt-driven compressor, and passes into a condenser mounted on the front of the radiator, where it loses heat and becomes liquid. The liquid passes through an expansion valve to an evaporator, where it changes from liquid under high pressure to gas under low pressure. This change is accompanied by a drop in temperature, which cools the evaporator. The refrigerant returns to the

TOOL TiP

Many car accessory shops sell one-shot air conditioning recharge aerosols. These generally contain refrigerant, compressor oil, leak sealer and system conditioner. Some also have a dye to help pinpoint leaks.

⚠️ *Warning: These products must only be used as directed by the manufacturer, and do not remove the need for regular maintenance.*

compressor, and the cycle begins again.

Air blown through the evaporator passes to the air distribution unit, where it is mixed with hot air blown through the heater matrix to achieve the desired temperature in the passenger compartment.

The heating side of the system works in the same way as on models without air conditioning (see Section 8).

The system is electronically-controlled and any problems with the system should be referred to a Nissan dealer.

Precautions

With an air conditioning system, it is necessary to observe special precautions whenever dealing with any part of the system, or its associated components. If for any reason the system must be disconnected, entrust this task to your Nissan dealer or a refrigeration engineer.

⚠️ *Warning: The refrigeration circuit contains a liquid refrigerant, and it is therefore dangerous to disconnect any part of the system without specialised knowledge and equipment. The refrigerant is potentially dangerous, and should only be handled by qualified persons. If it is splashed onto the skin, it can cause frostbite. It is not itself poisonous, but in the presence of a naked flame (including a cigarette) it forms a poisonous gas. Uncontrolled discharging of the refrigerant is dangerous, and potentially damaging to the environment.*

Do not operate the air conditioning system if it is known to be short of refrigerant, as this may damage the compressor.

11 Air conditioning system components – removal and refitting

⚠️ *Warning: Do not attempt to open the refrigerant circuit. Refer to the precautions given in Section 10.*

1 The only operation which can be carried out easily without discharging the refrigerant is renewal of the auxiliary (compressor) drivebelt which is described in Chapter 1. All other operations must be referred to a Nissan dealer or an air conditioning specialist.

2 If necessary for access to other components, the compressor can be unbolted and moved aside, *without disconnecting its flexible hoses*, after removing the drivebelt **(see illustration)**.

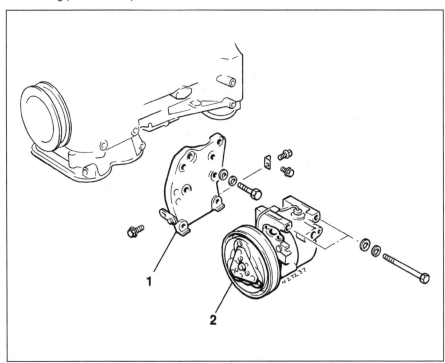

11.2 Air conditioning compressor mounting details – do not disconnect refrigerant lines

1 *Compressor mounting bracket*   2 *Compressor*

Chapter 4 Part A:
Fuel and exhaust systems

Contents

Degrees of difficulty

Easy, suitable for novice with little experience	**Fairly easy,** suitable for beginner with some experience	**Fairly difficult,** suitable for competent DIY mechanic	**Difficult,** suitable for experienced DIY mechanic	**Very difficult,** suitable for expert DIY or professional

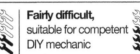

Specifications

General

System type . . . . Nissan Electronic Concentrated Control System (ECCS) multi-point injection

Fuel system data

For idle speed and mixture settings, refer to Chapter 1 Specifications

Fuel pump type . . . . Electric, immersed in tank
Fuel pump regulated constant pressure (approximate) . . . . 3.0 bars

Fuel system component test data

Fuel pump resistance (at fuel gauge sender unit connection) . . . . 0.2 to 5.0 ohms
Throttle potentiometer resistances:
  Meter connected between terminals A and B (see text Section 13):
    Throttle valve closed . . . . 0.5 kilohms
    Throttle valve partially open . . . . 0.5 to 4.0 kilohms
    Throttle valve fully open . . . . 4.0 kilohms
Fuel injector resistance . . . . 10 ohms
Auxiliary air control (AAC) valve resistance . . . . 10 ohms
Fast idle control device (FICD) solenoid valve:
  Apply 12 volts across solenoid terminals . . . . Solenoid should click, indicating correct operation
Power steering pressure switch:
  Steering wheel being turned . . . . Continuity between switch terminals
  Steering wheel stationary . . . . Open-circuit between switch terminals
Coolant temperature sensor resistances:
  At 20°C (68°F) . . . . 2.1 to 2.9 kilohms
  At 90°C (194°F) . . . . 0.24 to 0.26 kilohms
  At 110°C (230°F) . . . . 0.14 to 0.15 kilohms

Recommended fuel

Minimum octane rating . . . . 95 RON unleaded

Torque wrench settings

	Nm	lbf ft
Exhaust manifold nuts and bolts	27	20
Exhaust system fasteners:		
Catalytic converter support bracket bolts	18	13
Catalytic converter to manifold bolts	32	24
Front pipe-to-catalytic converter nuts	30	22
Intermediate pipe and tailpipe flange joint nuts	35	26
Fuel rail retaining bolts:		
Stage 1	10	7
Stage 2	18	13
Inlet manifold nuts and bolts	18	13
Fuel tank mounting bolts	30	22
Throttle housing retaining bolts:		
Stage 1	10	7
Stage 2	20	15

---

1 General information and precautions

The fuel system consists of a fuel tank mounted centrally under the car with an electric fuel pump immersed in it, a fuel filter, fuel feed and return lines. The fuel pump supplies fuel to the fuel rail which acts as a reservoir for the four fuel injectors which inject fuel into the inlet tracts. A fuel filter is incorporated in the feed line from the pump to the fuel rail to ensure that the fuel supplied to the injectors is clean.

Refer to Section 6 for further information on the operation of the fuel injection system, and Section 16 for information on the exhaust system.

⚠ **Warning: Many of the procedures in this Chapter require the removal of fuel lines and connections which may result in some fuel spillage. Before carrying out any operation on the fuel system refer to the precautions given in 'Safety first!' at the beginning of this Manual and follow them implicitly. Petrol is a highly dangerous and volatile liquid and the precautions necessary when handling it cannot be overstressed. Residual pressure will remain in the fuel lines long after the vehicle was last used, when disconnecting any fuel line, depressurise the fuel system as described in Section 7.**

2 Air cleaner assembly – removal and refitting

Removal

1 Release the retaining clips, then lift off the air cleaner housing lid along with its seal.

2 Lift out the filter element, noting which way around it is fitted.

3 Disconnect the crankcase ventilation hose from the left-hand end of the air cleaner housing (see illustration).

4 Undo and remove the retaining bolt, slacken the retaining clip, then free the air cleaner housing from the top of the throttle housing (see illustrations).

5 Disconnect the inlet duct from the left-hand end of the air cleaner housing (see illustration).

6 Disconnect the vacuum hose and remove the air cleaner housing (see illustration). Collect the rubber sealing ring if it remained in position on the throttle housing.

7 To remove the inlet duct, unscrew the lower mounting bracket bolt adjacent to the battery and the upper mounting bracket bolt on the inner wing (see illustrations).

8 Disengage the lower mounting pegs from the mounting plate rubber bushes and withdraw the inlet duct from the engine compartment (see illustration).

2.3 Disconnect the crankcase ventilation hose from the left-hand end of the air cleaner housing

2.4a Undo and remove the air cleaner retaining bolt . . .

2.4b . . . slacken the clip securing the air cleaner to the throttle housing . . .

2.4c . . . and free the air cleaner from the top of the throttle housing

2.5 Disconnect the inlet duct from the left-hand end of the air cleaner housing

Refitting

9 Refitting is a reversal of the relevant removal procedure, ensuring that all hoses are properly reconnected, and that all ducts are correctly seated and securely held by their retaining clips.

3 Accelerator cable –
removal, refitting
and adjustment

Removal

1 Remove the air cleaner assembly as described in Section 2.
2 Free the accelerator inner cable from the throttle cam on the throttle housing.
3 Slacken the outer cable locknut and adjuster nut, then free the outer cable from its mounting bracket **(see illustration)**.
4 Working back along the length of the cable, free it from any relevant retaining clips or ties, whilst noting its correct routing.
5 Working inside the vehicle, remove the lower facia panel from the driver's side of the facia, as described in Chapter 11, Section 29.
6 Reach up behind the facia, and detach the inner cable from the top of the accelerator pedal.
7 Undo the bolts securing the outer cable end fitting to the bulkhead, then withdraw the cable through the bulkhead and into the passenger compartment.

Refitting

8 From inside the vehicle, feed the cable through the bulkhead aperture.
9 When the end of the cable appears in the engine compartment, pull the cable through the bulkhead, and route it correctly around the engine compartment.
10 Return to the inside of the vehicle, and secure the outer cable end fitting to the bulkhead by securely tightening its retaining bolts. Clip the inner cable into position on the top of the accelerator pedal.
11 Make sure that the cable is securely retained, then refit the lower panel to the facia (see Chapter 11).
12 From within the engine compartment, work along the cable, and ensure that it is secured in position with all the relevant retaining clips and ties, and correctly routed.
13 Locate the outer cable in its mounting bracket, and reconnect the inner cable to the throttle cam. Adjust the cable as described below.

Adjustment

14 With the throttle cam resting against its stop, slacken the accelerator cable locknut, and position the adjuster nut so that only a slight amount of free play is present in the inner cable. Hold the adjuster nut stationary, and securely tighten the locknut.
15 Have an assistant depress the accelerator pedal, and check that the throttle cam opens

2.6 Disconnect the vacuum hose and remove the air cleaner housing

2.7b ... and the upper mounting bracket bolt on the inner wing (arrowed)

fully and returns smoothly to its stop. If necessary, readjust the cable as described above.
16 Refit the air cleaner assembly as described in Section 2 on completion.

4 Accelerator pedal –
removal and refitting

Removal

1 Remove the lower facia panel from the driver's side of the facia, as described in Chapter 11, Section 29.
2 Reach up behind the facia, and detach the inner cable from the top of the accelerator pedal.

3.3 Slacken the locknut (arrowed) and free the accelerator cable from its mounting bracket

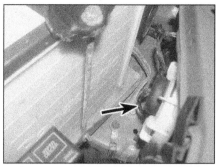

2.7a Unscrew the air inlet duct lower mounting bracket bolt (arrowed) ...

2.8 Disengage the lower mounting pegs from the rubber bushes and remove the air inlet duct

3 Undo the two mounting bolts securing the pedal mounting bracket to the bulkhead, and remove the pedal assembly from underneath the facia **(see illustration)**.
4 If necessary, remove the retaining clip from the end of the accelerator pedal pivot shaft, then slide the pedal out of position and recover the return spring from the mounting bracket.
5 Examine the mounting bracket and pedal pivot points for signs of wear, and renew as necessary.

Refitting

6 Refitting is a reversal of the removal procedure, applying a little multi-purpose grease to the pedal pivot shaft. On completion, adjust the accelerator cable as described in Section 3.

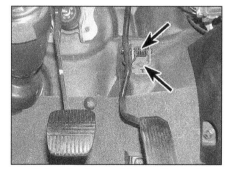

4.3 Accelerator pedal mounting bolts (arrowed)

5 Unleaded petrol – general information and usage

Note: *The information given in this Chapter is correct at the time of writing. If updated information is thought to be required, check with a Nissan dealer. If travelling abroad, consult one of the motoring organisations (or a similar authority) for advice on the petrols available, and their suitability for your vehicle.*

1 The fuel recommended by Nissan is given in the Specifications Section of this Chapter.

2 All Nissan Almera petrol engine models are designed to run on fuel with a minimum octane rating of 95 (RON). All models have a catalytic converter must be run on unleaded fuel **only**. Under no circumstances should leaded fuel or lead replacement petrol be used, as this may damage the catalyst.

6 Fuel injection system – general information

All models are fitted with a combined fuel injection/ignition (engine management) system, otherwise known as the Electronic Concentrated Control System (ECCS). Refer to Chapter 5B for information on the ignition side of the system, the fuel injection side of the system operates as follows.

The fuel pump, immersed in the fuel tank, supplies fuel from the fuel tank to the fuel rail, via a filter mounted on the engine compartment bulkhead. Fuel supply pressure is controlled by the pressure regulator, on the end of the fuel rail, which lifts to allow excess fuel to return to the tank when the optimum operating pressure of the fuel system is exceeded.

The electrical control system consists of the ECCS electronic control unit, along with the following sensors:

a) *Throttle potentiometer – informs the ECCS control unit of the throttle valve position, and the rate of throttle opening/closing.*

b) *Coolant temperature sensor – informs the ECCS control unit of engine temperature.*

c) *Airflow meter – informs the ECCS control unit of the mass and temperature of the air passing through the inlet duct.*

d) *Camshaft position sensor – housed in the distributor, the sensor informs the ECCS control unit of the engine speed and crankshaft position (see Chapter 5B for further information).*

e) *Vehicle speed sensor – informs the ECCS control unit of the vehicle speed.*

f) *Power steering and air conditioning system switches (where fitted) – informs the ECCS control unit if the system(s) are in operation, to allow it to adjust the idle speed to compensate for the extra load on the engine.*

g) *Exhaust gas sensor – informs the ECCS control unit of the oxygen content of the exhaust gases (see Part B of this Chapter for further information).*

All the above signals are analysed by the ECCS control unit. Based on this information, the ECCS control unit selects the response appropriate to those values, and controls the fuel injectors (varying their pulse width – the length of time each injector is held open – to provide a richer or weaker mixture, as appropriate). The mixture and idle speed are constantly varied by the ECCS control unit to provide the best settings for cranking, starting (with either a hot or cold engine) and engine warm-up, idle, cruising, and acceleration.

The ECCS control unit also has full control over the engine idle speed via the auxiliary air control valve. The valve, which is fitted to the throttle housing, controls the opening of an air passage which bypasses the throttle valve. When the throttle valve is closed, the ECCS control unit controls the opening of the valve, which regulates the amount of air which flows through the valve, and so controls the idle speed.

The throttle housing has a built-in fast idle facility, which is controlled by a thermostatic valve. When the engine is cold, the wax capsule in the valve (which is fitted to the throttle housing) is at its smallest, and the fast idle cam holds the throttle valve slightly open. As the engine warms-up, the wax capsule expands, forcing the valve plunger upwards, which in turn rotates the fast idle cam to the required position.

There is also a fast idle control solenoid valve which is controlled by the ECCS control unit. The solenoid valve controls the opening of an air passage which bypasses the throttle valve, and this is used to raise the idle speed when either the power steering and/or air conditioning systems are in operation. **Note:** *On some models, the valve may be fitted but will not be operational.*

The ECCS control unit also controls the exhaust and evaporative emission control systems, which are described in detail in Part B of this Chapter.

If there is an abnormality in any of the readings obtained from the sensors, the ECCS control unit switches to its back-up mode. If this happens, it ignores the abnormal sensor signal, and assumes a preprogrammed value which will allow the engine to continue running, albeit at reduced efficiency. If the ECCS control unit enters its back-up mode, the warning light on the instrument panel will come on, and the relevant fault code will be stored in the ECCS control unit memory.

If the warning light comes on, the vehicle should be taken to a Nissan dealer at the earliest opportunity. Once there, a complete test of the engine management system can be carried out, using a special electronic diagnostic test unit which is simply plugged into the system's diagnostic connector.

Note: *The ECCS control unit also has a self-diagnostic mode which can be accessed by the DIY mechanic. See Section 11 for further information.*

7 Fuel system – depressurisation

Note: *Refer to the warning note in Section 1 before proceeding.*

⚠ *Warning: The following procedure will merely relieve the pressure in the fuel system – remember that fuel will still be present in the system components, and take precautions accordingly before disconnecting any of them. Use clean rags wrapped around the connections to catch escaping fuel, and dispose of any fuel-soaked rags with care. Plug or tape over any open fuel lines, to prevent further loss of fuel or ingress of dirt.*

1 The fuel system referred to in this Section is defined as the tank-mounted fuel pump, the fuel filter, the fuel rail and injectors, the pressure regulator, and the metal pipes and flexible hoses of the fuel lines between these components. All these contain fuel which will be under pressure while the engine is running and/or while the ignition is switched on. The pressure will remain for some time after the ignition has been switched off, and must be relieved before any of these components are disturbed for servicing work.

2 Identify and remove the fuel pump fuse (typically fuse 17) from the passenger compartment fusebox – the fuses can also be identified from the label inside the fusebox cover, or from the wiring diagrams at the end of this manual.

3 Start the engine, and allow it to run until it stalls.

4 Try to start the engine at least twice more, to ensure that all residual pressure has been relieved.

5 Disconnect the battery negative terminal (refer to *Disconnecting the battery* in the Reference Chapter).

6 For safety, the fuel pump fuse should not be refitted until all work on the fuel system has been completed. If you refit the fuse now, **do not** switch on the ignition until completion of work.

8 Fuel gauge sender unit – removal and refitting

Note: *Refer to the warning note in Section 1 before proceeding.*

Removal

1 Depressurise the fuel system as described in Section 7.

2 Disconnect the battery negative terminal

(refer to *Disconnecting the battery* in the Reference Chapter).

3 To gain access to the sender unit, lift out the rear seat cushion, or fold the cushion forward, according to model.

4 Undo the retaining screws, and lift up the access cover to expose the sender unit.

5 Brush away any accumulated dust or dirt around the top of the sender unit so it does not drop into the fuel tank when the unit is removed **(see illustration)**.

6 Disconnect the wiring connectors from the top of the sender unit **(see illustration)**.

7 Mark the fuel feed and return hoses for identification purposes, then release the hose retaining clips **(see illustration)**. Disconnect both hoses from the top of the unit, and plug the hose ends.

8 Note the fitted position of the sender unit and if necessary make an alignment mark on the sender unit and fuel tank to ensure correct refitting.

9 Undo the six retaining screws and lift off the sender unit retaining plate.

10 Carefully lift the sender unit from the top of the fuel tank then, when sufficient clearance exists, disconnect the fuel pump wiring connector and the two fuel hoses from its base **(see illustration)**. Identify the hoses to ensure correct refitting

11 Remove the sender unit from the fuel tank taking care not to bend the float arm or to spill fuel onto the interior of the vehicle. Recover the rubber sealing ring from the tank aperture and discard it; a new one must be used on refitting.

Refitting

12 Refitting is a reversal of the removal procedure, noting the following points:

   a) *Fit a new rubber sealing ring to the fuel tank.*
   b) *Ensure that the fuel hoses are correctly reconnected to their original locations on the sender unit, and are securely held by their retaining clips.*
   c) *Align the marks made on the sender unit and fuel tank during removal and secure the unit with the retaining plate.*
   d) *Ensure that the fuel feed and return hoses are correctly reconnected, and are securely retained by their clips.*

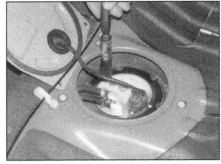

8.5 Brush away any accumulated dust or dirt around the top of the fuel gauge sender unit

8.7 Release the hose retaining clips and disconnect the fuel hoses from the sender unit

   e) *Prior to refitting the access cover, reconnect the battery, then start the engine and check the fuel hoses for signs of leaks.*
   f) *Refit the access cover with the arrow and front mark toward the front of the vehicle.*

9 Fuel pump – removal and refitting

Note: *Refer to the warning note in Section 1 before proceeding.*

Removal

1 Remove the fuel gauge sender unit as described in Section 8.

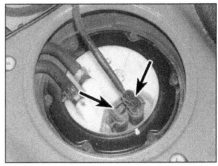

8.6 Disconnect the wiring connectors (arrowed) from the top of the sender unit

8.10 Disconnect the fuel pump wiring connector and the two fuel hoses from the base of the sender unit

2 Reach into the fuel tank, lift the front of the pump assembly upwards and slide it forward to disengage the retaining lugs. Withdraw the pump assembly from the fuel tank **(see illustration)**.

3 Disconnect the fuel feed hose and wiring connector from the end of the fuel pump **(see illustrations)**.

4 Release the four retaining lugs and separate the pump chamber cover from the base **(see illustrations)**.

5 Remove the fuel pump from the pump chamber base **(see illustration)**.

Refitting

6 Locate the fuel pump in the pump chamber base then clip the chamber cover into place.

7 Reconnect the wiring connector and fuel

9.2 Release the fuel pump from the fuel tank retaining lugs and withdraw the pump assembly

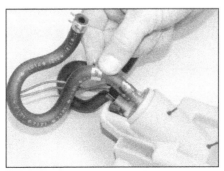

9.3a Disconnect the fuel feed hose . . .

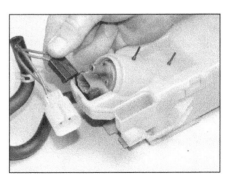

9.3b . . . and wiring connector from the end of the fuel pump

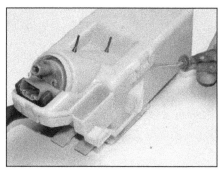

9.4a Release the four retaining lugs . . .

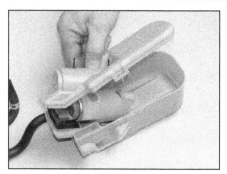

9.4b . . . and separate the fuel pump chamber cover from the base

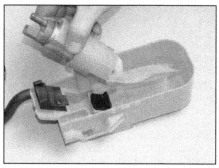

9.5 Remove the fuel pump from the pump chamber base

feed hose to the pump, ensuring that the wiring is routed around the left-hand side of the feed hose.

8 Place the pump assembly in the fuel tank, and engage the rear retaining lugs. Slide the assembly to the rear, push it down and engage the front retaining lug.

9 Refit the fuel gauge sender unit as described in Section 8.

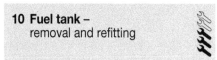

10 Fuel tank –
removal and refitting

Note: *Refer to the warning note in Section 1 before proceeding.*

Removal

1 Before removing the fuel tank, all fuel must be drained from the tank. Since a fuel tank drain plug is not provided, it is therefore preferable to carry out the removal operation when the tank is nearly empty.

2 Depressurise the fuel system as described in Section 7.

3 Disconnect the battery negative terminal (refer to *Disconnecting the battery* in the Reference Chapter), then syphon or hand-pump the remaining fuel from the tank.

4 For improved access, remove the relevant sections of the exhaust system as described in Section 16.

5 Disconnect the wiring connectors and fuel hoses from the fuel gauge sender unit, as described in paragraphs 1 to 7 of Section 8.

6 Working at the left-hand side of the rear of the fuel tank, release the retaining clips, then disconnect the filler neck vent pipe and main filler neck hose from the fuel tank.

7 Place a trolley jack with an interposed block of wood beneath the tank, then raise the jack until it is supporting the weight of the tank.

8 Working around the periphery of the fuel tank, slacken and remove the bolts securing the tank to the vehicle body.

9 Slowly lower the fuel tank out of position, disconnecting any other relevant vent pipes as they become accessible (where necessary). Remove the tank from underneath the vehicle, and recover the tank mounting rubbers, noting their correct fitted positions.

10 If the tank is contaminated with sediment or water, remove the sender unit and fuel pump (Sections 8 and 9) and swill the tank out with clean fuel. If any damage is evident, the tank should be renewed.

Refitting

11 Refitting is the reverse of the removal procedure, noting the following points:
a) *When lifting the tank back into position, reconnect all the relevant breather hoses, and take great care to ensure that none of the hoses become trapped between the tank and vehicle body. Tighten the fuel tank mounting bolts to the specified torque setting.*
b) *Ensure that all pipes and hoses are correctly routed, and securely held in position with their retaining clips.*
c) *On completion, refill the tank with fuel, and check for signs of leakage prior to taking the vehicle on the road.*

11 Fuel injection system –
fault diagnosis and adjustment

Fault diagnosis

General information

1 If a fault appears in the fuel injection/ignition system, first ensure that all the system wiring connectors are securely connected and free of corrosion. Then ensure that the fault is not due to poor maintenance –

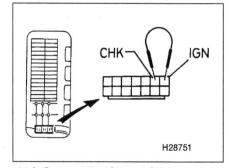

11.6 Connect the IGN and CHK terminals using a jumper wire

ie, check that the air cleaner filter element is clean, that the spark plugs are in good condition and correctly gapped, that the valve clearances are correctly adjusted, the cylinder compression pressures are correct, the ignition timing is correct, and that the emission control systems are operating correctly, referring to the appropriate parts of Chapters 1, 2, 4 and 5 for further information.

2 If these checks fail to reveal the cause of the problem, a quick check of the fuel injection/ignition circuits can be performed by setting the ECCS control unit to its self-diagnostic mode 2. In mode 2, the ECCS control unit will reveal any fault codes stored in its memory, using the engine check light in the instrument panel and the red LED on the right-hand side of the control unit.

3 Faults detected by ECCS control unit are stored in its memory, until the starter motor has been operated 50 times. If the fault is not detected again within this period, it will automatically be erased from the memory. Fault codes can also be erased from the memory by setting the control unit to self-diagnostic mode 2 and then switching it back to mode 1, as described below, or by leaving the battery disconnected for more than 24 hours.

Setting the self-diagnostic modes

4 Remove the passenger compartment fusebox cover, to gain access to the diagnostic connector which is clipped to the base of the fusebox.

5 Turn the ignition switch to the 'on' position, but do not start the engine. The ECCS control unit is now in self-diagnostic mode 1.

6 Using a spare piece of wire, connect the IGN terminal of the diagnostic connector to the CHK terminal **(see illustration)**. Keep the terminals connected for at least 2 seconds, then disconnect the wire. The ECCS control unit is now in self-diagnostic mode 2.

7 In mode 2, the control unit will reveal any fault codes stored in its memory.

8 The code is revealed using a series of long (0.6 second) and short (0.3 second) flashes of the instrument panel engine check light. The long flashes indicate the number of ten digits and the short flashes indicate the number of single digits. The long flashes, which indicate the first digit of the fault code, will be given out first, then after a gap of approximately 0.9 seconds, the short flashes, which indicate the second digit of the fault code, will follow. For example, five long flashes, followed by five short flashes would indicate fault code fifty five. There will be a gap of 2.1 seconds before any other codes are revealed. Once all codes have been revealed, the ECU will continuously run through the code(s) stored in its memory, revealing each one in turn with a gap of 2.1 seconds between each code. The fault codes are as follows.

Code	Faulty circuit
11	Camshaft position sensor circuit
12	Airflow meter circuit
13	Coolant temperature sensor circuit
21	Ignition signal circuit
43	Throttle potentiometer circuit
55	All circuits operating correctly

9 With the control unit in mode 2, if the engine is started it will automatically enter its exhaust gas sensor check mode. In this mode, the engine check light indicates the condition of the exhaust gases. When the light is illuminated, the exhaust gas mixture is lean, and when it is off, the mixture is rich. To check the sensor, with the control unit in self-diagnostic mode 2, start the engine and warm it up to normal operating temperature. Once it is warm, raise the engine speed to approximately 2000 rpm, and hold it there for approximately 2 minutes whilst observing the engine check light. If the exhaust gas sensor is functioning correctly, the light should flash on and off at least 5 times every 10 seconds.

10 When all the checks are complete, exit the self-diagnostic mode 2. If the engine has not been started, this can be achieved by reconnecting the IGN and CHK terminals of the diagnostic connector again for at least two seconds. If the engine is running, exit mode 2 by switching the ignition switch to the 'off' position and disconnecting the battery negative terminal.

11 If a more detailed check of the fuel injection/ignition system is required, take the vehicle to a Nissan dealer. They will have access to the special electronic diagnostic test unit which is plugged into the system's diagnostic connector, and can carry out a full check of the system components.

Adjustment

12 The only adjustment possible is that of the base idle speed. Refer to Chapter 1 for the adjustment procedures.

12 Throttle housing – removal and refitting

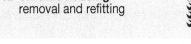

1 Disconnect the battery negative terminal (refer to *Disconnecting the battery* in the Reference Chapter).
2 Remove the air cleaner assembly as described in Section 2.
3 Depress the retaining clips and disconnect

the wiring connectors from the throttle potentiometer, the auxiliary air control valve, the airflow meter and the fast idle control device solenoid valve.
4 Free the accelerator inner cable from the throttle cam. Slacken the outer cable locknut and adjuster nut, then free the outer cable from its mounting bracket.
5 Make a note of the correct fitted positions of all the relevant vacuum pipes and breather hoses, to ensure that they are correctly positioned on refitting, then release the retaining clips (where fitted) and disconnect them from the throttle housing.
6 Release the retaining clips, disconnect the coolant hoses from each side of the throttle housing, and plug the hose ends. Work quickly, to minimise coolant loss.
7 Slacken and remove the bolts securing the throttle housing assembly to the inlet manifold, and remove it from the engine compartment **(see illustration)**. Remove the gasket and discard it; a new one must be used on refitting. Plug the inlet manifold port with a wad of clean cloth, to prevent the possible entry of foreign matter.

Refitting

8 Refitting is a reverse of the removal procedure, bearing in mind the following points:
 a) Ensure that the mating surfaces of the manifold and throttle housing are clean and dry, and fit a new gasket to the manifold. Fit the throttle housing then, working in a diagonal sequence, tighten the retaining bolts to the specified Stage 1 torque setting. Go around again in sequence, and tighten them to the specified Stage 2 torque setting.
 b) Ensure that all hoses are correctly reconnected and, where necessary, that their retaining clips are securely tightened.
 c) On completion, adjust the accelerator cable using the information given in Section 3. On automatic transmission models, also check the kickdown cable adjustment as described in Chapter 7B.

13 Fuel injection system components – removal and refitting

Fuel rail and injectors

Note: *Refer to the warning note in Section 1 before proceeding.*
Note: *If a faulty injector is suspected, before condemning the injector it is worth trying the effect of one of the proprietary injector-cleaning treatments.*
1 Depressurise the fuel system as described in Section 7.
2 Disconnect the battery negative terminal (refer to *Disconnecting the battery* in the Reference Chapter).

12.7 Throttle housing retaining bolts (arrowed)

3 Remove the air cleaner assembly as described in Section 2.
4 Disconnect the vacuum pipe from the fuel pressure regulator.
5 Slacken the retaining clips and disconnect the fuel feed and return hoses from the left-hand end of the fuel rail **(see illustration)**. Label the hoses if wished, to avoid confusion on refitting.
6 Depress the retaining tangs, and disconnect the wiring connectors from the four injectors.
7 Slacken and remove the fuel rail retaining bolts, then carefully ease the fuel rail and injector assembly out from the inlet manifold, and remove it from the vehicle. Recover the spacers fitted between the rail and manifold, and remove each injector seal from the manifold.
8 Undo the two retaining bolts, and remove the retaining plate from the relevant injector. Push the injector out of position, and recover the sealing rings. Repeat the procedure as required to remove any other injectors.
9 Discard the seals and sealing rings; new ones must be used on refitting.
10 Refitting is a reversal of the removal procedure, noting the following points:
 a) Fit new O-rings to all disturbed injectors, and fit new injector seals to the manifold.
 b) Apply a smear of engine oil to the O-rings to aid installation, then ease the injectors into the fuel rail.
 c) Fit a new seal to each injector, and ease the fuel rail assembly into position in the manifold. Fit the spacers between the rail and manifold then, working in a diagonal

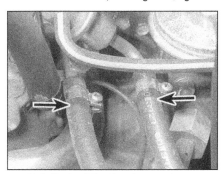

13.5 Fuel rail feed and return hoses (arrowed)

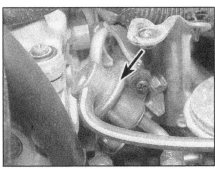

13.12 Fuel pressure regulator (arrowed) is mounted on the left-hand end of the fuel rail

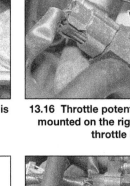

13.16 Throttle potentiometer (arrowed) is mounted on the right-hand side of the throttle housing

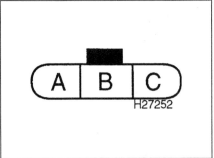

13.22 Throttle potentiometer wiring connector

sequence from the centre outwards, tighten the bolts to their specified Stage 1 torque setting. Go around again in sequence, and tighten them to the specified Stage 2 torque.

d) On completion, start the engine and check for fuel leaks.

Fuel pressure regulator

Note: *Refer to the warning note in Section 1 before proceeding.*

11 Depressurise the fuel system as described in Section 7, then disconnect the battery negative terminal (refer to *Disconnecting the battery* in the Reference Chapter).

12 Disconnect the vacuum hose from the fuel pressure regulator, which is mounted on the left-hand end of the fuel rail **(see illustration)**.

13 Slacken the retaining clips and disconnect the fuel hose from base of the regulator.

13.26 Auxiliary air control (AAC) valve wiring connector (1) and retaining screws (2)

14 Undo the two retaining bolts, and remove the regulator from the end of the fuel rail. Recover the sealing ring fitted to the regulator and discard it; a new one must be used on refitting.

15 Refitting is the reverse of removal, using a new sealing ring. On completion, start the engine and check for fuel leaks.

Throttle potentiometer

16 The throttle potentiometer is mounted onto the right-hand side of the throttle housing **(see illustration)**. Prior to removal, disconnect the battery negative terminal (refer to *Disconnecting the battery* in the Reference Chapter). Remove the air cleaner assembly as described in Section 2 to improve access.

17 Disconnect the wiring connector from the throttle potentiometer.

18 Using a dab of white paint or a suitable

13.31 Disconnect the wiring connector . . .

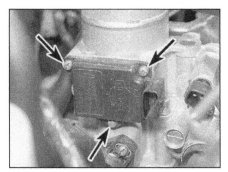

13.32 . . . then undo the retaining screws (arrowed) and remove the airflow meter

marker pen, make alignment marks between the potentiometer and the throttle housing.

19 Undo the two retaining screws, and remove the potentiometer.

20 On refitting, offer up the potentiometer, making sure its lever is correctly engaged with the throttle valve spindle lever.

21 Align the marks made prior to removal, and lightly tighten the retaining screws. Adjust the potentiometer as follows.

22 Connect a multi-meter between terminals A and B of the throttle potentiometer wiring plug socket **(see illustration)**. Set the multi-meter to the resistance scale and check that the potentiometer resistance readings are as given in the Specifications. If necessary, slacken the retaining screws and reposition the potentiometer until its resistances are as specified.

23 When the potentiometer is correctly positioned, securely tighten its retaining screws and reconnect the wiring connector.

24 Refit the air cleaner assembly and reconnect the battery negative terminal.

Auxiliary air control (AAC) valve

25 The auxiliary air control (AAC) valve is mounted onto the front of the throttle housing. Prior to removal, disconnect the battery negative terminal (refer to *Disconnecting the battery* in the Reference Chapter). Remove the air cleaner assembly as described in Section 2 to improve access.

26 Disconnect the wiring connector from the control valve **(see illustration)**.

27 Undo the retaining screws, then remove the valve from the throttle housing and recover the gasket.

28 If necessary, with the valve removed, separate the two halves of the valve by removing the retaining screws, then recover the sealing ring. Check that the valve plunger is free to move easily, and returns quickly under spring pressure. If not, the valve assembly must be renewed.

29 Refitting is the reverse of removal, using a new gasket.

Airflow meter

30 The airflow meter is mounted onto the right-hand side of the throttle housing. Prior to removal, disconnect the battery negative terminal (refer to *Disconnecting the battery* in the Reference Chapter). Remove the air cleaner assembly as described in Section 2 to improve access.

31 Disconnect the wiring connector from the airflow meter **(see illustration)**.

32 Undo the retaining screws, then remove the meter from the throttle housing **(see illustration)**. Recover its sealing ring (where fitted).

33 Refitting is the reverse of removal, using a new sealing ring (where applicable) and tightening its retaining screws securely.

Fast idle control device (FICD)

34 The fast idle control device (FICD) solenoid

valve is screwed into the front of the throttle housing **(see illustration)**. Prior to removal, disconnect the battery negative terminal (refer to *Disconnecting the battery* in the Reference Chapter). Remove the air cleaner assembly as described in Section 2 to improve access.

35 Disconnect the wiring connector from the solenoid.

36 Unscrew the solenoid valve from the body, and recover the plunger, spring and sealing washer.

37 Refitting is the reverse of removal, using a new sealing washer. Ensure that the solenoid plunger and spring are fitted in the correct order, and the correct way around.

Fast idle thermostatic valve

38 The fast idle thermostatic valve is fitted to the left-hand side of the throttle housing. Prior to removal, disconnect the battery negative terminal (refer to *Disconnecting the battery* in the Reference Chapter).

39 Remove the air cleaner assembly as described in Section 2.

40 Unscrew the nut and washer, then withdraw the fast idle cam retaining plate, spring, spacer and lever, noting each component's correct fitted position.

41 Undo the retaining screw, and remove the fast idle valve retaining plate. Lift the valve out of position, and recover its sealing ring.

42 Refitting is the reverse of removal. Use a new sealing ring, and ensure that all the fast idle cam components are refitted in their original positions.

Camshaft position sensor

43 The camshaft position sensor is an integral part of the distributor. Refer to Chapter 5B.

Coolant temperature sensor

44 Refer to Chapter 3.

ECCS control unit

45 The ECCS control unit is situated just in front of the centre console, mounted onto the transmission tunnel floor. Prior to removal, disconnect the battery negative terminal (refer to *Disconnecting the battery* in the Reference Chapter).

46 To gain access, remove the plastic cover fitted over the control unit.

47 Undo the retaining screws, and release the control unit from its mounting bracket. Disconnect the wiring connector(s), and remove the unit from the vehicle.

48 Refitting is the reverse of removal, ensuring that the wiring connector is securely reconnected.

Neutral switch

Manual transmission models

49 Refer to Chapter 7A.

Starter inhibitor/ reversing light switch

Automatic transmission models

50 Refer to Chapter 7B.

13.34 The fast idle control device (FICD) (arrowed) is screwed into the front of the throttle housing

System relays

51 Refer to Chapter 12.

Power steering idle-up switch

52 The power steering idle-up switch is screwed into the power steering feed pipe, in the right-hand rear corner of the engine compartment. Prior to removal, set the front wheels in the straight-ahead position, then disconnect the battery negative terminal (refer to *Disconnecting the battery* in the Reference Chapter).

53 Locate the switch and disconnect its wiring connector.

54 Unscrew the switch, recover its sealing washer, and plug the opening in the pipe. Work quickly, to minimise fluid loss and to prevent dirt entering the hydraulic system.

55 Refitting is the reverse of removal, using a new sealing washer. Check and if necessary top-up the power steering fluid reservoir as described in *Weekly checks*.

Air conditioning idle-up switch

56 The air conditioning switch is screwed into the air conditioning pipe, in the left-hand front corner of the engine compartment. Removal and refitting of the switch requires the air conditioning system to be discharged and recharged (see Chapter 3), and this should not be attempted by the home mechanic.

Vehicle speed sensor

57 The vehicle speed sensor is an integral

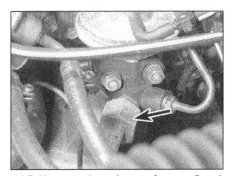

14.5 Unscrew the union nut (arrowed) and disconnect the EGR pipe from the valve

part of the speedometer drive assembly. Refer to Chapter 7A for removal and refitting details.

14 Inlet manifold – removal and refitting

Note: *Refer to the information in Section 1 for details of model identification.*

Removal

1 Drain the cooling system as described in Chapter 1.

2 Depressurise the fuel system as described in Section 7.

3 Carry out the operations described in paragraphs 1 to 6 of Section 12, and disconnect all components from the throttle housing.

4 Make a note of the correct fitted locations of all the relevant inlet manifold vacuum hose connections, and disconnect them from the manifold and (where necessary) from the associated vacuum valves. To avoid confusion on refitting, label each hose as it is disconnected.

5 Unscrew the union nut and free the EGR pipe linking the inlet and exhaust manifolds, from the control valve on the left-hand end of the inlet manifold **(see illustration)**.

6 Disconnect the wiring connectors from the injectors and the engine coolant temperature sensor. Free the wiring from any relevant retaining clips, and position it clear of the manifold. Where necessary, also undo the retaining bolt(s) and free the earth lead(s) from the manifold **(see illustration)**.

7 Undo the retaining bolts and remove the support bracket from the underside of the manifold.

8 Slacken the retaining clips and disconnect the fuel feed and return hoses from the left-hand end of the fuel rail. Plug the hose and rail unions, to minimise fuel loss and to prevent dirt entering the system.

9 Make a final check that all the necessary vacuum/breather hoses have been disconnected from the manifold then, working in the **reverse** of the tightening sequence **(see illustration 14.11)**, slacken and remove the manifold retaining nuts and bolts.

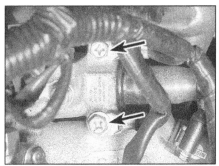

14.6 Undo the bolts, and free the earth leads (arrowed) from the right-hand end of the manifold

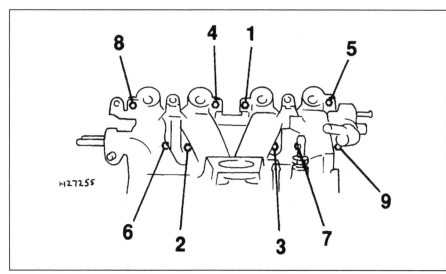

14.11 Inlet manifold nut and bolt tightening sequence

10 Manoeuvre the manifold away from the head, and out of the engine compartment. Remove the manifold gasket and discard it.

Refitting

11 Refitting is the reverse of the removal procedure, noting the following points:

a) Ensure that the manifold and cylinder head mating surfaces are clean and dry, and fit the new gasket to the head studs. Install the manifold, and tighten its retaining nuts and bolts to the specified torque in sequence **(see illustration)**.

b) Ensure that all relevant hoses are

reconnected to their original positions, and are securely held (where necessary) by their retaining clips.

c) Adjust the accelerator cable as described in Section 3.

d) On completion, refill the cooling system as described in Chapter 1.

15 Exhaust manifold – removal and refitting

Removal

1 Trace the wiring back from the exhaust gas sensor to its wiring connector, and disconnect it from the main wiring harness.

2 Slacken and remove the retaining screws, and remove the shroud from the top of the exhaust manifold **(see illustration)**.

3 Unscrew the union nut, and free the EGR pipe from the side of the manifold **(see illustration)**. Note that it may be necessary to slacken the inlet manifold end of the pipe, and the pipe mounting bolts (where fitted), in order to free the pipe from the manifold.

4 Firmly apply the handbrake, then jack up the front of the vehicle and support it securely on axle stands (see *Jacking and vehicle support*).

5 Remove the catalytic converter as described in Section 16.

6 Working in the **reverse** of the tightening sequence **(see illustration 15.8)**, slacken and remove the manifold retaining nuts and bolts.

7 Manoeuvre the manifold out of the engine compartment, and discard the manifold gasket(s).

Refitting

8 Refitting is the reverse of the removal procedure, noting the following points:

a) Examine all the exhaust manifold studs for signs of damage and corrosion; remove all traces of corrosion, and repair or renew any damaged studs.

b) Ensure that the manifold and cylinder head sealing faces are clean and flat, and fit the new manifold gasket(s). Tighten the manifold retaining nuts and bolts to the specified torque in sequence **(see illustration)**.

c) Refit the catalytic converter to the manifold, using the information given in Section 16.

15.2 Removing the exhaust manifold shroud

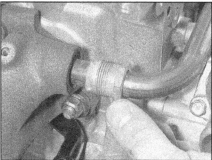

15.3 Unscrew the union nut, and free the EGR pipe from the side of the manifold

16 Exhaust system – general information, component removal and refitting

General information

1 The exhaust system consists of four sections – the catalytic converter, the front pipe, the intermediate pipe and silencer, and the tailpipe and silencer.

2 The system is suspended throughout its

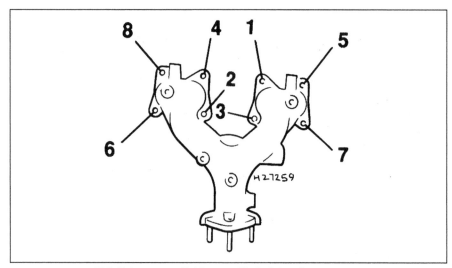

15.8 Exhaust manifold nut and bolt tightening sequence

entire length by rubber mountings, and all exhaust sections are joined by flanged joints which are secured together by nuts and/or bolts.

Removal

3 Each exhaust section can be removed individually or, alternatively, the complete system can be removed as a unit.

4 To remove the system or part of the system, firmly apply the handbrake, then jack up the front of the vehicle and support it securely on axle stands (see *Jacking and vehicle support*). Alternatively, position the car over an inspection pit, or on car ramps.

Front pipe

5 Undo the nuts securing the front pipe to the catalytic converter. Separate the front pipe from the converter and collect the gaskets.

6 Slacken and remove the two nuts/bolts securing the front pipe flange joint to the intermediate pipe. Withdraw the front pipe from underneath the vehicle, and recover the gasket from the joint.

Catalytic converter

7 Undo the nuts securing the front pipe to the catalytic converter. Separate the front pipe from the converter and collect the gaskets. Suitably support the disconnected front pipe so as not to place undue strain on the system mountings.

8 Undo the retaining bolts and remove the shroud from the front of the converter.

9 Undo the bolts securing the base of the converter to the engine support brackets.

10 From above, undo the bolts securing the catalytic converter to the manifold. Withdraw the converter and collect the gasket and sealing ring.

Intermediate pipe

11 Undo the nuts securing the front pipe and tailpipe flange joints to the intermediate pipe, then separate the joints and recover the gaskets.

12 Unhook the intermediate pipe from its mounting rubbers, then manoeuvre the pipe out from underneath the vehicle.

Tailpipe

13 Undo the two nuts securing the tailpipe to the intermediate pipe. Separate the joint and recover the gasket.

14 Unhook the tailpipe from its mounting rubbers, then manoeuvre the pipe over the rear axle beam and remove it from underneath the vehicle.

Complete system

15 If the complete system is to be removed, it will be necessary to remove the tailpipe first, due to the limited clearance around the rear axle beam. To do this proceed as described in paragraphs 13 and 14.

16 With the tailpipe removed, undo the nuts securing the front pipe to the catalytic converter. Separate the front pipe from the converter, and collect the gaskets.

17 With the aid of an assistant, free the system from all its mounting rubbers, and manoeuvre it out from underneath the vehicle.

Heat shield(s)

18 The heat shields are secured in position by a mixture of nuts, bolts and clamps. When an exhaust section is renewed, transfer any relevant heat shields from the original over to the new section before installing the exhaust section on the vehicle.

Refitting

19 Each section is refitted by a reverse of the removal sequence, noting the following points:

a) *Ensure that all traces of corrosion have been removed from the flanges, and renew all necessary gaskets.*

b) *Inspect the rubber mountings for signs of damage or deterioration, and renew as necessary.*

c) *Prior to tightening the exhaust system fasteners to the specified torque, ensure that all rubber mountings are correctly located, and that there is adequate clearance between the exhaust system and vehicle underbody/suspension components, etc.*

Notes

Chapter 4 Part B:
Emissions control systems

Contents

Degrees of difficulty

Easy, suitable for novice with little experience		Fairly easy, suitable for beginner with some experience		Fairly difficult, suitable for competent DIY mechanic		Difficult, suitable for experienced DIY mechanic		Very difficult, suitable for expert DIY or professional	

Specifications

Torque wrench setting	Nm	lbf ft
Exhaust gas sensor to manifold	45	32

1 General information

All Nissan Almera models covered in this manual use unleaded petrol, and have various features built into the fuel/exhaust system to help minimise harmful emissions.

In addition to the catalytic converter, a crankcase emissions control system, evaporative emissions control system and exhaust gas recirculation (EGR) system are also used.

The various emissions control systems operate as follows.

Catalytic converter

To minimise the amount of pollutants which escape into the atmosphere, all models are fitted with a catalytic converter in the exhaust system. The system is of the 'closed-loop' type, in which an exhaust gas sensor provides the ECCS control unit constant feedback, enabling the unit to adjust the mixture to provide the best possible conditions for the converter to operate.

The sensor's tip is sensitive to oxygen, and sends the control unit a varying voltage depending on the amount of oxygen in the exhaust gases; if the intake air/fuel mixture is too rich, the sensor sends a high-voltage signal. The voltage falls as the mixture weakens. Peak conversion efficiency of all major pollutants occurs if the intake air/fuel mixture is maintained at the chemically-correct ratio for the complete combustion of petrol – 14.7 parts (by weight) of air to 1 part of fuel (the 'stoichiometric' ratio). The sensor output voltage alters in a large step at this point, the control unit using the signal change as a reference point, and correcting the intake air/fuel mixture accordingly by altering the fuel injector pulse width (injector opening time). The sensor has a built-in heating element (controlled by the control unit), to quickly bring the sensor's tip to an efficient operating temperature.

Crankcase emissions

To reduce the emission of unburned hydrocarbons from the crankcase into the atmosphere, the engine is sealed, and the blow-by gases and oil vapour are drawn from inside the crankcase, through the PCV valve, into the inlet tract, to be burned by the engine during normal combustion.

Under conditions of high manifold vacuum, the gases will be sucked positively out of the crankcase. Under conditions of low manifold vacuum, the gases are forced out of the crankcase by the (relatively) higher crankcase pressure; if the engine is worn, the raised crankcase pressure (due to increased blow-by) will cause some of the flow to return under all manifold conditions.

Evaporative emissions control

To minimise the escape of unburned hydrocarbons into the atmosphere, an evaporative emissions control system is fitted. The fuel tank filler cap is sealed, and a carbon canister collects the petrol vapours generated in the tank when the car is parked. It stores them until the vapours can be cleared into the inlet tract when the engine is running.

The system is controlled by the ECCS control unit via a solenoid control valve; the same solenoid control valve also operates the EGR system. To ensure that the engine runs correctly when it is cold and/or idling, and to protect the catalytic converter from the effects of an over-rich mixture, the solenoid control valve is not opened by the control unit until the engine has warmed-up and is under load. Once these conditions are met, the valve is modulated on and off to allow the stored vapour to pass into the inlet tract.

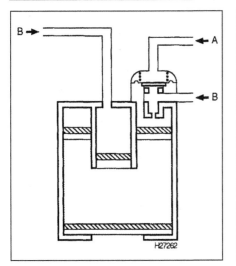

2.3 Carbon canister hose union identification

For A and B, refer to text

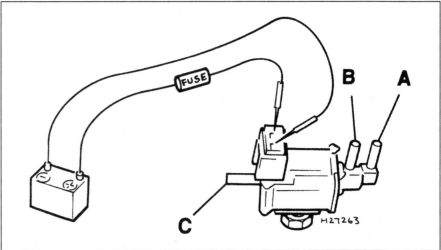

2.6 Solenoid control valve test details

For A, B and C, refer to text

Exhaust gas recirculation (EGR)

This system reduces the amount of unburnt hydrocarbons in the exhaust gases before the gases reach the catalytic converter. This is achieved by taking some of the exhaust gases from the exhaust manifold and recirculating them back into the inlet manifold, via a pipe linking the two, where they are burned again during normal combustion. The exhaust gas recirculation (EGR) valve is fitted to the inlet manifold end of the pipe.

The system is controlled by the ECCS control unit, via a solenoid control valve and the back-pressure transducer (BPT) valve; the solenoid valve also operates the evaporative emissions control system. When the engine is cold, the ECCS control unit keeps the solenoid control valve closed, cutting off the vacuum supply to the EGR valve. When the engine reaches operating temperature, the ECCS control unit opens the solenoid valve, allowing the vacuum supply to act on the EGR valve, via the BPT valve. The BPT valve is sensitive to the exhaust gas back-pressure, and regulates the EGR valve on and off accordingly. When the back-pressure is high, the BPT valve closes, allowing the vacuum supply to act on the EGR valve, opening the valve. When the

2.8 Carbon canister assembly

back-pressure drops, the BPT valve opens, cutting off the vacuum supply to the EGR valve, and so closing the valve.

2 Emissions control systems – testing and component renewal

Crankcase emissions control

1 This system requires no attention, other than to check that the hose(s) are clear and undamaged, and to renew the PCV filter at the intervals given in Chapter 1.

Evaporative emissions control

Testing

2 If the system is thought to be faulty, disconnect the hoses from the carbon canister and solenoid control valve, and check that they are clear by blowing through them. The canister is located in the left-hand rear corner of the engine compartment, and the solenoid control valve is mounted onto the rear of the inlet manifold.

3 The carbon canister and valve can be tested as follows **(see illustration)**. Disconnect the hoses from the canister, and blow down port A of the canister; there should be no sign of leakage. Apply vacuum to port A then blow down each of the ports B in turn. Both ports should be clear, and should freely pass air. If the checks do not give the expected results, the canister and valve assembly is faulty and must be renewed.

4 The solenoid control valve can be checked as follows **(see illustration 2.6)**. If necessary, remove the valve as described in paragraphs 13 to 16 to improve access.

5 Blow down port A, and check that no air flows out of port B, but that air flows out of port C.

6 Connect a fused 12 volt supply to the

solenoid valve **(see illustration)**. With the voltage applied, blow down port A, and check that air passes through the valve and flows out of port B, but not out of port C.

7 If the solenoid control valve does not perform as expected, it is faulty and must be renewed.

Carbon canister renewal

8 The carbon canister is located in the left-hand rear corner of the engine compartment **(see illustration)**.

9 Make a note of the correct fitted location of each hose on the canister. To avoid the possibility of connecting the hoses incorrectly on refitting, make identification marks between each hose and its canister union.

10 Release the retaining clips (where fitted) and disconnect the hoses from the top of the canister.

11 Free the canister from its mounting bracket, and remove it from the engine compartment.

12 Refitting is a reverse of the removal procedure, ensuring that the hoses are correctly reconnected.

Solenoid control valve renewal

13 The solenoid control valve is mounted onto the rear of the inlet manifold. To improve access, remove the air cleaner assembly as described in Part A of this Chapter.

14 Depress the retaining clip and disconnect the wiring connector from the valve.

15 Make a note of the correct fitted location of each hose on the valve. To avoid the possibility of connecting the hoses incorrectly on refitting, make identification marks between each hose and its valve union.

16 Release the retaining clips (where fitted) then disconnect the hoses from the valve, and free the valve from its mounting bracket.

17 Refitting is a reverse of the removal procedure, ensuring that the hoses are correctly reconnected.

2.23 EGR valve details

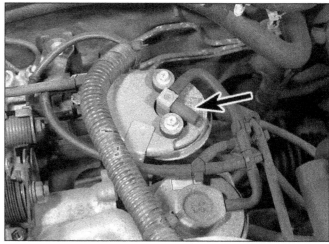

2.28 BPT valve location (arrowed) on the inlet manifold

Exhaust gas recirculation (EGR)

Testing

18 If the system is thought to be faulty, disconnect the hoses from the EGR valve, the back-pressure transducer (BPT) valve and solenoid control valve, and check that they are clear by blowing through them. If all is well, reconnect all hoses.

19 To check the operation of the EGR valve, disconnect the vacuum hose from the top of the valve, and fit a length of hose to the valve union. Suck on the hose end; check that the valve diaphragm is pulled up, and returns quickly when the vacuum is released. This can be checked by placing a finger lightly against the underside of the valve where the movement of the diaphragm can be felt. If the valve operation is sticky or the diaphragm does not move at all, the EGR valve must be renewed.

20 The BPT valve can be checked as follows. Warm the engine up to normal operating temperature, and disconnect the vacuum hose from the EGR valve. Place a finger over the end of the disconnected hose, and rev the engine in short bursts. As the engine speed (and exhaust gas pressure) increases, a vacuum should be felt in the pipe. As the engine speed falls, the vacuum should be switched off by the BPT valve. If this is not the case, the BPT valve is faulty and should be renewed.

21 The solenoid control valve is the same valve that controls the evaporative emissions control system, and can be tested as described previously in paragraphs 4 to 7.

EGR valve renewal

22 Remove the air cleaner assembly as described in Part A of this Chapter.

23 Disconnect the vacuum hose from the EGR valve, which is mounted on the left-hand end of the inlet manifold **(see illustration)**. On some models, access can be further

improved by first removing the BPT valve (see below).

24 Slacken the union nuts, and disconnect the EGR pipe and BPT valve pipe from the EGR valve.

25 Unscrew the two retaining nuts/bolts and washers, and remove the valve from the manifold. Remove the gasket and discard it.

26 Refitting is the reverse of removal, using a new gasket.

Back-pressure transducer (BPT) valve renewal

27 Remove the air cleaner assembly as described in Part A of this Chapter.

28 Disconnect the vacuum hose from the top of the BPT valve, which is mounted on the left-hand side of the inlet manifold **(see illustration)**.

29 Undo the two retaining bolts and remove the valve from the manifold, disconnecting it from the exhaust gas pipe.

30 Refitting is the reverse of removal.

Solenoid control valve renewal

31 Refer to paragraphs 13 to 17 of this Section.

Exhaust emissions control

Testing

32 If the CO level at the tailpipe is too high, the operation of the exhaust gas sensor should be tested using the ECCS control unit self-diagnosis facility as described in Part A of this Chapter. Detailed testing of the sensor and catalytic converter must be left to a Nissan dealer.

Catalytic converter renewal

33 Refer to Part A of this Chapter.

Exhaust gas sensor renewal

Note: The exhaust gas sensor is delicate, and it will not work if it is dropped or knocked, if its power supply is disrupted, or if any cleaning materials are used on it.

34 Trace the wiring back from the exhaust gas sensor, which is screwed into the exhaust manifold **(see illustration)**. Disconnect the wiring connector, and free the wiring from any relevant retaining clips or ties.

35 Unscrew the sensor, and remove it along with its sealing washer.

36 Refitting is a reverse of the removal procedure, using a new sealing washer. Prior to installing the sensor, apply a smear of high-temperature grease to the sensor threads. Ensure that the sensor is securely tightened. Check that the wiring is correctly routed, and in no danger of contacting either the exhaust system or the engine.

3 Catalytic converter – general information and precautions

The catalytic converter is a reliable and simple device which needs no maintenance in itself, but there are some facts of which an owner should be aware if the converter is to function properly for its full service life.

a) DO NOT use leaded petrol, or lead replacement petrol, in a car with a

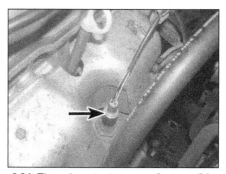

2.34 The exhaust gas sensor (arrowed) is screwed into the top of the exhaust manifold

catalytic converter – the lead will coat the precious metals, reducing their converting efficiency, and will eventually destroy the converter.

b) Always keep the ignition and fuel systems well-maintained in accordance with the manufacturer's schedule (see Chapter 1).

c) If the engine develops a misfire, do not drive the car at all (or at least as little as possible) until the fault is cured.

d) DO NOT push- or tow-start the car – this will soak the catalytic converter in unburned fuel, causing it to overheat when the engine does start.

e) DO NOT switch off the ignition at high engine speeds – ie, do not 'blip' the throttle immediately before switching off.

f) DO NOT use fuel or engine oil additives – these may contain substances harmful to the catalytic converter.

g) DO NOT continue to use the car if the engine burns oil to the extent of leaving a visible trail of blue smoke.

h) Remember that the catalytic converter operates at very high temperatures. DO NOT, therefore, park the car in dry undergrowth, over long grass, or over piles of dead leaves, after a long run.

i) Remember that the catalytic converter is FRAGILE – do not strike it with tools during servicing work.

j) The catalytic converter, used on a well-maintained and well-driven car, should last for between 50 000 and 100 000 miles, but if the converter is no longer effective, it must be renewed.

k) In some cases, a sulphurous smell (like that of rotten eggs) may be noticed from the exhaust. This is common to many catalytic converters when new, however, after a few thousand miles, the problem should disappear.

Chapter 5 Part A:
Starting and charging systems

Contents

Degrees of difficulty

Easy, suitable for novice with little experience	Fairly easy, suitable for beginner with some experience	Fairly difficult, suitable for competent DIY mechanic	Difficult, suitable for experienced DIY mechanic 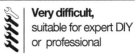	Very difficult, suitable for expert DIY or professional

Specifications

System type . . . . 12 volt, negative earth

Battery
Type . . . . Low-maintenance or maintenance-free, depending on model
Charge condition:
   Poor . . . . 12.5 volts
   Normal . . . . 12.6 volts
   Good . . . . 12.7 volts

Alternator
Type . . . . Mitsubishi or Hitachi
Rating . . . . 70 amp

Starter motor
Type . . . . Mitsubishi or Hitachi

Torque wrench settings	Nm	lbf ft
Alternator lower mounting bolts	44	33
Alternator upper mounting (adjustment link) bolt	19	14
Starter motor mounting bolts	36	27

1 General information and precautions

General information

The engine electrical system consists mainly of the charging and starting systems. Because of their engine-related functions, these components are covered separately from the body electrical devices such as the lights, instruments, etc (which are covered in Chapter 12). Information on the ignition system is covered in Part B of this Chapter.

The electrical system is of the 12 volt negative earth type.

The battery is of the low-maintenance or 'maintenance-free' (sealed for life) type, and is charged by the alternator, which is belt-driven from the crankshaft pulley.

The starter motor is of the pre-engaged type, incorporating an integral solenoid. On starting, the solenoid moves the drive pinion into engagement with the flywheel ring gear before the starter motor is energised. Once the engine has started, a one-way clutch prevents the motor armature being driven by the engine until the pinion disengages from the flywheel.

Precautions

Further details of the various systems are given in the relevant Sections of this Chapter. While some repair procedures are given, the usual course of action is to renew the component concerned. The owner whose interest extends beyond mere component renewal should obtain a copy of the *Automotive*

Electrical & Electronic Systems Manual, available from the publishers of this manual.

It is necessary to take extra care when working on the electrical system, to avoid damage to semi-conductor devices (diodes and transistors), and to avoid the risk of personal injury. In addition to the precautions given in *Safety first!* at the beginning of this manual, observe the following when working on the system:

Always remove rings, watches, etc, before working on the electrical system. Even with the battery disconnected, capacitive discharge could occur if a component's live terminal is earthed through a metal object. This could cause a shock or nasty burn.

Do not reverse the battery connections. Components such as the alternator, ECCS control unit, or any other components having semi-conductor circuitry could be irreparably damaged.

If the engine is being started using jump leads and a slave battery, connect the batteries *positive-to-positive* and *negative-to-negative* (see *Jump starting*). This also applies when connecting a battery charger.

Never disconnect the battery terminals, the alternator, any electrical wiring, or any test instruments, when the engine is running.

Do not allow the engine to turn the alternator when the alternator is not connected.

Never 'test' for alternator output by 'flashing' the output lead to earth.

Never use an ohmmeter of the type incorporating a hand-cranked generator for circuit or continuity testing.

Always ensure that the battery negative terminal is disconnected when working on the electrical system.

Before using electric-arc welding equipment on the car, disconnect the battery, alternator and components such as electronic control units, to protect them from the risk of damage.

Several systems fitted to the vehicle require battery power to be available at all times, either to ensure their continued operation (such as the clock) or to maintain security codes which would be wiped if the battery were to be disconnected. To ensure that there are no unforeseen consequences of this action, refer to *Disconnecting the battery* in the Reference Chapter of this manual for further information.

2 Battery –
checking, testing and charging

Checking

Standard and
low-maintenance battery

1 In addition to the checks described in *Weekly checks* at the start of this manual, the battery electrolyte level should also be periodically checked as follows.

Batteries with a translucent casing

2 On this type of battery, the electrolyte level is visible through the casing. Make sure that the level in each cell is between the UPPER and LOWER level marks on the side of the battery casing.

3 If topping-up is necessary, remove the cell cap(s) and top-up the relevant cell to the UPPER level marking using only distilled water. **Note:** *Do not use ordinary tap water, as this will damage the battery.* Refit the cell cap(s), ensuring each one is securely fitted, and mop-up any spilt water.

Batteries with a solid
(non-translucent) casing

4 On batteries where it is not possible to see the electrolyte level through the casing, the level is checked via the cell filler cap apertures. Remove the cap from each battery cell and, looking down through cap apertures, check that the electrolyte level is up to the base of the aperture neck.

5 If topping-up is necessary, top-up the relevant cell to the base of the neck using only distilled water. **Note:** *Do not use ordinary tap water, as this will damage the battery.* Refit the cell cap(s), ensuring each one is securely fitted, and mop-up any spilt water.

Testing

Standard and
low-maintenance battery

6 If the vehicle covers a small annual mileage, it is worthwhile checking the specific gravity of the electrolyte every three months, to determine the state of charge of the battery. Use a hydrometer to make the check, and compare the results with the following table. Note that the specific gravity readings assume an electrolyte temperature of 15°C (60°F); for every 10°C (18°F) below 15°C (60°F), subtract 0.007. For every 10°C (18°F) above 15°C (60°F), add 0.007. However, for convenience, the temperatures quoted in the following table are **ambient** (outdoor air) temperatures, above or below 25°C (77°F):

	Above 25°C (77°F)	Below 25°C (77°F)
Fully-charged	1.210 to 1.230	1.270 to 1.290
70% charged	1.170 to 1.190	1.230 to 1.250
Discharged	1.050 to 1.070	1.110 to 1.130

7 If the battery condition is suspect, first check the specific gravity of electrolyte in each cell. A variation of 0.040 or more between any cells indicates loss of electrolyte, or deterioration of the internal plates.

8 If the specific gravity variation is 0.040 or more, a new battery should be fitted. If the cell variation is satisfactory but the battery is discharged, it should be charged as described later in this Section.

Maintenance-free battery

9 In cases where a 'sealed for life' maintenance-free battery is fitted, topping-up and testing of the electrolyte in each cell is not possible. The condition of the battery can therefore only be tested using a battery condition indicator or a voltmeter.

10 One type of maintenance-free battery which may be fitted is the 'Delco' type maintenance-free battery, with a built-in charge condition indicator. The indicator is located in the top of the battery casing, and indicates the condition of the battery from its colour. If the indicator shows green, then the battery is in a good state of charge. If the indicator turns darker, eventually to black, then the battery requires charging, as described later in this Section. If the indicator shows clear/yellow, then the electrolyte level in the battery is too low to allow further use, and the battery should be renewed. **Do not** attempt to charge, load or jump start a battery when the indicator shows clear/yellow.

11 If testing the battery using a voltmeter, connect the voltmeter across the battery, and compare the result with those given in the Specifications under 'charge condition'. The test is only accurate if the battery has not been subjected to any kind of charge for the previous six hours. If this is not the case, switch on the headlights for 30 seconds, then wait four to five minutes before testing the battery after switching off the headlights. All other electrical circuits must be switched off, so check that the doors and tailgate are fully shut when making the test.

12 If the voltage reading is less than 12.2 volts, then the battery is discharged, whilst a reading of 12.2 to 12.4 volts indicates a partially-discharged condition.

13 If the battery is to be charged, remove it from the vehicle (Section 3) and charge it as described later in this Section.

Charging

Standard and
low-maintenance battery

Note: *The following is intended as a guide only. Always refer to the manufacturer's recommendations (often printed on a label attached to the battery) before charging a battery.*

14 Charge the battery at a rate of 3.5 to 4 amps, and continue to charge the battery at this rate until no further rise in specific gravity is noted over a four-hour period.

15 Alternatively, a trickle charger charging at the rate of 1.5 amps can safely be used overnight.

16 Specially rapid 'boost' charges which are claimed to restore the power of the battery in 1 to 2 hours are not recommended, as they can cause serious damage to the battery plates through overheating.

17 While charging the battery, note that the temperature of the electrolyte should never exceed 37.8°C (100°F).

Maintenance-free battery

Note: *The following is intended as a guide only. Always refer to the manufacturer's recommendations (often printed on a label attached to the battery) before charging a battery.*

18 This battery type takes considerably longer to fully recharge than the standard type, the time taken being dependent on the extent of discharge, but it can take anything up to three days.

19 A constant-voltage type charger is required, to be set, when connected, to 13.9 to 14.9 volts, with a charger current below 25 amps. Using this method, the battery should be usable within three hours, giving a voltage reading of 12.5 volts, but this is for a partially-discharged battery and, as mentioned, full charging can take considerably longer.

20 If the battery is to be charged from a fully-discharged state (condition reading less than 12.2 volts), have it recharged by your Nissan dealer or local automotive electrician, as the charge rate is higher, and constant supervision during charging is necessary.

3 Battery – removal and refitting

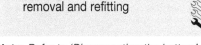

Note: *Refer to 'Disconnecting the battery' in the Reference Chapter before proceeding.*

Removal

1 The battery is located on the left-hand side of the engine compartment.
2 Slacken the clamp nut/bolt, and disconnect the clamp from the battery negative terminal.
3 Remove the insulation cover (where fitted) and disconnect the positive clamp in the same way.
4 Unscrew the nuts, and remove the battery retaining clamp.
5 Lift the battery out of the engine compartment and, where necessary, remove the plastic battery tray. If necessary, the battery mounting bracket can also be unbolted and removed from the engine compartment.

Refitting

6 Refitting is a reversal of removal, but smear petroleum jelly on the terminals when reconnecting the leads, and always reconnect the positive lead first, and the negative lead last.

4 Charging system – testing

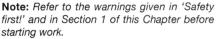

Note: *Refer to the warnings given in 'Safety first!' and in Section 1 of this Chapter before starting work.*

1 If the ignition/no-charge warning light fails to come on when the ignition is switched on, first check the alternator wiring connections for security. If satisfactory, check that the warning light bulb has not blown, and that the bulbholder is secure in its location in the instrument panel. If the light still fails to come on, check the continuity of the warning light feed wire from the alternator to the bulbholder. If all is satisfactory, the alternator is at fault, and should be taken to an auto-electrician for testing and repair.
2 If the ignition warning light comes on when the engine is running, stop the engine as soon as possible. Check that the drivebelt is correctly tensioned (see Chapter 1), that the drivebelt is not contaminated (with oil or water, for example), and that the alternator connections are secure. If all is so far satisfactory, the alternator should be renewed or taken to an auto-electrician for testing and repair.
3 If the alternator output is suspect, even though the warning light functions correctly, the regulated voltage may be checked as follows.
4 Connect a voltmeter across the battery terminals, and start the engine.

5 Increase the engine speed until the voltmeter reading remains steady; the reading should be approximately 12 to 13 volts, and no more than 14 volts.
6 Switch on as many electrical accessories (eg, the headlights, heated rear window and heater blower) as possible, and check that the alternator maintains the regulated voltage at around 13 to 14 volts.
7 If the regulated voltage is not as stated, the fault may be due to worn brushes, weak brush springs, a faulty voltage regulator, a faulty diode, a severed phase winding, or worn or damaged slip-rings. The alternator should taken to an auto-electrician for testing and repair.

5 Alternator – removal and refitting

Removal

1 Disconnect the battery negative terminal (refer to *Disconnecting the battery* in the Reference Chapter).
2 Slacken the auxiliary drivebelt as described in Chapter 1, and disengage it from the alternator pulley.
3 Remove the rubber covers (where fitted) from the alternator terminals, then unscrew the retaining nut(s) and disconnect the wiring from the rear of the alternator. Where necessary, also undo the retaining screw and disconnect the earth lead.
4 Unscrew the alternator upper and lower mounting bolts and washers, then manoeuvre the alternator away from its mounting brackets and out of position. Recover the spacer (where fitted) from between the alternator and the adjustment link.

Refitting

5 Refitting is a reversal of removal, tensioning the auxiliary drivebelt as described in Chapter 1, and ensuring that the alternator mountings are tightened to the specified torque.

6 Alternator – testing and overhaul

If the alternator is thought to be suspect, it should be removed from the vehicle and taken to an auto-electrician for testing. Most auto-electricians will be able to supply fit new parts at reasonable cost. However, check on the cost of repairs before proceeding as it may prove more economical to obtain a new or exchange alternator.

7 Starting system – testing

Note: *Refer to the precautions given in 'Safety*

first!' and in Section 1 of this Chapter before starting work.

1 If the starter motor fails to operate when the ignition key is turned to the appropriate position, the following may be to blame:
  a) The battery is faulty.
  b) The electrical connections between the switch, solenoid, battery and starter motor are somewhere failing to pass the necessary current from the battery through the starter to earth.
  c) The solenoid is faulty.
  d) The starter motor is mechanically or electrically defective.
2 To check the battery, switch on the headlights. If they dim after a few seconds, this indicates that the battery is discharged – recharge (see Section 2) or renew the battery. If the headlights glow brightly, operate the ignition switch and observe the lights. If they dim, then this indicates that current is reaching the starter motor, therefore the fault must lie in the starter motor. If the lights continue to glow brightly (and no clicking sound can be heard from the starter motor solenoid), this indicates that there is a fault in the circuit or solenoid – see following paragraphs. If the starter motor turns slowly when operated, but the battery is in good condition, then this indicates that either the starter motor is faulty, or there is considerable resistance somewhere in the circuit.
3 If a fault in the circuit is suspected, disconnect the battery leads (including the earth connection to the body), the starter/solenoid wiring and the engine/transmission earth strap (refer to *Disconnecting the battery* in the Reference Chapter of this manual). Thoroughly clean the connections, reconnect the leads and wiring, then use a voltmeter or test light to check that full battery voltage is available at the battery positive lead connection to the solenoid, and that the earth is sound. Smear petroleum jelly around the battery terminals to prevent corrosion – corroded connections are amongst the most frequent causes of electrical system faults.
4 If the battery and all connections are in good condition, check the circuit by disconnecting the wire from the solenoid blade terminal. Connect a voltmeter or test light between the wire end and a good earth (such as the battery negative terminal), and check that the wire is live when the ignition switch is turned to the 'start' position. If it is, then the circuit is sound – if not, the circuit wiring can be checked as described in Chapter 12.
5 The solenoid contacts can be checked by connecting a voltmeter or test light between the battery positive feed connection on the starter side of the solenoid, and earth. When the ignition switch is turned to the 'start' position, there should be a reading or lighted bulb, as applicable. If there is no reading or lighted bulb, the solenoid is faulty and should be renewed.

8.3 Disconnect the wiring (arrowed) from the starter motor

6 If the circuit and solenoid are proved sound, the fault must lie in the starter motor. In this event, it may be possible to have the starter motor overhauled by a specialist, but check on the availability and cost of spares before proceeding, as it may prove more economical to obtain a new or exchange motor.

8 Starter motor – removal and refitting

Removal

Manual transmission models

1 Disconnect the battery negative terminal (refer to *Disconnecting the battery* in the Reference Chapter).
2 Remove the air cleaner assembly as described in Chapter 4A.
3 Slacken and remove the retaining nut, and disconnect the main battery cable from the starter motor solenoid. Also disconnect the wiring connector from the solenoid **(see illustration)**.
4 Unscrew the starter motor mounting bolts and withdraw the starter motor from the transmission bellhousing.

Automatic transmission models

5 Disconnect the battery negative terminal (refer to *Disconnecting the battery* in the Reference Chapter).
6 Remove the air cleaner assembly as described in Chapter 4A.
7 So that access to the motor can be gained both from above and below, firmly apply the handbrake, then jack up the front of the vehicle and support it securely on axle stands (see *Jacking and vehicle support*).
8 Remove the exhaust front pipe as described in Chapter 4A.
9 To improve access to the starter motor, undo the bolts securing the inlet manifold support bracket(s), and move the bracket(s) clear of the starter motor. Note that it is not necessary to remove the brackets completely.
10 Slacken and remove the retaining nut, and disconnect the main battery cable from the starter motor solenoid. Also disconnect the wiring connector from the solenoid.
11 Unscrew the starter motor mounting bolts, supporting the motor as the bolts are withdrawn, and manoeuvre the starter motor out from its location.

Refitting

12 Refitting is a reversal of removal.

9 Starter motor – testing and overhaul

If the starter motor is thought to be suspect, it should be removed from the vehicle and taken to an auto-electrician for testing. Most auto-electricians will be able to supply fit new parts at reasonable cost. However, check on the cost of repairs before proceeding as it may prove more economical to obtain a new or exchange motor.

10 Ignition switch – removal and refitting

The ignition switch is integral with the steering column lock, and can be removed as described in Chapter 10.

11 Oil pressure warning light switch – removal and refitting

Removal

1 The switch is located at the rear of the cylinder block, towards its right-hand end. Access to the switch is improved if the vehicle is jacked up and supported on axle stands (see *Jacking and Vehicle Support*), so that the switch can be reached from underneath.
2 Disconnect the battery negative terminal (refer to *Disconnecting the battery* in the Reference Chapter).
3 Remove the protective sleeve from the wiring plug (where applicable), then disconnect the wiring from the switch.
4 Unscrew the switch and recover the sealing washer (where fitted). Be prepared for oil spillage. If the switch is to be left removed from the engine for any length of time, plug the hole to prevent excessive oil loss.

Refitting

5 Where the switch was fitted with a sealing washer, examine the sealing washer for signs of damage or deterioration, and if necessary renew it. Where no sealing washer was fitted, clean the switch and apply a smear of sealant to its threads.
6 Refit the switch, tightening it securely, and reconnect the wiring connector.
7 Lower the vehicle to the ground, then check and if necessary, top-up the engine oil as described in *Weekly checks*.

Chapter 5 Part B:
Ignition system

Contents

Degrees of difficulty

Easy, suitable for novice with little experience	**Fairly easy,** suitable for beginner with some experience	**Fairly difficult,** suitable for competent DIY mechanic	**Difficult,** suitable for experienced DIY mechanic	**Very difficult,** suitable for expert DIY or professional

Specifications

General

System type	Breakerless electronic ignition controlled by ECCS control unit
Firing order	1-3-4-2 (No 1 cylinder at timing chain end)
Ignition timing (at idle speed)	10° ± 2° BTDC

Torque wrench setting	Nm	lbf ft
Distributor mounting bolts	8	6

---

1 Ignition system – general information

The ignition system is integrated with the fuel injection system, to form a combined engine management system which is controlled by the ECCS control unit (see Chapter 4A for further information on the fuel injection side of the system).

The distributor contains a camshaft position sensor, which informs the control unit of engine speed and crankshaft position.

Based on this information, and the information received from its other sensors, the ECCS control unit then calculates the correct ignition timing setting, and switches the power transistor unit on and off accordingly. This causes a high voltage to be induced in the coil secondary (HT) windings, which then travels down the HT lead to the distributor and onto the relevant spark plug.

The ignition coil and power transistor (together with the camshaft position sensor) are integral with the distributor body and cannot be individually removed.

2 Ignition system – testing

Warning: Due to the high voltages produced by the electronic ignition system, extreme care must be taken when working on the system with the ignition switched on. Persons with surgically-implanted cardiac pacemaker devices should keep well clear of the ignition circuits, components and test equipment.

3.2 Undo the retaining screws and lift off the distributor cap

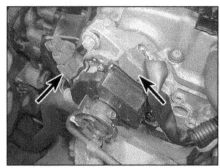

3.3 On early models, disconnect the two wiring connectors (arrowed) on the distributor body

3.4 Undo the two mounting bolts, and withdraw the distributor from the cylinder head

1 If a fault appears in the ignition system, first ensure that the fault is not due to a poor electrical connection or poor maintenance; ie, check that the air cleaner filter element is clean, that the spark plugs are in good condition and correctly gapped, that the engine breather hoses are clear and undamaged, referring to Chapter 1 for further information. Also check that the accelerator cable is correctly adjusted, as described in Chapter 4A. If the engine is running very roughly, check the compression pressures and the valve clearances, as described in Chapter 2A.

2 The only specific ignition system checks which can be carried out by the home mechanic are those described in Chapter 1, relating to the spark plugs. If necessary, the system wiring and wiring connectors can be checked as described in Chapter 12, ensuring that the control unit wiring connector(s) have first been disconnected.

3 Alternatively, a quick check of the complete system can be carried out using the control unit self-diagnosis mode (see Chapter 4A).

4 If the above checks fail to reveal the cause of the problem, the vehicle should be taken to a suitably-equipped Nissan dealer for testing.

3 Distributor – removal and refitting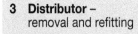

Removal

1 Disconnect the battery negative terminal (refer to *Disconnecting the battery* in the Reference Chapter).

2 Slacken and remove the distributor cap retaining screws **(see illustration)**. Lift off the cap, position it clear of the distributor body, and recover the cap seal.

3 Disconnect the wiring connector(s) from the distributor body. On early models there is a large connector on the top of the distributor and a smaller connector on the side of the unit **(see illustration)**. On later models, only the single large connector is used.

4 Check the cylinder head and distributor flange for signs of alignment marks. If no marks are visible, using a scriber or suitable

marker pen, mark the relationship of the distributor body to the cylinder head. Slacken and remove the two mounting bolts, and withdraw the distributor from the cylinder head **(see illustration)**. Remove the O-ring from the end of the distributor body and discard it; a new one must be used on refitting.

Refitting

5 Lubricate the new O-ring with a smear of engine oil, and fit it to the groove in the distributor body. Examine the distributor cap seal for wear or damage, and renew if necessary.

6 Align the distributor rotor shaft drive coupling key with the slots in the camshaft end, noting that the slots are offset to ensure that the distributor can only be fitted in one position. Carefully insert the distributor into the cylinder head, whilst rotating the rotor arm slightly to ensure that the coupling is correctly engaged.

7 Align the marks noted or made on removal, and install the distributor retaining bolts, tightening them lightly only.

8 Ensure that the seal is correctly located in its groove, then refit the cap assembly to the distributor and tighten its retaining screws securely.

9 Reconnect the distributor wiring connector(s) and the battery negative terminal.

10 Check and, if necessary, adjust the ignition timing as described in Section 4, then tighten the distributor mounting bolts to the specified torque.

4.2 Crankshaft pulley TDC notch aligned with timing chain cover pointer (arrowed)

4 Ignition timing – checking and adjustment

Checking

1 To check the ignition timing, a stroboscopic timing light will be required.

2 The timing marks are in the form of notches on the crankshaft pulley rim, which align with a pointer on the timing chain cover. The notches are spaced at intervals of 5°, and go from 20° before top dead centre (BTDC) to 5° after top dead centre (ATDC). The TDC mark is highlighted with paint to aid identification **(see illustration)**.

3 Start the engine, warm it up to normal operating temperature, and then switch off.

4 Disconnect the wiring connector from the throttle potentiometer (see Chapter 4A).

5 Connect the timing light to No 1 cylinder (nearest the timing chain) plug lead as described in the timing light manufacturer's instructions.

6 Start the engine, allowing it to idle and point the timing light at the crankshaft pulley. The relevant timing mark should be aligned with the pointer on the timing chain cover (see Specifications for the correct timing setting).

Adjustment

7 If adjustment is necessary, slacken the two distributor mounting bolts, then slowly rotate the distributor body as required until the crankshaft pulley marks are correctly positioned.

⚠️ *Warning: At all times, avoid touching the HT leads, and keep loose clothing, long hair, etc, well away from the moving parts of the engine. Once the marks are correctly aligned, hold the distributor stationary, and tighten its mounting bolts to the specified torque. Recheck that the timing marks are still correctly aligned and, if necessary, repeat the adjustment procedure.*

8 When the ignition timing is correctly set, stop the engine and disconnect the timing light.

9 Reconnect the throttle potentiometer wiring connector.

Chapter 6
Clutch

Contents

Degrees of difficulty

Easy, suitable for novice with little experience	Fairly easy, suitable for beginner with some experience	Fairly difficult, suitable for competent DIY mechanic	Difficult, suitable for experienced DIY mechanic	Very difficult, suitable for expert DIY or professional 

Specifications

Type .... Single dry plate with diaphragm spring. Cable-operated release mechanism

Adjustment data
Clutch pedal height:
  Right-hand drive models .... 159.0 to 169.0 mm
  Left-hand drive models .... 153.0 to 163.0 mm
Clutch pedal free play (measured at pedal pad) .... 11.0 to 15.0 mm

Friction disc
Diameter:
  1.4 litre models .... 180.0 mm
  1.6 litre models .... 190.0 mm
Friction material thickness (new) .... 8.0 to 8.4 mm
Minimum friction material-to-rivet head depth .... 0.3 mm
Maximum friction disc run-out:
  1.4 litre models (measured 85.0 mm out from plate centre) .... 1.0 mm
  1.6 litre models (measured 95.0 mm out from the plate centre) ..... 1.0 mm

Torque wrench settings

	Nm	lbf ft
Clutch cable retainer-to-bulkhead nuts	10	7
Clutch pedal height adjustment bolt locknut	19	14
Clutch pedal pivot bolt	19	14
Pressure plate retaining bolts	25	18

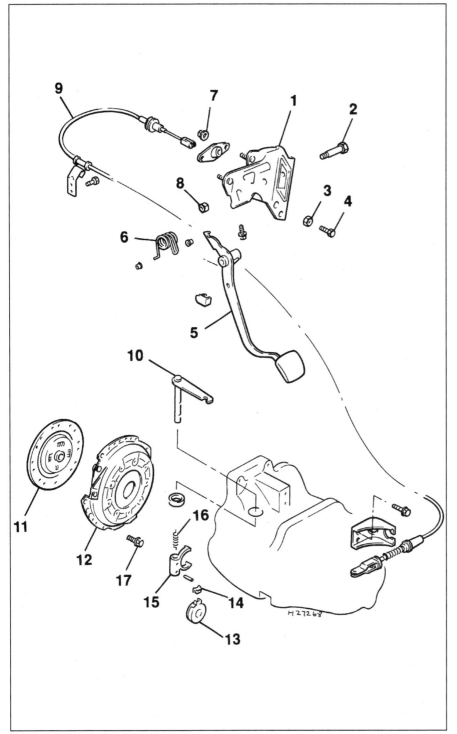

1.1 Exploded view of the clutch and associated components

1 Pedal mounting bracket	9 Cable
2 Pedal pivot bolt	10 Release lever
3 Pedal height adjustment bolt	11 Friction disc
locknut	12 Pressure plate
4 Pedal height adjustment bolt	13 Release bearing
5 Clutch pedal	14 Retaining clip
6 Assist spring	15 Release fork
7 Cable-to-bulkhead nut	16 Return spring
8 Pedal pivot bolt nut	17 Pressure plate bolt

1 General information

The clutch consists of a friction disc, a pressure plate assembly, a release bearing and the release mechanism. All of these components are contained in the large cast-aluminium alloy bellhousing, and sandwiched between the engine and the transmission. The release mechanism is mechanical, being operated by a cable **(see illustration).**

The friction disc is fitted between the engine flywheel and the clutch pressure plate, and is allowed to slide on the transmission input shaft splines. It consists of two circular facings of friction material riveted in position to provide the clutch bearing surface, and a spring-cushioned hub to damp out transmission shocks.

The pressure plate assembly is bolted to the engine flywheel, and is located by three dowel pins. When the engine is running, drive is transmitted from the crankshaft via the flywheel to the friction disc (these components being clamped securely together by the pressure plate assembly), and from the friction disc to the transmission input shaft.

To interrupt the drive, the spring pressure must be relaxed. This is achieved by a sealed release bearing fitted concentrically around the transmission input shaft; when the driver depresses the clutch pedal, the release bearing is pressed against the fingers at the centre of the diaphragm spring. Since the spring is held by rivets between two annular fulcrum rings, the pressure at its centre causes it to deform, so that it flattens and thus releases the clamping force it exerts, at its periphery, on the pressure plate.

Depressing the clutch pedal pulls the clutch inner cable, and this in turn rotates the release fork by acting on the lever at the fork's upper end, above the bellhousing. The fork itself is clipped to the left of the release bearing.

As the friction disc facings wear, the pressure plate moves towards the flywheel; this causes the diaphragm spring fingers to push against the release bearing, thus reducing the clearance which must be present in the mechanism. To ensure correct operation, the clutch cable must be regularly adjusted.

2 Clutch – adjustment

1 The clutch adjustment is checked by first by setting the clutch pedal height, and then by adjusting the pedal free play.
2 Peel back the carpet from underneath the clutch pedal, and ensure that there are no obstructions between the pedal and floor panel. Measure the distance from the centre of the clutch pedal pad to the floor **(see illustration)**. **Note:** *This measurement can be*

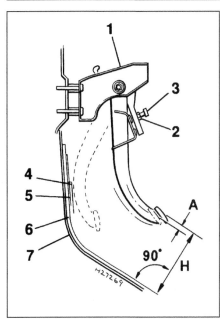

2.2 Clutch pedal height adjustment details

Dimension A – pedal free play measurement
Dimension H – pedal height measurement

1 Pedal mounting bracket	4 Carpet
2 Locknut	5 Insulator sheet
3 Pedal height adjustment bolt	6 Insulator sheet
	7 Floor panel

taken with the carpet in position, so long as the thickness of the carpet is added onto the pedal height measurement. The pedal height should be within the range given in the Specifications at the start of this Chapter.

3 If height adjustment is necessary, reach up behind the facia, and slacken the pedal height adjustment bolt locknut **(see illustration)**. If necessary, remove the lower facia panel (see Chapter 11, Section 29) to improve access to the bolt. Position the bolt as required, so that the pedal height is correctly set, then tighten the locknut to the specified torque.

4 With the pedal height correctly set, check the pedal free play as follows.

5 Slowly depress the clutch pedal, and measure the distance that the clutch pedal pad travels from the at-rest position to the point where resistance is felt **(see illustration 2.2)**. This is the pedal free play, and should be within the range given in the Specifications.

6 If free play adjustment is necessary, working within the engine compartment, locate the clutch release lever, which is situated on the top of the transmission. Slacken the cable locknut, then rotate the knurled adjusting nut until the release lever travel from the at-rest position to the position where resistance is felt is approximately 2.5 to 3.5 mm **(see illustration)**. Recheck the clutch pedal free play as described in paragraph 5, and readjust if necessary. Once the pedal free play is correctly set, securely tighten the cable locknut.

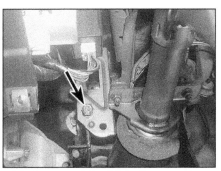

2.3 Clutch pedal height adjustment bolt and locknut (arrowed)

7 With the pedal height and free play correctly adjusted, refit all components removed for access.

3 Clutch cable – removal and refitting

Removal

1 Remove the air cleaner assembly as described in Chapter 4A.

2 Locate the clutch release lever which is situated on the top of the transmission, then slacken the locknut and knurled adjuster nut situated on the cable end fitting.

3 Release the inner cable end fitting from the release lever, and the outer cable fitting from its mounting bracket on the transmission **(see illustration)**.

4 Working inside the vehicle, remove the lower facia panel from the driver's side of the facia as described in Chapter 11, Section 29. Unhook the clutch inner cable from the top of the clutch pedal.

5 Return to the engine compartment, and withdraw the cable from the engine compartment bulkhead. If necessary, undo the two nuts securing the cable retainer to the bulkhead, and remove the retainer along with the cable.

6 Work back along the cable, releasing it from any relevant retaining clips and guides, and noting its correct routing, and remove it from the vehicle.

7 Examine the cable, looking for worn end

3.3 Detach the clutch inner cable from the release lever, then free the outer cable from its mounting bracket

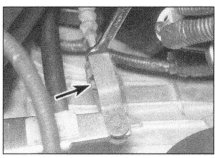

2.6 Slacken the locknut, and adjust the clutch cable by rotating the knurled adjusting nut (arrowed)

fittings or a damaged outer casing, and for signs of fraying of the inner cable. Check the cable's operation; the inner cable should move smoothly and easily through the outer casing. Remember that a cable that appears serviceable when tested off the car may well be much heavier in operation, when compressed into its working position. Renew the cable if it shows any signs of excessive wear or damage. If the cable has seen several years' service, it would be best to renew it anyway, as a precaution against it breaking in service.

Refitting

8 Apply a thin smear of multi-purpose grease to the cable end fittings, then pass the cable through the engine compartment bulkhead. Locate the cable retainer on its studs (where removed), and tighten its retaining nuts to the specified torque.

9 From inside the vehicle, hook the inner cable over the clutch pedal end, and check that it is securely retained.

10 Work along the cable, ensuring that it is correctly routed, and retained by all the relevant retaining clips and guides. Clip the outer cable into its mounting bracket on the transmission.

11 Hook the inner cable end fitting over the end of the clutch release lever, and adjust the clutch pedal settings as described in Section 2. If a new cable has been fitted, fully depress and release the clutch pedal approximately 50 times to prestretch the cable before carrying out the adjustment.

12 On completion, refit the air cleaner assembly as described in Chapter 4A.

4 Clutch pedal – removal and refitting

Note: *Access to the pedal and spring is very poor, and can only be significantly improved by removing the complete facia as described in Chapter 11, Section 29.*

Removal

1 Working as described in Section 2, slacken the clutch cable locknut and knurled adjuster nut, to obtain maximum free play in the cable.

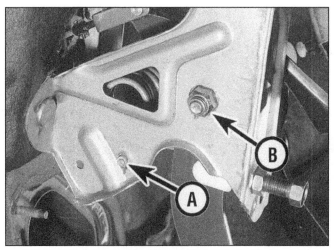

4.3 Clutch pedal assist spring end (A) and pivot bolt nut (B) (shown with facia removed)

5.2 Prior to removal, make alignment marks between the clutch pressure plate and flywheel

2 Working inside the vehicle, remove the lower facia panel from the driver's side of the facia (see Chapter 11, Section 29). Unhook the clutch inner cable from the top of the clutch pedal.

3 Carefully unhook the ends of the clutch pedal assist spring from its locations in the pedal mounting bracket, and recover the spring seats **(see illustration)**.

4 Slacken and remove the nut and pivot bolt, then withdraw the clutch pedal and assist spring from the mounting bracket. If necessary, the pedal and spring can then be separated.

5 Carefully clean all components, and renew any that are worn or damaged. Check the bearing surfaces of the pivot bushes and bolt with particular care; the bushes can be renewed separately if worn.

Refitting

6 Press the pivot bushes into the pedal bore, then apply a smear of multi-purpose grease to their bearing surfaces. Ensure that the clutch pedal and assist spring are correctly mated.

7 Refit the spring seats to the pedal mounting bracket, and install the pedal assembly. Refit the pivot bolt and nut, and tighten it to the specified torque setting.

8 Locate each end of the assist spring in its relevant seat.

9 Hook the clutch cable onto the end of the pedal, then adjust the clutch pedal settings as described in Section 2.

5 Clutch assembly –
removal, inspection and refitting

⚠️ **Warning: Dust created by clutch wear and deposited on the clutch components may contain asbestos, which is a health hazard. DO NOT blow it out with compressed air, or inhale any of it. DO NOT use petrol or petroleum-based solvents to clean off the dust. Brake system cleaner or methylated spirit should be used to flush the dust into a suitable receptacle. After the clutch components are wiped clean with rags, dispose of the contaminated rags and cleaner in a sealed, marked container.**

Note: *Although some friction materials may no longer contain asbestos, it is safest to assume that they DO, and to take precautions accordingly*

Removal

1 Unless the complete engine/transmission is to be removed from the car, and separated for major overhaul (see Chapter 2B), the clutch can be reached by removing the transmission as described in Chapter 7A.

2 Before disturbing the clutch, use a dab of quick-drying paint or a marker pen to mark the relationship of the pressure plate assembly to the flywheel **(see illustration)**.

3 Working in a diagonal sequence, slacken the pressure plate bolts by half a turn at a time, until the spring pressure is released and the bolts can be unscrewed by hand.

4 Prise the pressure plate assembly off its locating dowels, and collect the friction disc, noting which way round the friction disc is fitted.

Inspection

Note: *Due to the amount of work necessary to remove and refit clutch components, it is usually considered good practice to renew the clutch friction disc, pressure plate assembly and release bearing as a matched set, even if only one of these is actually worn enough to require renewal.*

5 Remove the clutch assembly.

6 When cleaning clutch components, first read the warning at the beginning of this Section. Remove the dust only as described – working with dampened cloths will help to keep dust levels to a minimum. Wherever possible, work in a well-ventilated atmosphere.

7 Check the friction disc facings for signs of wear, damage or oil contamination. If the friction material is cracked, burnt, scored or damaged, or if it is contaminated with oil or grease (shown by shiny black patches), the friction disc must be renewed. Measure the depth of the rivets below the friction material surface. If depth of any rivet is equal to, or less than, the service limit given in the Specifications, then the friction disc must be renewed.

8 If the friction material is still serviceable, check that the centre boss splines are unworn, that the torsion springs are in good condition and securely fastened, and that all the rivets are tightly fastened. If excessive wear or damage is found, the friction disc must be renewed.

9 If the friction material is fouled with oil, this must be due to an oil leak from the crankshaft left-hand oil seal, from the sump-to-cylinder block joint, or from the transmission input shaft. Renew the seal or repair the joint, as appropriate, as described in Chapter 2A or 7A before installing the new friction disc, or the new disc will quickly go the same way.

10 Check the pressure plate assembly for obvious signs of wear or damage; shake it to check for loose rivets, or worn or damaged fulcrum rings. Check that the drive straps securing the pressure plate to the cover do not show signs (such as a deep yellow or blue discoloration) of overheating. If the diaphragm spring is worn or damaged, or if its pressure is in any way suspect, the pressure plate assembly should be renewed.

11 Examine the machined bearing surfaces of the pressure plate and of the flywheel; they should be clean, completely flat, and free from scratches or scoring. If either is discoloured from excessive heat, or shows signs of cracks, it should be renewed; however, minor damage of this nature can sometimes be polished away using emery paper.

12 Check that the release bearing contact surface rotates smoothly and easily, with no

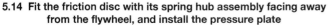

5.14 Fit the friction disc with its spring hub assembly facing away from the flywheel, and install the pressure plate

5.17 Using a clutch-aligning tool to centralise the friction disc

sign of noise or roughness, and that the surface itself is smooth and unworn, with no signs of cracks, pitting or scoring. If there is any doubt about its condition, the bearing must be renewed

Refitting

13 On reassembly, ensure that the bearing surfaces of the flywheel and pressure plate are completely clean, smooth, and free from oil or grease. Use solvent to remove any protective grease from new components.
14 Fit the friction disc so that its spring hub assembly faces away from the flywheel; there may also be a marking showing which way round the plate is to be refitted **(see illustration)**.
15 Refit the pressure plate assembly, aligning the marks made on dismantling (if the original pressure plate is re-used), and locating the pressure plate on its three locating dowels. Fit the pressure plate bolts, but tighten them only finger-tight so that the friction disc can still be moved.
16 The friction disc must now be centralised, so that when the transmission is refitted, its input shaft will pass through the splines at the centre of the friction disc.
17 Centralisation can be achieved by passing a screwdriver or other long bar through the friction disc, and into the hole in the crankshaft. The friction disc can then be moved around until it is centred on the crankshaft hole. Alternatively, a clutch-aligning tool can be used to eliminate the guesswork **(see illustration)**. These can be obtained from most accessory shops, or can be made up from a length of metal rod or wooden dowel which fits closely inside the crankshaft hole, and has insulating tape wound around it to match the diameter of the friction disc splined hole.
18 When the friction disc is centralised, tighten the pressure plate bolts evenly and in a diagonal sequence to the specified torque setting.

19 Apply a thin smear of high-melting point grease to the splines of the friction disc and the transmission input shaft, also to the release bearing bore and release fork shaft.
20 Refit the transmission as described in Chapter 7A.

6 Clutch release mechanism – removal, inspection and refitting

Note: *Refer to the warning concerning the dangers of asbestos dust at the beginning of Section 5.*

Removal

1 Unless the complete engine/transmission is to be removed from the car, and separated for major overhaul (see Chapter 2B), the clutch release mechanism can be reached by removing the transmission as described in Chapter 7A.
2 Lift the retaining clips, then slide the release bearing from the fork and remove it from the transmission. Note the correct fitted direction of each clip on the bearing.
3 Rotate the release fork then, using a hammer and suitable punch, drive out the roll pins securing the release fork to the shaft (drive out the small inner pins first, then remove the outer pins) **(see illustration)**. Discard the roll pins; new ones should be used on refitting.
4 Note the correct fitted position of the return spring, then withdraw the release lever; recover the release fork and spring from the transmission housing.

Inspection

5 Check the release mechanism, renewing any component which is worn or damaged. Carefully check all bearing surfaces and points of contact.
6 Inspect the bush and dust seal fitted to the

transmission housing for signs of damage and deterioration, and renew if necessary. The old bush can be tapped out of position, and the new one installed using a hammer and suitable tubular socket.
7 When checking the release bearing itself, note that it is often considered worthwhile to renew it as a matter of course, given that a significant amount of work is required to gain access to it. Check that the contact surface rotates smoothly and easily, with no sign of noise or roughness. Also check that the surface itself is smooth and unworn, with no signs of cracks, pitting or scoring. If there is any doubt about its condition, the bearing must be renewed.

Refitting

8 Apply a smear of high-melting point grease to the shaft pivot points and the contact surfaces of the release fork.
9 Slide the release lever partially into position in the transmission.
10 Offer up the release fork and spring, aligning them with the lever shaft, and push the release lever fully into position.
11 Align the release fork holes with the holes in the lever shaft, and drive the two new roll pins into position.

6.3 The clutch release fork is secured to the release lever shaft by two dual-roll pin arrangements (arrowed)

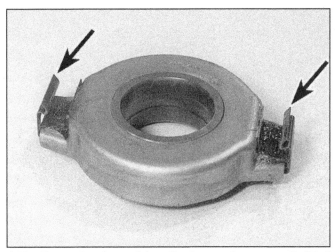

6.12a Ensure that the retaining clips (arrowed) are correctly fitted to the release bearing . . .

6.12b . . . then fit the bearing to the transmission, making sure the clips are correctly engaged with the release fork ends

12 Fit the retaining clips (where removed) to the release bearing, making sure that they are fitted the correct way round. Apply a smear of high-melting point grease to the contact surfaces of the bearing and input shaft, then slide the bearing along the shaft and clip it onto the release fork **(see illustrations)**.

13 Check the operation of the release mechanism, ensuring that it moves smoothly, and returns easily under the pressure of the return spring, then refit the transmission as described in Chapter 7A.

Chapter 7 Part A:
Manual transmission

Contents

Degrees of difficulty

| Easy, suitable for novice with little experience | Fairly easy, suitable for beginner with some experience | Fairly difficult, suitable for competent DIY mechanic | Difficult, suitable for experienced DIY mechanic | Very difficult, suitable for expert DIY or professional |

Specifications

General
Type . . . . Manual, five forward speeds and reverse. Synchromesh on all forward speeds

Designation:
1.4 litre models RS5F30A
1.6 litre models RS5F31A

Torque wrench settings	Nm	lbf ft
Engine-to-transmission fixing bolts:		
Bolts less than 30 mm long	19	14
Bolts 30 mm long, and longer	35	26
Gearchange linkage components:		
Selector rod front pivot bolt	16	12
Selector rod rear pivot bolt	20	15
Support rod-to-transmission bolt	35	26
Support rod-to-gear lever retaining plate nuts	14	10
Support rod rear mounting bracket nuts	14	10
Support rod-to-rear mounting bracket nut	20	15
Gaiter retaining plate nuts	5	4
Left-hand engine/transmission mounting:		
Through-bolt	49	36
Mounting-to-transmission bolts	49	36
Neutral switch	19	14
Oil drain plug	29	21
Oil filler/level plug	15	11
Rear engine/transmission mounting:		
Through-bolt	69	51
Mounting bracket retaining bolts	80	59
Reversing light switch	25	18

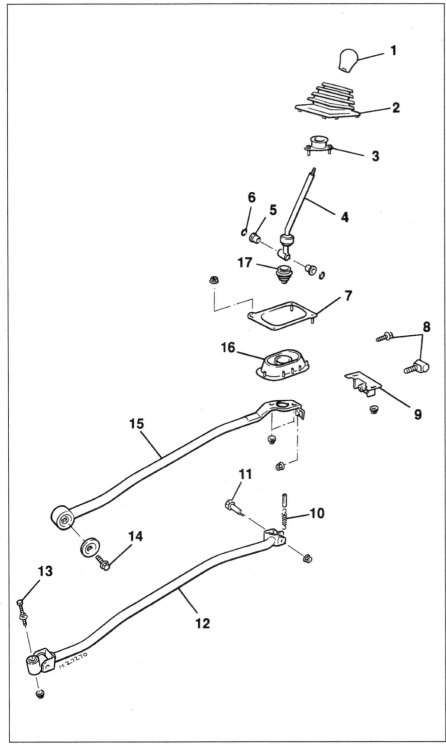

2.3 Gear lever and linkage components

1 Gear lever knob	8 Support rod damper/	12 Selector rod
2 Main gaiter	bolt	13 Selector rod-to-
3 Gear lever retaining	9 Support rod rear	transmission pivot bolt
plate	mounting bracket	14 Support rod-to-
4 Gear lever	10 Selector rod return	transmission bolt
5 Pivot bush	spring	15 Support rod
6 O-ring	11 Selector rod-to-lever	16 Gaiter
7 Gaiter retaining plate	pivot bolt	17 Gaiter

1 General information

The transmission is contained in a cast-aluminium alloy casing bolted to the left-hand end of the engine, and consists of the gearbox and final drive differential, often called a transaxle.

Drive is transmitted from the crankshaft via the clutch to the input shaft, which has a splined extension to accept the clutch friction disc, and rotates in sealed ball-bearings. From the input shaft, drive is transmitted to the output shaft, which rotates in a roller bearing at its right-hand end, and a sealed ball-bearing at its left-hand end. From the output shaft, the drive is transmitted to the differential crownwheel, which rotates with the differential case and planetary gears, thus driving the sun gears and driveshafts. The rotation of the planetary gears on their shaft allows the inner roadwheel to rotate at a slower speed than the outer roadwheel when the car is cornering.

The input and output shafts are arranged side-by-side, parallel to the crankshaft and driveshafts, so that their gear pinion teeth are in constant mesh. In the neutral position, the output shaft gear pinions rotate freely, so that drive cannot be transmitted to the crownwheel.

Gear selection is via a floor-mounted lever and selector rod mechanism. The selector rod causes the appropriate selector fork to move its respective synchro-sleeve along the shaft, to lock the gear pinion to the synchro-hub. Since the synchro-hubs are splined to the output shaft, this locks the pinion to the shaft so that drive can be transmitted. To ensure that gearchanging can be made quickly and quietly, a synchromesh system is fitted to all forward gears, consisting of baulk rings and spring-loaded fingers, as well as the gear pinions and synchro-hubs; the synchromesh cones are formed on the mating faces of the baulk rings and gear pinions.

2 Gearchange linkage – removal and refitting

Removal

1 Remove the centre console as described in Chapter 11.

2 Firmly apply the handbrake, then jack up the front of the vehicle and support it securely on axle stands (see *Jacking and vehicle support*).

3 From underneath the vehicle, using a pair of pliers, unhook the spring securing the rear of the selector rod to the support rod **(see illustration)**.

4 Slacken and remove the nuts, and withdraw the pivot bolts securing the selector rod to the

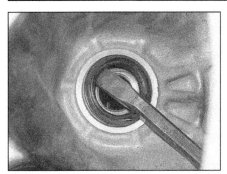

3.4 Using a large flat-bladed screwdriver to lever out a driveshaft oil seal

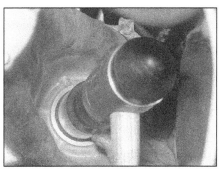

3.5 Tap the new seal into position using a suitable tubular drift

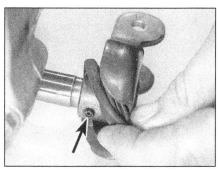

3.10 Selector rod end fitting is retained by a dual-roll pin arrangement (arrowed)

gear lever and transmission. Remove the selector rod from underneath the vehicle, and recover the O-rings and bushes from the base of the gear lever.

5 Unscrew the bolt and washer securing the support rod to the transmission housing. Slacken and remove the nuts securing the rear of the support rod and the support rod bracket to the underside of the vehicle, and manoeuvre the rod and bracket assembly out of position. If necessary, undo the retaining nut, and separate the support rod and mounting bracket.

6 From inside the vehicle, lift out the gear lever and retaining plate, and recover the small rubber gaiter from the base of the lever.

7 If necessary, slacken and remove the retaining nuts, then lift off the retaining plate and remove the main gear lever gaiter from the vehicle.

8 Inspect all the linkage components for signs of wear or damage, paying particular attention to the pivot bushes and selector rod universal joint, and renew worn components as necessary. Renew the rubber gaiters if they are split or badly perished.

Refitting

9 Refitting is a reversal of the removal procedure, applying a smear of multi-purpose grease to the gear lever pivot ball and bushes, and to the selector rod rear pivot bolt. Tighten all the linkage nuts and bolts to their specified torque settings (where given).

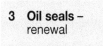

3 Oil seals – renewal

Driveshaft oil seal

1 Firmly apply the handbrake, then jack up the front of the vehicle and support it securely on axle stands (see *Jacking and vehicle support*). Remove the appropriate front roadwheel.

2 Drain the transmission oil as described in Chapter 1.

3 Working as described in Chapter 8, free the inner end of the driveshaft from the transmission, and place it clear of the seal, noting

that there is no need to unscrew the driveshaft retaining nut; the driveshaft can be left secured to the hub. Support the driveshaft, to avoid placing any strain on the driveshaft joints or gaiters.

4 Carefully prise the oil seal out of the transmission using a large flat-bladed screwdriver **(see illustration)**.

5 Remove all traces of dirt from the area around the oil seal aperture, then apply a smear of grease to the outer lip of the new oil seal, and locate it in its aperture. Drive the seal squarely into position, using a suitable tubular drift (such as a socket) which bears only on the hard outer edge of the seal **(see illustration)**. Drive the seal into position until it seats against its locating shoulder.

6 Refit the driveshaft as described in Chapter 8.

7 Refill the transmission with the specified type and quantity of oil, as described in Chapter 1.

Input shaft oil seal

8 To renew the input shaft seal, the transmission must be dismantled. This task should therefore be entrusted to a Nissan dealer.

Selector shaft oil seal

9 Firmly apply the handbrake, then jack up the front of the vehicle and support it securely on axle stands (see *Jacking and vehicle support*). Drain the transmission oil as described in Chapter 1, or be prepared for some oil loss as the seal is removed.

10 Slide back the transmission selector rod rubber gaiter, to reveal the roll pin **(see illustration)**.

11 Using a hammer and suitable punch, tap out the roll pin (tap out the small inner pin first, followed by the larger outer pin) then free the selector rod from the transmission. Discard the roll pin; a new one should be used on refitting.

12 Remove the rubber gaiter, then carefully lever the selector shaft oil seal out of position and slide it off the shaft.

13 Before fitting a new seal, check the selector shaft's seal rubbing surface for signs of burrs, scratches or other damage which may have caused the seal to fail in the first

place. It may be possible to polish away minor faults of this sort using fine abrasive paper, but more serious defects will require the renewal of the selector shaft.

14 Apply a smear of grease to the new seal's outer edge and sealing lip, then carefully slide the seal along the selector rod. Press the seal fully into position in the transmission housing.

15 Refit the rubber gaiter, making sure that it is correctly seated on the seal.

16 Engage the selector rod with the shaft. Align the pin holes, and tap the new roll pins into position.

17 Lower the vehicle to the ground, and top-up/refill (as applicable) the transmission oil as described in Chapter 1.

4 Neutral switch – testing, removal and refitting

Testing

1 A neutral switch is fitted to all models and is one of the many engine management system switches and sensors (see the relevant Part of Chapter 4 for further information). The switch is screwed into the base of the transmission on its right-hand side **(see illustration)**. **Note:** *Do not confuse the neutral switch with the reversing light switch, which is screwed in to the left-hand side of the transmission.*

2 To test the switch, trace the wiring back from the switch to its connector, which is on

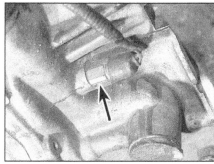

4.1 The neutral switch is screwed into the right-hand side of the transmission (arrowed)

top of the transmission. Disconnect the wiring connector, and use a multi-meter (set to the resistance function) or a battery-and-bulb test circuit to check that there is continuity between the switch terminals only when the transmission is in neutral. If this is not the case, and there are no obvious breaks or other damage to the wires, the switch is faulty and must be renewed.

Removal

3 Firmly apply the handbrake, then jack up the front of the vehicle and support it securely on axle stands (see *Jacking and vehicle support*). Drain the transmission oil as described in Chapter 1, or be prepared for some oil loss as the switch is removed.
4 Trace the switch wiring back to its connector, and disconnect it from the main harness.
5 Unscrew the switch from the transmission, and remove it. Plug the transmission housing aperture, to minimise oil loss (if the transmission has not been drained) and to prevent dirt entry.

Refitting

6 Remove all traces of sealant from the threads of the switch, and apply a smear of fresh sealant to them.
7 Remove the plug from the switch aperture in the transmission, and screw the switch into position.
8 Tighten the switch to the specified torque, and reconnect the wiring connector.
9 Lower the vehicle to the ground, and top-up/refill the transmission oil (as applicable) as described in Chapter 1.

---

5 Reversing light switch – testing, removal and refitting

Testing

1 The reversing light circuit is controlled by a plunger-type switch that is screwed into the base of the transmission on its left-hand side **(see illustration)**. **Note:** *Do not confuse the reversing light switch with the neutral switch, which is screwed in to the right-hand side of*

5.1 The reversing light switch (arrowed) is screwed into the left-hand side of the transmission

the transmission. If a fault develops in the circuit, first ensure that the circuit fuse has not blown.
2 To test the switch, trace the wiring back from the switch to its connector, which is on top of the transmission. Disconnect the wiring connector, and use a multi-meter (set to the resistance function) or a battery-and-bulb test circuit to check that there is continuity between the switch terminals only when reverse gear is selected. If this is not the case, and there are no obvious breaks or other damage to the wires, the switch is faulty and must be renewed.

Removal

3 Firmly apply the handbrake, then jack up the front of the vehicle and support it securely on axle stands (see *Jacking and vehicle support*). Drain the transmission oil as described in Chapter 1, or be prepared for some oil loss as the switch is removed.
4 Trace the switch wiring back to its connector, and disconnect it from the main harness.
5 Unscrew the switch from the transmission, and remove it. Plug the transmission housing aperture, to minimise oil loss (if the transmission has not been drained), and to prevent dirt entry.

Refitting

6 Remove all traces of sealant from the threads of the switch, and apply a smear of fresh sealant to them.
7 Remove the plug from the switch aperture in the transmission, and screw the switch into position.
8 Tighten the switch to the specified torque, then reconnect the wiring connector and check the operation of the circuit.
9 Lower the vehicle to the ground, and top-up/refill the transmission oil (as applicable) as described in Chapter 1.

---

6 Speedometer drive – removal and refitting

Removal

1 Firmly apply the handbrake, then jack up the front of the vehicle and support it securely on axle stands (see *Jacking and vehicle support*). The speedometer drive is situated on the rear of the transmission housing, next to the inner end of the right-hand driveshaft.
2 Disconnect the speedometer sensor wiring connector from the speedometer drive unit.
3 Slacken and remove the retaining bolt, and withdraw the speedometer drive and driven pinion assembly from the transmission housing, along with its O-ring.
4 If necessary, tap out the retaining pin, then slide the pinion out of the housing and recover the thrustwasher. The oil seal can then be removed from the housing.

5 Examine all components for signs of damage, and renew if necessary. Renew the housing O-ring as a matter of course. Note that the thrustwasher is available in various thicknesses – measure the thickness of the original, and quote this when ordering a new one.
6 If the driven pinion is worn or damaged, also examine the drive pinion in the transmission housing for signs of wear or damage. To renew the drive pinion, the transmission must be dismantled and the differential gear removed. This task should therefore be entrusted to a Nissan dealer.

Refitting

7 Where necessary, press the new seal into the housing. Apply a smear of multi-purpose grease to the driven pinion shaft, and slide on the thrustwasher. Insert the driven pinion into the housing, and secure it in position with the retaining pin.
8 Fit a new O-ring to the speedometer drive, and refit it to the transmission, ensuring that the drive and driven pinions are correctly engaged. Securely tighten the housing retaining bolt.
9 Reconnect the speedometer sensor wiring connector, then lower the vehicle to the ground.

---

7 Manual transmission – removal and refitting

Removal

1 Firmly apply the handbrake, then jack up the front of the vehicle and support it securely on axle stands (see *Jacking and vehicle support*). Remove both front roadwheels. Undo the retaining screws, and remove the plastic undershields from beneath the engine/transmission, and the covers from underneath both wheel arches.
2 Drain the transmission oil as described in Chapter 1, then refit the drain and filler/level plugs and tighten them to their specified torque settings.
3 Remove the air cleaner assembly as described in Chapter 4A.
4 Remove the battery and starter motor as described in Chapter 5A.
5 Remove the exhaust system front pipe as described in Chapter 4A.
6 Slacken the clutch cable locknut and knurled adjuster nut, then free the cable end fitting from the release lever, and the outer cable fitting from its mounting bracket **(see illustration)**.
7 Disconnect the wiring connectors from the reversing light switch and the neutral switch. Disconnect the earth lead(s) from the top of the transmission housing **(see illustrations)**. Free the wiring from any relevant retaining clips, and position it clear of the transmission.
8 Slacken and remove the nut and pivot bolt

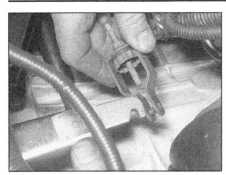

7.6 Free the clutch cable from the release lever, and position it clear of the transmission

7.7a Disconnect the reversing light switch wiring connector . . .

7.7b . . . and neutral switch connector, then detach the earth lead(s) from the transmission housing (arrowed)

securing the gearchange linkage selector rod to the transmission, and the bolt and washer securing the support rod in position. Free both rods, and position them clear of the transmission **(see illustrations)**.

9 Disconnect the breather pipe from the top of the transmission housing **(see illustration)**.

10 Disconnect the speedometer sensor wiring connector from the speedometer drive unit.

11 Working as described in Chapter 8, free the inner end of each driveshaft from the transmission, and place the ends clear of the transmission. Note that there is no need to unscrew the driveshaft retaining nuts – each driveshaft can be left secured to the hub. Support the driveshafts, to avoid placing any strain on the driveshaft joints or gaiters.

12 Place a jack with interposed block of wood beneath the engine, to take the weight of the engine. Alternatively, attach a hoist or support bar to the engine and take the weight of the engine.

13 Place a jack and block of wood beneath the transmission, and raise the jack to take the weight of the transmission.

14 Slacken and remove the through-bolt from the left-hand engine/transmission mounting. Undo the three bolts securing the mounting to the transmission, and manoeuvre the mounting out of position.

15 Slacken and remove the through-bolt from the rear engine/transmission mounting. Undo the bolts securing the mounting bracket in position, and manoeuvre it away from the engine/transmission.

16 With the jack positioned beneath the transmission taking the weight, slacken and remove the remaining bolts securing the transmission housing to the engine. Note the correct fitted positions of each bolt (and the relevant brackets) as they are removed, to use as a reference on refitting – the bolts are of different lengths. Note that it may be necessary to raise the transmission slightly to gain access to the lower bolts.

17 Make a final check that all necessary components have been disconnected, and are positioned clear of the transmission so that they will not hinder the removal procedure.

18 Move the trolley jack and transmission to the left to free it from its locating dowels. Keep the transmission fully supported until the input shaft is free of the engine.

19 Once the transmission is free, lower the jack and manoeuvre the unit out from under the car. If they are loose, remove the locating dowels from the transmission or engine, and keep them in a safe place.

Refitting

20 The transmission is refitted by a reversal of the removal procedure, bearing in mind the following points:

a) *Apply a little high-melting point grease to the splines of the transmission input shaft. Do not apply too much, otherwise there is a possibility of the grease contaminating the clutch friction disc.*

7.8a Unscrew the bolt and recover the washer securing the gearchange linkage support rod to the mounting bracket

7.8c . . . and free the selector rod from the transmission

b) *Ensure that the locating dowels are correctly positioned prior to installation.*
c) *Insert the transmission-to-engine bolts into their original locations, as noted on removal. Tighten all nuts and bolts to the specified torque (where given).*
d) *Renew the driveshaft oil seals using the information given in Section 3.*
e) *On completion, refill the transmission with the specified type and quantity of lubricant as described in Chapter 1.*

8 Manual transmission overhaul – general information

Overhauling a manual transmission is a difficult and involved job for the DIY home

7.8b Unscrew the nut, then withdraw the pivot bolt . . .

7.9 Disconnect the breather hose from the top of the transmission

mechanic. In addition to dismantling and reassembling many small parts, clearances must be precisely measured and, if necessary, changed by selecting shims and spacers. Internal transmission components are also often difficult to obtain, and in many instances, extremely expensive. Because of this, if the transmission develops a fault or becomes noisy, the best course of action is to have the unit overhauled by a specialist repairer, or to obtain an exchange reconditioned unit.

Nevertheless, it is not impossible for the more experienced mechanic to overhaul the transmission, if the special tools are available, and the job is done in a deliberate step-by-step manner so that nothing is overlooked.

The tools necessary for an overhaul include internal and external circlip pliers, bearing pullers, a slide hammer, a set of pin punches, a dial test indicator, and possibly a hydraulic press. In addition, a large, sturdy workbench and a vice will be required.

During dismantling of the transmission, make careful notes of how each component is fitted, to make reassembly easier and accurate.

Before dismantling the transmission, it will help if you have some idea which area is malfunctioning. Certain problems can be closely related to specific areas in the transmission, which can make component examination and renewal easier. Refer to the *Fault finding* Section in the Reference Chapter for more information.

Chapter 7 Part B:
Automatic transmission

Contents

Degrees of difficulty

Easy, suitable for novice with little experience	Fairly easy, suitable for beginner with some experience	Fairly difficult, suitable for competent DIY mechanic	Difficult, suitable for experienced DIY mechanic 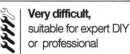	Very difficult, suitable for expert DIY or professional

Specifications

General

Type	Automatic, four forward speeds and reverse
Designation	RL4F03A

Torque wrench settings

	Nm	lbf ft
Drain plug	35	26
Engine-to-transmission fixing bolts:		
Bolts less than 30 mm long	19	14
Bolts 30 mm long, and longer	35	26
Left-hand engine/transmission mounting:		
Through-bolt	49	36
Mounting-to-transmission bolts	49	36
Rear engine/transmission mounting:		
Through-bolt	69	51
Mounting bracket retaining bolts	80	59
Selector cable-to-transmission lever nut	21	15
Torque converter-to-driveplate bolts	51	38

1 General information

A four-speed fully-automatic transmission is available as an option on 1.6 litre models. The transmission consists of a torque converter, an epicyclic geartrain, and hydraulically-operated clutches and brakes.

The torque converter provides a fluid coupling between engine and transmission which acts as an automatic clutch, and also provides a degree of torque multiplication when accelerating.

The epicyclic geartrain provides either of the four forward or one reverse gear ratios, according to which of its component parts are held stationary, or are allowed to turn. The components of the geartrain are held or

released by brakes and clutches which are activated by a hydraulic control unit. A fluid pump within the transmission provides the necessary hydraulic pressure to operate the brakes and clutches.

Driver control of the transmission is by a six-position selector lever. The transmission has a 'drive' option, and a 'hold' facility on the first two gear ratios. The drive option D provides automatic changing throughout the range of all four gear ratios, and is the one to select for normal driving. An automatic kickdown facility will automatically shift the transmission down a gear if the accelerator pedal is fully depressed. The first and second gear 'hold' options are very similar, but they limit the gear ratios available – ie, when the selector lever is in the 2 position, only the first two ratios can be selected, and in the 1 position, only the first ratio can be selected. The lower ratios can be used to provide

engine braking whilst travelling down steep gradients. Note, however, that the transmission should never be shifted down into position 2 whilst the vehicle is travelling at 68 mph (110 km/h) or more, or into position 1 whilst travelling at 56 mph (90 km/h) or more. An overdrive switch button is also provided on the shift lever. For normal driving the switch should be in the 'on' position, however if engine braking is required when driving up and down long slopes, the switch should be 'off'.

Due to the complexity of the automatic transmission, any repair or overhaul work must be left to a Nissan dealer with the necessary special equipment for fault diagnosis and repair. The contents of the following Sections are therefore confined to supplying general information, and any service information and instructions that can be used by the owner.

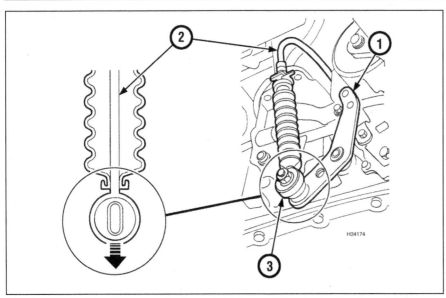

2.4 Selector cable adjustment details. Pull cable end fitting in the direction of the arrow with the specified force

1 Transmission selector lever	2 Selector cable
	3 Cable retaining nut

2 Selector cable – adjustment

1 Position the selector lever firmly against its detent mechanism in the P position.
2 Firmly apply the handbrake, then jack up the front of the vehicle and support it securely on axle stands (see *Jacking and vehicle support*).
3 Working underneath the vehicle, slacken the nut securing the cable end fitting to the transmission selector lever.
4 Hook a spring balance onto the end of the cable, and pull the cable downwards with a force of 6.9 N (0.7 kg/1.5 lbs) **(see illustration)**.
5 From this position, remove the spring balance, and move the cable end fitting backwards by 1.0 mm. Hold the cable in this position, and tighten its retaining nut to the specified torque.
6 Check the operation of the selector lever, ensuring that it moves smoothly and easily, without any sign of the cable binding. If not, repeat the operations in paragraphs 3 to 5.
7 Once the selector lever is operating correctly, apply multi-purpose grease to the contact surfaces of the transmission selector lever and cable, and lower the vehicle to the ground.

3 Selector cable – removal and refitting

Removal

1 Firmly apply the handbrake, then jack up the front of the vehicle and support it securely on axle stands (see *Jacking and vehicle support*). Position the selector lever in the P position.
2 Remove the centre console as described in Chapter 11.
3 Unscrew the nut securing the selector cable to the lever. Slide out the clip securing the outer cable to its mounting bracket, and free the selector cable from the selector lever **(see illustration)**.
4 From underneath the vehicle, undo the two bolts securing the cable to the vehicle body.
5 Withdraw the rear end of the cable from the vehicle. Work along the cable, freeing it from any relevant ties and clips, whilst noting its correct routing.
6 Slacken and remove the nut and washer securing the cable to the transmission selector lever. Slide out the clip securing the outer cable to its mounting bracket, then free the cable from the transmission bracket and remove it from underneath the vehicle. Note that the transmission selector lever must not be disturbed until the cable is refitted. As a precaution, mark the position of the lever in relation to the transmission housing.
7 Examine the cable, looking for worn end fittings or a damaged outer casing. Check for signs of fraying of the inner cable. Check the cable's operation; the inner cable should move smoothly and easily through the outer casing. Renew the cable if it shows any signs of excessive wear or of damage.

Refitting

8 Feed the cable up into position from underneath the vehicle. Refit the two bolts securing the cable to the floor, and tighten them securely.
9 From inside the vehicle, connect the cable to the selector lever, and locate the outer cable into its mounting bracket. Tighten the cable retaining nut, and secure the outer cable in position with the retaining clip.
10 From underneath the vehicle, work along the cable, securing it in position with all the relevant clips and ties, whilst ensuring that it is correctly routed. Feed the cable over the top of the transmission and engage it with the transmission bracket.
11 Ensuring that the selector lever is in the P position, and that the transmission selector lever is still in the Park position, hook the inner cable onto the transmission selector lever. Locate the outer cable into position in its mounting bracket, and secure it with the retaining clip.
12 Hook a spring balance onto the end of the cable, and pull the cable downwards with a force of 6.9 N (0.7 kg/1.5 lbs).

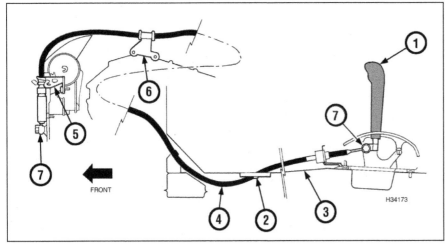

3.3 Selector cable fixing details

1 Selector lever	3 Vehicle floorpan	6 Mounting bracket
2 Selector cable-to-vehicle body fixing	4 Selector cable	7 Selector cable retaining nut
	5 Mounting bracket	

13 From this position, remove the spring balance, and move the cable end fitting backwards by 1.0 mm. Hold the cable in this position, and tighten its retaining nut to the specified torque setting.

14 Check the operation of the selector lever, ensuring that it moves smoothly and easily, without any sign of the cable binding. If not, slacken the cable retaining nut and repeat the operations in paragraphs 12 and 13.

15 Once the selector lever is operating correctly, apply multi-purpose grease to the contact surfaces of the selector levers and cable, and lower the vehicle to the ground.

16 Refit the centre console as described in Chapter 11.

---

4 Selector lever assembly – removal and refitting

Removal

1 Firmly apply the handbrake, then jack up the front of the vehicle and support it securely on axle stands (see *Jacking and vehicle support*). Position the selector lever in the P position.

2 Remove the centre console as described in Chapter 11.

3 Unscrew the nut securing the selector cable to the lever. Slide out the clip securing the outer cable to its mounting bracket, and free the selector cable from the selector lever **(see illustration)**.

4 Disconnect the wiring from the overdrive switch and gear indicator light bulb.

5 Undo the two retaining screws, depress the detent button, and slide the handle off the top of the selector lever. If necessary, the detent button and spring can then be withdrawn from the handle, as can the overdrive switch.

6 Undo the nuts, accessed both from inside and underneath the vehicle, securing the selector lever to the floor, and lift the lever assembly out of the vehicle. Do not attempt to dismantle the selector lever components; the lever assembly is a sealed unit, with no individual components being available separately.

Refitting

7 Refit the selector lever assembly to the vehicle, and securely tighten its retaining nuts.

8 Clip the overdrive switch, detent button and spring back into position in the selector lever handle (where removed).

9 Ensure that the switch wiring is correctly routed down through the handle, and refit the handle to the selector lever. Securely tighten its retaining screws, then check the operation of the lever detent mechanism.

10 Connect the overdrive switch and gear indicator light wiring connector.

11 Connect the cable to the selector lever, and clip the outer cable into its mounting bracket. Tighten the cable retaining nut and secure the outer cable in position with the retaining clip.

12 Adjust the selector cable as described in Section 2, then refit the centre console as described in Chapter 11.

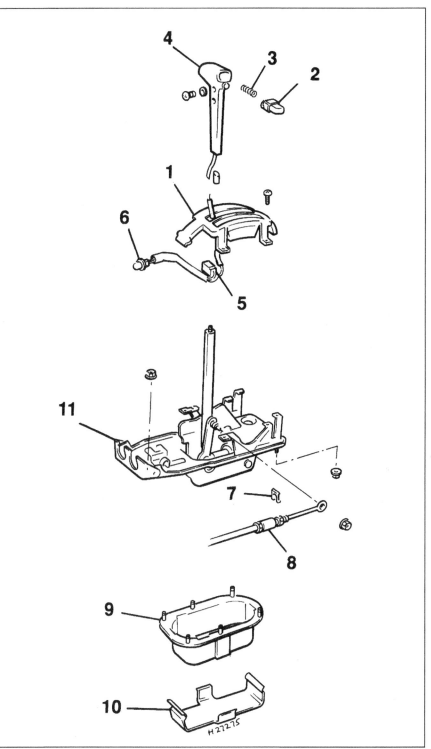

4.3 Selector lever components

1 *Gear indicator panel*	5 *Wiring connector*	9 *Dust cover*
2 *Detent button*	6 *Gear indicator light bulb*	10 *Dust cover support*
3 *Spring*	7 *Cable clip*	11 *Selector lever*
4 *Selector lever handle*	8 *Selector cable*	*assembly*

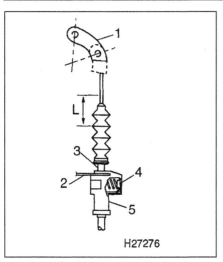

5.4 Adjust the kickdown cable so dimension L is as specified

1 Throttle cam
2 Mounting bracket
3 Outer cable tube
4 Adjuster
5 Kickdown cable casing
L = 40 to 42 mm

5 Kickdown cable – adjustment

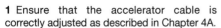

1 Ensure that the accelerator cable is correctly adjusted as described in Chapter 4A.
2 Depress the adjuster to release the outer cable locking mechanism, and move the outer cable fully towards the mounting bracket. Release the adjuster, so that the outer cable is locked in position.
3 Using a dab of white paint or a suitable marker pen, make a mark on the kickdown inner cable. This mark can then be used for measuring the cable travel.
4 Position a ruler next to the mark on the inner cable, then quickly move the throttle cam from its fully closed position to its fully open position, whilst measuring the travel of the kickdown cable. This should be within the range quoted **(see illustration)**.
5 If adjustment is necessary, depress the adjuster, then reposition the outer cable and repeat the procedure described in paragraph 4.
6 Repeat the procedure as necessary, until the kickdown cable travel is within the specified range.

6 Kickdown cable – removal and refitting

Renewal of the kickdown cable is complex task, which should be entrusted to a Nissan dealer. To detach the cable at the transmission end first requires the removal of the hydraulic control valve assembly, which is a task that should not be undertaken by the home mechanic.

7 Speedometer drive – removal and refitting

Refer to Chapter 7A.

8 Oil seals – renewal

Driveshaft seals

1 Refer to the information given in Chapter 7A, noting that the seals must be correctly positioned **(see illustration)**.

Selector shaft seal

2 Renewal of the selector shaft seal is complex task, requiring much dismantling of the transmission, and should therefore be entrusted to a Nissan dealer.

9 Starter inhibitor/ reversing light switch – testing, removal and refitting

Testing

1 The starter inhibitor/reversing light switch is a dual-function switch which is screwed onto the front of the transmission housing. The inhibitor function of the switch ensures that the engine can only be started whilst the selector lever is in either the N or P positions, therefore preventing the engine being started whilst the transmission is in gear. If at any time it is noted that the engine can be started whilst the selector lever is in any position other than P or N, then it is likely that the inhibitor function of the switch is incorrectly adjusted or faulty. The switch also performs the function of the reversing light switch, illuminating the reversing lights whenever the selector lever is in the R position. If either function of the switch is faulty, the switch must be tested as follows.
2 To improve access to the switch, firmly apply the handbrake, then jack up the front of the vehicle and support it securely on axle stands (see *Jacking and vehicle support*).
3 Disconnect the wiring connector from the switch.
4 Measure the resistances between the various terminals of the switch with the selector lever in the P, N and R positions **(see illustration)**. If the switch is functioning correctly, there should be continuity between terminals 1 and 2 in the P and N positions, and continuity between terminals 3 and 4 in the R position.
5 If this is not the case, slacken the switch retaining bolts, and adjust the switch as described in paragraphs 10 and 11. If the switch still fails to function properly after adjustment, it is faulty and must be renewed.

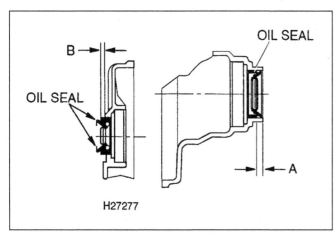

8.1 Correct fitted positions of driveshaft seals

Right-hand side seal (A) = 5.5 to 6.5 mm
Left-hand side seal (B) = 0.5 mm or less

9.4 Starter/inhibitor switch terminal identification. Test switch as described in text

Removal

6 To improve access to the switch, firmly apply the handbrake, then jack up the front of the vehicle and support it securely on axle stands (see *Jacking and vehicle support*).
7 Disconnect the wiring connector from the switch.
8 Slacken and remove the three retaining bolts, then free the switch from the transmission selector mechanism and remove it from the vehicle.

Refitting

9 Manoeuvre the switch into position, and engage it with the selector mechanism. Install the switch retaining bolts, tightening them loosely at this stage.
10 Position the selector lever in the N position, and obtain a 4 mm diameter twist drill or rod.
11 Insert the drill/rod through the hole in the top of the selector lever, and through the hole in the top of the switch lever **(see illustration)**. With the holes aligned, securely tighten the switch retaining bolts, then withdraw the drill/rod.
12 Reconnect the switch wiring, and check the operation of the switch.

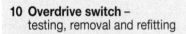

10 Overdrive switch –
testing, removal and refitting

Testing

1 Remove the centre console as described in Chapter 11.
2 Disconnect the wiring from the overdrive switch and gear indicator light.
3 Measure the resistance between the two outer terminals of the overdrive switch connector. If the switch is functioning correctly, there should be continuity between the terminals only when the switch is 'off' (switch button released). If not, the switch must be renewed.

Removal

4 Remove the centre console as described in Chapter 11.
5 Disconnect the wiring from the overdrive switch and gear indicator light.
6 Undo the two retaining screws, depress the detent button, and slide the handle off the top of the selector lever.
7 Unclip the overdrive switch, and remove it from the handle.

Refitting

8 Clip the overdrive switch back into position in the selector lever handle.
9 Ensure that the switch wiring is correctly routed down through the handle, and refit the handle to the selector lever. Securely tighten its retaining screws, then check the operation of the lever detent mechanism.
10 Connect the overdrive switch and gear indicator light wiring, then refit the centre console as described in Chapter 11.

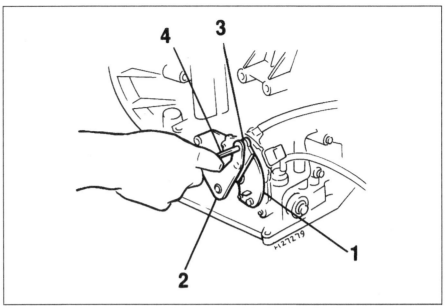

9.11 Adjust the switch using a drill/rod as described in text

1 *Starter/inhibitor switch*
2 *Transmission selector lever*
3 *Adjustment holes*
4 *4 mm diameter drill/rod*

11 Automatic transmission –
removal and refitting

Removal

1 Firmly apply the handbrake, then jack up the front of the vehicle and support it securely on axle stands (see *Jacking and vehicle support*). Position the selector lever in the N (neutral) position and remove both front roadwheels.
2 Drain the transmission fluid as described in Chapter 1, then refit the drain plug and tighten it to the specified torque.
3 Remove the battery and battery tray as described in Chapter 5A.
4 Remove the air cleaner assembly as described in Chapter 4A.
5 Disconnect the wiring connectors from the starter inhibitor/reversing light switch and the transmission solenoid wiring.
6 Detach the kickdown inner cable from the throttle body cam. Slacken the outer cable locknuts, and unbolt the cable mounting bracket. Release the kickdown cable from any relevant retaining clips, so that it is free to be removed with the transmission.
7 Slacken and remove the nut and washer securing the selector cable to the transmission selector lever. Slide out the clip securing the outer cable to its mounting bracket, then free the cable from the transmission.
8 Using a hose clamp or similar, clamp the fluid cooler hoses and disconnect both hoses from the transmission – be prepared for some spillage. Wipe up any spilt fluid immediately.
9 Remove the driveshafts as described in Chapter 8.

10 Remove the exhaust system front pipe as described in Chapter 4A.
11 Remove the starter motor as described in Chapter 5A.
12 Undo the retaining bolts, and remove the cover plate from the sump flange to gain access to the torque converter retaining bolts. Slacken and remove the visible bolt then, using a socket and extension bar to rotate the crankshaft pulley, undo the remaining bolts securing the torque converter to the driveplate as they become accessible. There are four bolts in total.
13 Place a jack with interposed block of wood beneath the engine, to take the weight of the engine. Alternatively, attach a hoist or support bar to the engine lifting eyes, and take the weight of the engine.
14 Place a jack and block of wood beneath the transmission, and raise the jack to take the weight of the transmission.
15 Slacken and remove the through-bolt from the rear engine/transmission mounting. Undo the bolts securing the mounting bracket in position, and manoeuvre it away from the engine/transmission..
16 Slacken and remove the through-bolt from the left-hand engine/transmission mounting. Undo the three bolts securing the mounting to the transmission, and manoeuvre the mounting out of position. Recover the rubbers from each side of the mounting bracket.
17 To ensure that the torque converter does not fall out as the transmission is removed, secure it in position using a length of metal strip bolted to one of the starter motor bolt holes.
18 With the jack positioned beneath the transmission taking the weight, slacken and

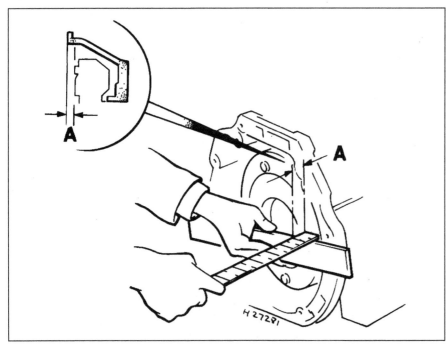

11.22 Prior to refitting, ensure that the distance (A) from the torque converter bolt holes to the transmission mating surface is at least 21.1 mm, indicating that the converter is correctly seated

remove the remaining bolts securing the transmission housing to the engine. Note the correct fitted positions of each bolt (and any relevant brackets) as they are removed, to use as a reference on refitting.

19 Make a final check that all necessary components have been disconnected, and are positioned clear of the transmission so that they will not hinder the removal procedure.

20 With the bolts removed, move the trolley jack and transmission to the left, to free it from its locating dowels.

21 Once the transmission is completely free from the engine, lower the jack and manoeuvre the unit out from under the car. If they are loose, remove the locating dowels from the transmission or engine, and keep them in a safe place.

Refitting

22 The transmission is refitted by a reversal of the removal procedure, bearing in mind the following points:

a) *Prior to installing the transmission, ensure that the torque converter is correctly engaged with the transmission. This can be checked by measuring the distance from the converter mounting bolt holes to the transmission mating surface; if the converter is correctly seated, this distance will be at least 21.1 mm* **(see illustration)**.

b) *Ensure that the locating dowels are correctly positioned prior to installation.*

c) *Tighten all nuts and bolts to the specified torque (where given).*

d) *Prior to refitting the driveshafts, renew the driveshaft seals using the information given in Section 8.*

e) *Adjust the selector cable and kickdown cable as described in Sections 2 and 5 of this Chapter.*

f) *On completion, refill the transmission with the specified type and quantity of lubricant, as described in Chapter 1.*

12 Automatic transmission overhaul –
general information

In the event of a fault occurring on the transmission, it is first necessary to determine whether it is of an electrical, mechanical or hydraulic nature, and to do this, special test equipment is required. It is therefore essential to have the work carried out by a Nissan dealer if a transmission fault is suspected.

Do not remove the transmission from the car for possible repair before professional fault diagnosis has been carried out, since most tests require the transmission to be in the vehicle.

Chapter 8
Driveshafts

Contents

Degrees of difficulty

Easy, suitable for novice with little experience	**Fairly easy,** suitable for beginner with some experience	**Fairly difficult,** suitable for competent DIY mechanic	**Difficult,** suitable for experienced DIY mechanic	**Very difficult,** suitable for expert DIY or professional

Specifications

General

Type	Unequal-length, solid steel shafts, splined to inner and outer constant velocity joints

Overhaul

Lubricant type	Nissan grease supplied with gaiter repair kit
Lubricant quantity:	
Outer CV joint:	
1.4 litre models	75 to 85 g
1.6 litre models	115 to 125 g
Inner CV joint:	
1.4 litre models	110 to 120 g
1.6 litre models	155 to 165 g
Outer joint gaiter setting dimension:	
1.4 litre models	90.5 to 92.5 mm
1.6 litre models	96.0 to 98.0 mm
Inner joint gaiter setting dimension:	
1.4 litre models	95.5 to 97.5 mm
1.6 litre models	101.5 to 103.5 mm

Torque wrench settings

	Nm	lbf ft
Driveshaft nut*	250	185
Hub carrier-to-suspension strut retaining bolt nuts*	124	92
Roadwheel nuts	110	81

*New nut(s) must be used.

2.3a Remove the split-pin . . .

2.3b . . . then, on early models, withdraw the locking plate . . .

2.3c . . . and the spacer

1 General information

Drive is transmitted from the differential to the front wheels by means of two solid steel driveshafts of unequal length.

Both driveshafts are splined at their outer ends, to accept the wheel hubs, and are threaded so that each hub can be fastened to the driveshaft by a large nut. The inner end of each driveshaft is splined, to accept the differential sun gear.

Constant velocity (CV) joints are fitted to

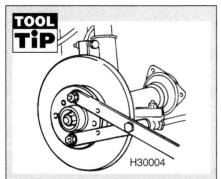

TOOL TiP

H30004

A tool to hold the front hub stationary whilst the driveshaft nut is slackened can be fabricated from two lengths of steel strip (one long, one short) and a nut and bolt; the nut and bolt forming the pivot of a forked tool

each end of the driveshafts, to ensure the smooth and efficient transmission of power at all suspension and steering angles. The outer constant velocity joints are of the ball-and-cage type, and the inner joints are of the tripod type.

2 Driveshafts – removal and refitting

Note: *A new driveshaft nut, driveshaft nut split-pin, driveshaft inner joint circlip, and new hub carrier-to-suspension strut retaining bolt nuts must be used on refitting.*

Removal

1 Firmly apply the handbrake, then jack up the front of the vehicle and support it securely on axle stands (see *Jacking and vehicle support*). Remove the appropriate roadwheel(s).
2 To reduce spillage when the inner end of the driveshaft is withdrawn from the transmission, drain the transmission oil/fluid as described in Chapter 1.
3 Remove the split-pin from the outer end of the driveshaft, then withdraw the castellated locking plate and the spacer. Note that the castellated locking plate and spacer are not fitted to later models. Discard the split-pin – a new one must be used on refitting **(see illustrations)**.
4 The front hub must now be held stationary in order to loosen the driveshaft nut. Ideally, the hub should be held by a suitable tool bolted into place using two of the roadwheel

nuts **(see Tool Tip)**. Alternatively, have an assistant firmly apply the brake pedal to prevent the hub from rotating. Using a socket and extension bar, slacken and remove the driveshaft nut, then remove the washer **(see illustration)**. Note that a new driveshaft nut must be used on refitting.

⚠️ **Warning: Take care, the nut is very tight.**

5 Extract the retaining clip and release the brake hydraulic hose from the support bracket on the suspension strut **(see illustration)**.
6 Undo the two nuts from the bolts securing the hub carrier to the suspension strut **(see illustration)**. Discard the nuts – new nuts must be used on refitting. Withdraw the bolts and release the upper end of the hub carrier from the strut.
7 Temporarily refit the driveshaft nut to the end of the driveshaft, to prevent damage to the driveshaft threads. Using a soft-faced mallet, carefully tap the driveshaft to free it from the hub carrier. If the shaft is a tight fit, a suitable puller can be used to force the end of the shaft from the hub.
8 Once the driveshaft is free, remove the driveshaft nut, tip the hub carrier outwards, and fully withdraw the outer end of the driveshaft from the hub **(see illustration)**. On models with ABS, take care not to strain the ABS wheel sensor wiring during this operation – if necessary, unscrew the bolt securing the sensor to the hub carrier, withdraw the sensor and position it clear of the work area.
9 Proceed as follows, according to transmission type.

2.4 Remove the driveshaft nut and the washer

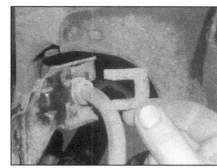

2.5 Extract the clip and release the brake hydraulic hose from the suspension strut

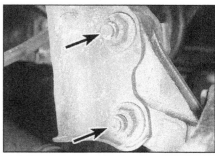

2.6 Undo the two nuts (arrowed) from the bolts securing the hub carrier to the suspension strut

Manual transmission models

10 Separate the relevant driveshaft from the transmission, using a suitable lever inserted between the casing of the inner constant velocity joint and the transmission casing **(see illustration)**. Prise out the driveshaft until the retaining circlip compresses into its groove, and is released from the differential sun gear. Discard the circlip – a new clip must be used on refitting.

11 Withdraw the driveshaft assembly from under the vehicle.

Automatic transmission right-hand shaft

12 The procedure is as described for manual transmission models in paragraphs 10 and 11, but take care not to damage the oil seal when prising the driveshaft from the transmission.

Automatic transmission left-hand shaft

13 To remove the left-hand driveshaft, the right-hand driveshaft **must** be removed first, as described previously.

14 Working through the aperture in the right-hand side of the differential created by removal of the right-hand driveshaft, pass a suitable long-bladed screwdriver or a similar drift into the differential until it rests against the inner end of the left-hand driveshaft. Take care not to damage the planet gear shaft or the sun gear **(see illustration)**.

15 Tap the end of the driveshaft until the retaining circlip is compressed into its groove, and is released from the differential sun gear. Discard the circlip – a new clip must be used on refitting.

16 Withdraw the driveshaft assembly from under the vehicle.

Refitting

17 Before installing a driveshaft, examine the driveshaft oil seal in the transmission for signs of damage or deterioration and, if necessary, renew it, referring to Chapter 7A or 7B for further information (it is advisable to renew the seal as a matter of course).

18 Thoroughly clean the driveshaft splines, and the apertures in the transmission and hub assembly. Apply a thin film of grease to the oil seal lips, and to the driveshaft splines and shoulders. Check that all driveshaft gaiter clips are securely fastened.

19 Note that the circlip at the inner end of the driveshaft **must** be renewed on refitting.

20 When refitting a driveshaft, great care must be taken to prevent damage to the driveshaft oil seals. Nissan specify the use of special tools which guide the shafts through the seal lips on refitting. Provided that the seal lips and the shaft ends are lightly greased, and that care is taken on refitting, these tools should not be necessary.

21 Insert the inner end of the driveshaft into the transmission, taking care not to damage the oil seal.

22 Engage the driveshaft splines with those

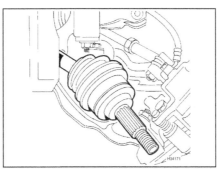

2.8 Tip the hub carrier outwards, and withdraw the outer end of the driveshaft from the hub

2.10 Release the driveshaft inner end using a suitable lever inserted between the inner joint and transmission casing – manual transmission models

of the differential sun gear, and press the driveshaft into place until the retaining crimp engages correctly behind (inboard of) the sun gear.

23 Grasp the inner joint body firmly, and check that the circlip is correctly engaged by attempting to pull the driveshaft from the transmission.

24 Apply a thin film of grease to the outer driveshaft joint splines, then engage the outer end of the driveshaft with the hub, ensuring that the splines engage correctly.

25 Refit the washer and a new driveshaft nut, but do not tighten the nut fully at this stage.

26 Engage the upper end of the hub carrier with the suspension strut and insert the two retaining bolts with their bolt heads toward the rear of the vehicle. Fit new nuts to the bolts, and tighten to the specified torque.

27 Locate the brake hydraulic hose in the suspension strut support bracket and secure the hose with the retaining clip.

28 If removed, refit the ABS wheel sensor and tighten its retaining bolt securely.

29 Hold the front hub stationary as during removal, then tighten the new driveshaft nut to the specified torque.

30 On early models, fit the spacer and the

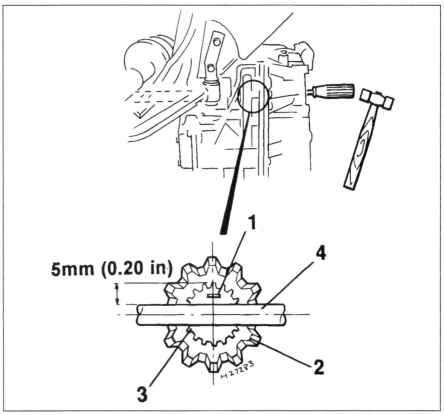

2.14 Using a screwdriver to release the inner end of the left-hand driveshaft from the differential – automatic transmission models

| 1 Screwdriver tip | 2 Sun gear | 3 Driveshaft | 4 Planet gear shaft |

3.2 Release the rubber gaiter large retaining clip by cutting through it using a hacksaw

3.5 Sharply strike the edge of the outer joint to drive it off the end of the driveshaft

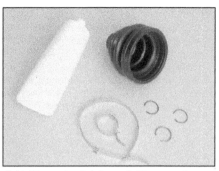

3.11 The outer joint repair kit comprises a new gaiter, retaining clips, circlips and the correct type of grease

castellated locking plate, ensuring that the grooves in the locking plate are aligned with the split-pin hole in the end of the driveshaft. On all models, fit a new split-pin and bend over the split-pin legs.

31 Refit the roadwheel(s), and lower the vehicle to the ground.

32 Refill the transmission with oil/fluid as described in Chapter 1.

3 Driveshaft rubber gaiters – renewal

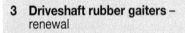

Outer joint

1 Remove the driveshaft as described in Section 2.

2 Release the rubber gaiter large retaining

clip by cutting through it using a junior hacksaw **(see illustration)**. Spread the clip and remove it from the gaiter.

3 Fold the gaiter back to expose the outer constant velocity joint then scoop out the excess grease.

4 If the original joint is to be re-used, make alignment marks between the joint and the driveshaft.

5 Mount the driveshaft in a vice and, using a mallet, sharply strike the edge of the outer joint to drive it off the end of the shaft **(see illustration)**. The joint is retained on the driveshaft by a circlip, and striking the joint in this manner forces the circlip into its groove, so allowing the joint to slide off.

6 Cut off the small retaining clip then slide the old gaiter off the end of the driveshaft.

7 Remove the circlip from the groove in the driveshaft splines, and discard it. A new circlip

must be fitted on reassembly.

8 With the constant velocity joint removed from the driveshaft, thoroughly clean the joint using paraffin, or a suitable solvent, and dry it thoroughly. Carry out a visual inspection of the joint.

9 Move the inner splined driving member from side-to-side, to expose each ball in turn at the top of its track. Examine the balls for cracks, flat spots, or signs of surface pitting.

10 Inspect the ball tracks on the inner and outer members. If the tracks have widened, the balls will no longer be a tight fit. At the same time, check the ball cage windows for wear or cracking between the windows.

11 If any of the constant velocity joint components are found to be worn or damaged, it will be necessary to renew the complete joint assembly as the internal parts are not available separately. If the joint is in satisfactory condition, obtain a repair kit consisting of a new gaiter, circlips, retaining clips, and the correct type of grease **(see illustration)**.

12 Commence reassembly by sliding the smaller gaiter securing clip onto the driveshaft, followed by the gaiter **(see illustrations)**.

13 Fit a new joint retaining circlip to the groove in the end of the shaft **(see illustration)**.

14 Fit the outer joint to the shaft, and engage it with the shaft splines. If the original joint is re-used, align the previously made marks on the joint and the end of the driveshaft.

15 Temporarily refit the driveshaft nut to protect the joint threaded end, and use a mallet to tap the joint onto the shaft until the circlip engages correctly behind the joint cage.

16 Pack the joint with the correct amount of the specified grease (supplied with the gaiter kit), then twist the joint to ensure that all the recesses are filled **(see illustration)**.

17 Check that the smaller end of the gaiter is located in the driveshaft groove, then slide the gaiter onto the outer joint.

18 Check that the gaiter does not swell or deform when positioned at the setting dimension given in the Specifications **(see illustration)**.

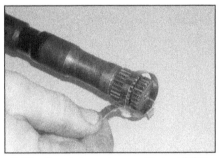

3.12a Commence reassembly by sliding the smaller gaiter securing clip onto the driveshaft . . .

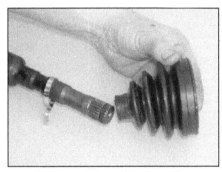

3.12b . . . followed by the gaiter

3.13 Fit a new joint retaining circlip to the groove in the end of the shaft

3.16 Pack the joint with the correct amount of the grease supplied with the kit, then twist the joint to fill all recesses

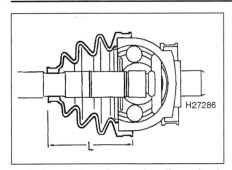

3.18 Outer joint gaiter setting dimension L

See Specifications for dimension L

3.19a Fit the new outer gaiter securing clip . . .

3.19b . . . bend back the end of the clip . . .

3.19c . . . and squeeze closed the tags to secure

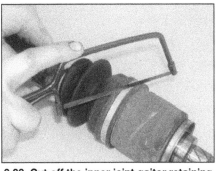

3.23 Cut off the inner joint gaiter retaining clips using a hacksaw

3.25 Withdraw the inner joint body from the spider

19 Fit the new outer gaiter large securing clip, bend back the end of the clip, and squeeze closed the tags to secure the clip **(see illustrations)**.
20 Slide the smaller securing clip over the gaiter, and tighten and secure it as described previously.
21 Refit the driveshaft as described in Section 2.

Inner joint

22 Remove the driveshaft as described in Section 2.
23 Release the rubber gaiter retaining clips by cutting through them using a junior hacksaw **(see illustration)**. Slide the gaiter off the joint towards the middle of the driveshaft.
24 If the original joint is to be re-used, make alignment marks between the joint body and the driveshaft.
25 Withdraw the joint body from the spider, then wipe off the excess grease from the spider and the end of the driveshaft **(see illustration)**.
26 If the original spider is to be re-used, mark the relationship of the spider and the driveshaft, then using circlip pliers, remove the circlip securing the spider to the end of the shaft **(see illustration)**. Discard the circlip – a new one should be used on refitting.
27 Withdraw the spider and the gaiter from the end of the shaft **(see illustrations)**.
28 Thoroughly clean the constant velocity joint components and the end of the driveshaft using paraffin, or a suitable solvent, and dry thoroughly. Carry out a visual

inspection of the joint. If any of the joint components are worn, the spider assembly or the joint body can be renewed separately as complete units, but no other spare parts are

available. If the joint is in satisfactory condition, obtain a repair kit consisting of a new gaiter, circlips, retaining clips, and the correct type of grease **(see illustration)**.

3.26 Remove the circlip securing the spider to the end of the driveshaft

3.27a Withdraw the spider . . .

3.27b . . . and the gaiter from the end of the shaft

3.28 The inner joint repair kit comprises a new gaiter, retaining clips, circlips and the correct type of grease

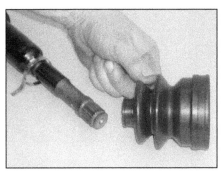

3.29 Commence reassembly by sliding the smaller gaiter securing clip onto the shaft, followed by the gaiter

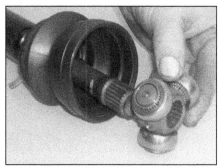

3.30 Refit the spider, aligning the marks made previously if the original spider is being used

3.31 Fit a new circlip to secure the spider to the driveshaft

3.32 Pack the correct amount of the grease supplied with the kit, around the spider rollers and into the joint body

3.33 Fit the joint body over the spider, aligning the marks made previously if the original body is being used

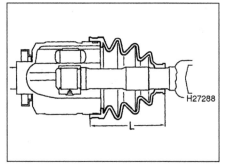

3.35 Inner joint gaiter setting dimension L

See Specifications for dimension L

29 Commence reassembly by sliding the smaller gaiter securing clip onto the driveshaft, followed by the gaiter **(see illustration)**.

30 Refit the spider, aligning the marks made previously if the original spider is being used **(see illustration)**.

31 Fit a new circlip to secure the spider to the driveshaft **(see illustration)**.

32 Pack the correct amount of the specified grease (supplied with the gaiter kit), around the spider rollers and into the joint body **(see illustration)**.

33 Fit the joint body over the spider. If the original body is being refitted, align the marks made between the body and the driveshaft before removal **(see illustration)**.

34 Slide the gaiter onto the joint body, expelling any trapped air, then fit the new large securing clip, and tighten it. Bend back the end of the clip, and secure it under the tags **(see illustrations 3.19b and 3.19c)**.

35 Position the gaiter at the setting dimension given in the Specifications **(see illustration)**. Check that the smaller end of the gaiter is located in the driveshaft groove, and that the gaiter is not stretched or deformed when at this length.

36 Locate the smaller gaiter securing clip on the end of the gaiter, and secure it as described previously.

37 Refit the driveshaft as described in Section 2.

4 Driveshaft overhaul – general information

1 If any of the checks described in Chapter 1 reveal wear in any driveshaft joint, first remove the roadwheel trim or centre cap (as appropriate).

2 Check that the driveshaft nut is correctly tightened; if in doubt, remove the split-pin and, where fitted, the castellated locking plate and spacer. Check that the nut is tightened to the specified torque, then refit the spacer and locking plate (where applicable), and a new split-pin. Refit the roadwheel trim or centre cap (as applicable), and repeat the check on the remaining driveshaft nut.

3 Road test the vehicle, and listen for a metallic clicking from the front as the vehicle is driven slowly in a circle on full-lock. If a clicking noise is heard, this indicates wear in the outer constant velocity joint.

4 If vibration, consistent with roadspeed, is felt through the car when accelerating, there is a possibility of wear in the inner constant velocity joints.

5 To check the joints for wear, remove the driveshafts, then dismantle them as described in Section 3. If any wear or free play is found, the relevant joint, or joint components, must be renewed.

Chapter 9
Braking system

Contents

Degrees of difficulty

| **Easy,** suitable for novice with little experience | **Fairly easy,** suitable for beginner with some experience | **Fairly difficult,** suitable for competent DIY mechanic | **Difficult,** suitable for experienced DIY mechanic | **Very difficult,** suitable for expert DIY or professional |

Specifications

General

System type .... Dual hydraulic circuit. Anti-lock braking system fitted to certain models. Front ventilated disc brakes on all models. Rear drum or rear disc brakes according to model. Vacuum servo-assistance on all models. Cable-operated handbrake acting on rear wheels

Front brakes

Type .... Ventilated disc, with single-piston sliding caliper
Disc diameter .... 232.0 mm

Disc thickness:

	Non-ABS	**ABS**
New	18.0 mm	20.0 mm
Minimum	16.0 mm	18.0 mm

Maximum disc run-out .... 0.07 mm
Minimum pad friction material thickness .... 2.0 mm

Rear disc brakes

Type .... Solid disc with single-piston sliding caliper
Disc diameter .... 234.0 mm
Disc thickness:
New .... 7.0 mm
Minimum thickness .... 6.0 mm
Maximum disc run-out .... 0.07 mm
Minimum pad friction material thickness .... 1.5 mm

Rear drum brakes

Type .... Drum, with leading and trailing shoes operated by twin-piston wheel cylinder

Drum inner diameter:
New .... 180.0 mm
Maximum diameter after machining .... 181.0 mm
Maximum out-of-round .... 0.03 mm
Minimum shoe lining thickness .... 1.5 mm

Brake pedal

Free height:
   Right-hand drive models:
      Manual transmission .... 155.0 to 165.0 mm
      Automatic transmission .... 164.0 to 174.0 mm
   Left-hand-drive models:
      Manual transmission .... 148.0 to 158.0 mm
      Automatic transmission .... 157.0 to 167.0 mm
Pedal free play .... 0.3 to 1.0 mm

Vacuum servo

Pushrod length (models without ABS only) .... 125.0 mm

Handbrake

Number of handbrake clicks required to operate warning light .... 0 to 1 click

Torque wrench settings

	Nm	lbf ft
ABS wheel sensor securing bolts:		
Front sensor	20	15
Rear sensor	30	22
Brake fluid hose union banjo bolts	18	13
Brake servo/pedal bracket securing nuts	15	11
Front brake caliper guide pin bolts	25	18
Front brake caliper mounting bracket bolts	60	44
Handbrake lever mounting bolts	18	13
Master cylinder securing nuts	15	11
Rear brake backplate bolts (drum brake models)	50	37
Rear brake caliper guide pin bolts	25	18
Rear brake caliper mounting bracket bolts	45	33
Rear brake caliper handbrake lever cam nut	25	18
Rear wheel cylinder mounting bolts	10	7

1 General information

The braking system is of the servo-assisted, dual-circuit hydraulic type. The arrangement of the hydraulic system is such that each circuit operates one front and one rear brake from a tandem master cylinder. Under normal circumstances, both circuits operate in unison. However, in the event of hydraulic failure in one circuit, full braking force will still be available at two diagonally-opposite wheels.

All models are fitted with front disc brakes and either rear drum brakes or rear disc brakes.

The front disc brakes are actuated by single-piston sliding type calipers, which ensure that equal pressure is applied to each disc pad.

The rear drum brakes incorporate leading and trailing shoes, which are actuated by twin-piston wheel cylinders. A self-adjust mechanism is incorporated, to automatically compensate for brake shoe wear. As the brake shoe linings wear, the footbrake operation automatically operates the adjuster mechanism, which effectively lengthens the shoe adjuster strut and repositions the brake shoes to reduce the lining-to-drum clearance. The mechanical handbrake linkage operates the brake shoes via a lever attached to the trailing brake shoe.

The rear disc brakes are actuated by single-piston sliding type calipers, and incorporate a mechanical handbrake mechanism.

To prevent the possibility of the rear wheels locking before the front wheels under heavy braking, pressure regulating valves are incorporated in the hydraulic circuit to the rear brakes. On models with conventional (non-ABS) braking systems, the valves are an integral part of the brake master cylinder. On models with ABS the valves are located in a separate unit mounted on the engine compartment bulkhead. The valves are of the pressure-sensitive type which regulate rear brake hydraulic pressure proportionally, depending on the pressure applied to the front brakes.

Note: *When servicing any part of the system, work carefully and methodically; also observe scrupulous cleanliness when overhauling any part of the hydraulic system. Always renew components (in axle sets, where applicable) if in doubt about their condition, and use only genuine Nissan replacement parts, or at least those of known good quality. Note the warnings given in 'Safety first!' and at relevant points in this Chapter concerning the dangers of asbestos dust and hydraulic fluid.*

2 Hydraulic system – bleeding

⚠ *Warning: Brake hydraulic fluid is poisonous; wash off immediately and thoroughly in the case of skin contact, and seek immediate medical advice if any fluid is swallowed, or gets into the eyes. Certain types of hydraulic fluid are inflammable, and may ignite when allowed into contact with hot components. When servicing any hydraulic system, it is safest to assume that the fluid IS inflammable, and to take precautions against the risk of fire as though it is petrol that is being handled. Hydraulic fluid is also an effective paint stripper, and will attack plastics; if any is spilt, it should be washed off immediately, using copious quantities of fresh water. Finally, it is hygroscopic (it absorbs moisture from the air) – old fluid may be contaminated and unfit for further use. When topping-up or renewing the fluid, always use the recommended type, and ensure that it comes from a freshly-opened sealed container.*

General

1 The correct operation of any hydraulic system is only possible after removing all air from the components and circuit; and this is achieved by bleeding the system.

2 During the bleeding procedure, add only clean, unused brake hydraulic fluid of the recommended type; never re-use fluid that has already been bled from the system. Ensure that sufficient fluid is available before starting work.

3 If there is any possibility of incorrect fluid being already in the system, the brake components and circuit must be flushed completely with uncontaminated, correct fluid, and new seals should be fitted throughout the system.

4 If hydraulic fluid has been lost from the system, or air has entered because of a leak, ensure that the fault is cured before proceeding further.

5 Park the vehicle on level ground, switch off the engine and select first or reverse gear (or P), then chock the wheels and release the handbrake.

6 Check that all pipes and hoses are secure, unions tight and bleed screws closed. Remove the dust caps (where applicable), and clean any dirt from around the bleed screws.

7 Unscrew the master cylinder reservoir cap, and top the master cylinder reservoir up to the MAX level line; refit the cap loosely. Remember to maintain the fluid level at least above the MIN level line throughout the procedure, otherwise there is a risk of further air entering the system.

8 There are a number of one-man, do-it-yourself brake bleeding kits currently available from motor accessory shops. It is recommended that one of these kits is used whenever possible, as they greatly simplify the bleeding operation, and also reduce the risk of expelled air and fluid being drawn back into the system. If such a kit is not available, the basic (two-man) method must be used, which is described in detail below.

9 If a kit is to be used, prepare the vehicle as described previously, and follow the kit manufacturer's instructions, as the procedure may vary slightly according to the type being used; generally, they are as outlined below in the relevant sub-section.

10 Whichever method is used, the same sequence must be followed (paragraphs 11 and 12) to ensure that the removal of all air from the system.

Bleeding sequence

11 If the system has been only partially disconnected, and suitable precautions were taken to minimise fluid loss, it should be necessary to bleed only that part of the system (ie, the primary or secondary circuit).

12 If the complete system is to be bled, then it should be done working in the following sequence:

Right-hand drive models

a) Left-hand rear wheel.
b) Right-hand front wheel.
c) Right-hand rear wheel.
d) Left-hand front wheel.

Left-hand drive models

a) Right-hand rear wheel.
b) Left-hand front wheel.
c) Left-hand rear wheel.
d) Right-hand front wheel.

Bleeding

Caution: On models equipped with ABS, switch off the ignition and disconnect the battery negative terminal (refer to

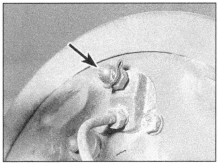

2.14a Bleed screw dust cap (arrowed) – rear drum brake

'Disconnecting the battery' in the Reference Chapter), before carrying out the bleeding procedure.

Basic (two-man) method

13 Collect a clean glass jar, a suitable length of plastic or rubber tubing which is a tight fit over the bleed screw, and a ring spanner to fit the screw. The help of an assistant will also be required.

14 Remove the dust cap from the first screw in the sequence (if not already done) **(see illustrations)**. Fit a suitable spanner and tube to the screw, place the other end of the tube in the jar, and pour in sufficient fluid to cover the end of the tube.

15 Ensure that the master cylinder reservoir fluid level is maintained at least above the MIN level line throughout the procedure.

16 Have the assistant fully depress the brake pedal several times to build-up pressure, then maintain it on the final downstroke.

17 While pedal pressure is maintained, unscrew the bleed screw (approximately one turn) and allow the compressed fluid and air to flow into the jar. The assistant should maintain pedal pressure, following the pedal down to the floor if necessary, and should not release the pedal until instructed to do so. When the flow stops, tighten the bleed screw again, have the assistant release the pedal slowly, and recheck the reservoir fluid level.

18 Repeat the steps given in paragraphs 16 and 17 until the fluid emerging from the bleed screw is free from air bubbles. If the master cylinder has been drained and refilled, and air is being bled from the first screw in the sequence, allow approximately five seconds between cycles for the master cylinder passages to refill.

19 When no more air bubbles appear, tighten the bleed screw securely, remove the tube and spanner, and refit the dust cap (where applicable). Do not overtighten the bleed screw.

20 Repeat the procedure on the remaining screws in the sequence, until all air is removed from the system, and the brake pedal feels firm again.

Using a one-way valve kit

21 As their name implies, these kits consist of a length of tubing with a one-way valve fitted, to prevent expelled air and fluid being

2.14b Removing the dust cap from a front caliper bleed screw

drawn back into the system; some kits include a translucent container, which can be positioned so that the air bubbles can be more easily seen flowing from the end of the tube.

22 The kit is connected to the bleed screw, which is then opened. The user returns to the driver's seat, depresses the brake pedal with a smooth, steady stroke, and slowly releases it; this is repeated until the expelled fluid is clear of air bubbles **(see illustration)**.

23 Note that these kits simplify work so much that it is easy to forget the master cylinder reservoir fluid level; ensure that this is maintained at least above the MIN level line at all times.

Using a pressure-bleeding kit

24 These kits are usually operated by the reservoir of pressurised air contained in the spare tyre. However, note that it will probably be necessary to reduce the pressure to a lower level than normal; refer to the instructions supplied with the kit.

25 By connecting a pressurised, fluid-filled container to the master cylinder reservoir, bleeding can be carried out simply by opening each screw in turn (in the specified sequence), and allowing the fluid to flow out until no more air bubbles can be seen in the expelled fluid.

26 This method has the advantage that the large reservoir of fluid provides an additional safeguard against air being drawn into the system during bleeding.

27 Pressure-bleeding is particularly effective when bleeding 'difficult' systems, or when bleeding the complete system at the time of routine fluid renewal.

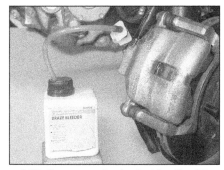

2.22 One-way valve brake bleeding kit connected to a front caliper

3.2a Brake pipe union-to-bracket securing clip (arrowed)

All methods

28 When bleeding is complete, and firm pedal feel is restored, wash off any spilt fluid, tighten the bleed screws securely, and refit their dust caps (where applicable).

29 Check the hydraulic fluid level in the master cylinder reservoir, and top up if necessary.

30 Discard any hydraulic fluid that has been bled from the system; it will not be fit for re-use.

31 Check the feel of the brake pedal. If it feels at all spongy, air must still be present in the system, and further bleeding is required. Failure to bleed satisfactorily after a reasonable repetition of the bleeding procedure may be due to worn master cylinder seals.

3 Hydraulic pipes and hoses – renewal

Note: *Before starting work, refer to the note at the beginning of Section 2 concerning the dangers of hydraulic fluid.*

1 If any pipe or hose is to be renewed, minimise fluid loss by first removing the master cylinder reservoir cap, then tighten the cap down onto a piece of polythene to obtain an airtight seal. Alternatively, flexible hoses can be sealed, if required, using a proprietary brake hose clamp; metal brake pipe unions can be plugged (if care is taken not to allow dirt into the system) or capped immediately they are disconnected. Place a wad of rag

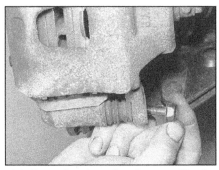

4.3 Removing the caliper lower guide pin bolt

3.2b Removing a front brake hose-to-bracket securing clip

under any union that is to be disconnected, to catch any spilt fluid.

2 If a flexible hose is to be disconnected, unscrew the brake pipe union nut before removing the spring clip which secures the hose to its mounting bracket **(see illustrations)**.

3 To unscrew the union nuts, it is preferable to obtain a brake pipe spanner of the correct size; these are available from most large motor accessory shops. Failing this, a close-fitting open-ended spanner will be required, though if the nuts are tight or corroded, their flats may be rounded-off if the spanner slips. In such a case, a self-locking wrench is often the only way to unscrew a stubborn union, but it follows that the pipe and the damaged nuts must be renewed on reassembly. Always clean a union and surrounding area before disconnecting it. If disconnecting a component with more than one union, make a careful note of the connections before disturbing any of them.

4 If a brake pipe is to be renewed, it can be obtained, cut to length and with the union nuts and end flares in place, from Nissan dealers. All that is then necessary is to bend it to shape, following the line of the original, before fitting it to the vehicle. Alternatively, most motor accessory shops can make up brake pipes from kits, but this requires very careful measurement of the original, to ensure that the replacement is of the correct length. The safest answer is usually to take the original to the shop as a pattern.

5 On refitting, do not overtighten the union nuts. It is not necessary to exercise brute force to obtain a sound joint.

4.4 Pivot the caliper body upwards to expose the brake pads

6 Ensure that the pipes and hoses are correctly routed, with no kinks, and that they are secured in the clips or brackets provided. After fitting, remove the polythene from the reservoir, and bleed the hydraulic system as described in Section 2. Wash off any spilt fluid, and check carefully for fluid leaks.

4 Front brake pads – renewal

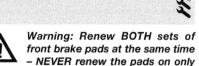

⚠ **Warning: Renew BOTH sets of front brake pads at the same time – NEVER renew the pads on only one wheel, as uneven braking may result.**

⚠ **Warning: Note that the dust created by wear of the pads may contain asbestos, which is a health hazard. Never blow it out with compressed air, and don't inhale any of it. An approved filtering mask should be worn when working on the brakes. DO NOT use petrol or petroleum-based solvents to clean brake parts; use brake cleaner or methylated spirit only.**

1 Firmly apply the handbrake, then jack up the front of the vehicle and support it securely on axle stands (see *Jacking and vehicle support*). Remove the front roadwheels.

2 Working on one side of the vehicle, push the caliper piston into its bore by pulling the caliper outwards.

3 Unscrew the caliper lower guide pin bolt (if necessary, use a slim open-ended spanner to counterhold the head of the guide pin), then remove the bolt **(see illustration)**.

4 Pivot the caliper body upwards to expose the brake pads **(see illustration)**. **Do not** depress the brake pedal until the caliper is refitted. Take care not to strain the brake fluid hose.

5 Note the locations and orientation of the shims fitted to the rear of each pad, and the anti-rattle clips fitted to the top and bottom of the pads. On early models, an upper and lower brake pad return spring is also fitted.

6 Withdraw the return springs (where fitted), then lift out the pads and shims, followed by the anti-rattle clips **(see illustrations)**.

7 Separate the shims from the brake pads, noting that there are two shims fitted to the inboard pad **(see illustration)**.

8 First measure the thickness of each brake pad's friction material. If either pad is worn at any point to the specified minimum thickness or less, all four pads must be renewed. Also, the pads should be renewed if any are fouled with oil or grease; there is no satisfactory way of degreasing friction material, once contaminated. If any of the brake pads are worn unevenly, or are fouled with oil or grease, trace and rectify the cause before reassembly. New brake pads and shim/clip kits are available from Nissan dealers. Do not be tempted to swap brake pads over to compensate for uneven wear.

9 If the brake pads are still serviceable,

4.6a Lift out the brake pads and shims . . .

4.6b . . . followed by the anti-rattle clips

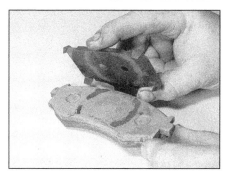

4.7 Separate the shims from the brake pads

carefully clean them using a clean, fine wire brush or similar, paying particular attention to the sides and back of the metal backing. Where applicable, clean out the grooves in the friction material, and pick out any large embedded particles of dirt or debris.

10 Clean the anti-rattle clips, shims, and the brake pad locations in the caliper body/ mounting bracket.

11 Prior to fitting the pads, check that the guide pins are free to slide easily in the caliper body/mounting bracket, and check that the rubber guide pin gaiters are undamaged. Brush the dust and dirt from the caliper and piston, but *do not* inhale it, as it is a health hazard. Inspect the dust seal around the piston for damage, and the piston for evidence of fluid leaks, corrosion or damage. If attention to any of these components is necessary, refer to Section 9.

12 If new brake pads are to be fitted, the caliper piston must be pushed back into the cylinder, to make room for them. Either use a G-clamp or similar tool, or use suitable pieces of wood as levers. Provided that the master cylinder reservoir has not been overfilled with hydraulic fluid, there should be no spillage, but keep a careful watch on the fluid level while retracting the piston. If the fluid level rises above the MAX level line at any time, the surplus should be syphoned off or ejected via a plastic tube connected to the bleed screw (see Section 2).

 Warning: Do not syphon the fluid by mouth, as it is poisonous; use a syringe or an old poultry baster.

13 Apply a little anti-squeal brake grease to the contact surfaces of the pad backing plates and the shims, but take great care not to allow any grease onto the pad friction linings. Similarly, apply brake grease to the contact surfaces of the anti-rattle clips – again take care not to apply excess grease, which may contaminate the pads.

14 Refit the anti-rattle clips to the caliper mounting bracket, then refit the pads and shims, in the positions noted before removal, ensuring that the pad friction material is against the disc.

15 Pivot the caliper back into position, over the pads and mounting bracket.

16 Refit the caliper lower guide pin bolt and tighten it to the specified torque.

17 Check that the caliper body slides smoothly on the guide pins.

18 Repeat the procedure on the remaining front caliper.

19 With both sets of front brake pads refitted, depress the brake pedal repeatedly until the pads are pressed into firm contact with the brake disc, and normal pedal pressure is restored.

20 Refit the roadwheels, and lower the vehicle to the ground.

21 Finally, check the brake hydraulic fluid level as described in *Weekly checks*.

22 Note that new pads will not give full braking efficiency until they have bedded-in. Be prepared for this, and avoid hard braking as far as possible for the first hundred miles or so after pad renewal.

5 Rear brake pads – renewal

⚠ *Warning: Renew BOTH sets of rear brake pads at the same time – NEVER renew the pads on only one wheel, as uneven braking may result.*

⚠ *Warning: Before starting work, refer to the warning given at the beginning of Section 4, concerning the dangers of asbestos dust.*

1 Chock the front wheels, then jack up the rear of the car and support it on axle stands (see *Jacking and vehicle support*). Remove the rear roadwheels, and release the handbrake fully.

2 Working on one side of the vehicle, extract the spring clip securing the handbrake cable to the bracket on the brake caliper, then release the cable from the bracket, and disconnect the end of the cable from the lever on the caliper **(see illustrations)**. Alternatively, undo the bracket retaining bolt and release the bracket and cable from the caliper.

3 Extract the spring clip securing the brake hydraulic hose to its mounting bracket on the rear axle beam. Withdraw the hose from the bracket.

4 Unscrew the caliper lower guide pin bolt **(see illustration)**. If necessary, use a slim open-ended spanner to counterhold the head of the guide pin.

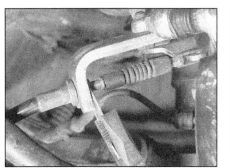

5.2a Extract the spring clip . . .

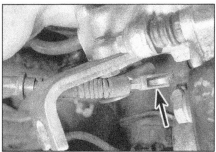

5.2b . . . then release the handbrake cable from the bracket and the lever on the caliper (arrowed)

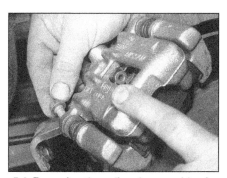

5.4 Removing the caliper lower guide pin bolt

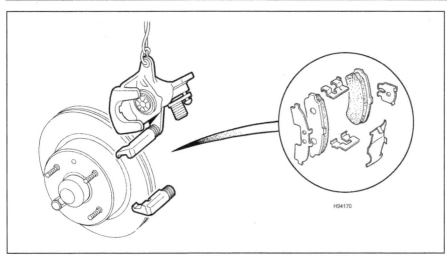

5.6 Rear brake pads, shims and anti-rattle clip arrangement

5 Pivot the caliper body upwards to expose the brake pads. **Do not** depress the brake pedal until the caliper is refitted. Take care not to strain the brake fluid hose.

6 Note the locations and orientation of the shims fitted to the rear of each pad and the anti-rattle clips fitted to the top and bottom of the pads **(see illustration)**.

7 Withdraw the pads, shims and the anti-rattle clips.

8 First measure the thickness of each brake pad's friction material. If either pad is worn at any point to the specified minimum thickness or less, all four pads must be renewed. Also, the pads should be renewed if any are fouled with oil or grease; there is no satisfactory way of degreasing friction material, once contaminated. If any of the brake pads are worn unevenly, or are fouled with oil or grease, trace and rectify the cause before reassembly. New brake pads and shim/spring kits are available from Nissan dealers. Do not be tempted to swap brake pads over to compensate for uneven wear.

9 If the brake pads are still serviceable, carefully clean them using a clean, fine wire brush or similar, paying particular attention to the sides and back of the metal backing. Where applicable, clean out the grooves in the friction material, and pick out any large

embedded particles of dirt or debris. Carefully clean the pad locations in the caliper body/mounting bracket.

10 Prior to fitting the pads, check that the guide pins are free to slide easily in the caliper body/mounting bracket, and check that the rubber guide pin gaiters are undamaged. Brush the dust and dirt from the caliper and piston, but *do not* inhale it, as it is a health hazard. Inspect the dust seal around the piston for damage, and the piston for evidence of fluid leaks, corrosion or damage. If attention to any of these components is necessary, refer to Section 10.

11 If new brake pads are to be fitted, the caliper piston must be pushed back into the cylinder, to make room for them. To retract the piston, ideally a suitable retractor tool should be used (available from motor factors). Alternatively, engage a suitable pair of long-nosed pliers with the notches in the piston, then turn the piston clockwise (viewed from the outboard end of the caliper), whilst simultaneously pushing it into the caliper **(see illustration)**. Take care not to damage the surface of the piston, or twist the dust seal. Provided that the master cylinder reservoir has not been overfilled with hydraulic fluid, there should be no spillage, but keep a careful watch on the fluid level while retracting the piston. If the fluid level rises above the MAX level line at any time, the surplus should be syphoned off or ejected via a plastic tube connected to the bleed screw (see Section 2).

⚠️ *Warning: Do not syphon the fluid by mouth, as it is poisonous; use a syringe or an old poultry baster.*

12 Apply a little anti-squeal brake grease to the contact surfaces of the pad backing plates and the shims, but take great care not to allow any grease onto the pad friction linings. Similarly, apply brake grease to the contact surfaces of the anti-rattle clips – again take care not to apply excess grease, which may contaminate the pads. Note that new pads may be supplied with an anti-squeal

coating on their contact faces, in which case there is no need to apply brake grease.

13 Refit the anti-rattle clips to the caliper mounting bracket, then refit the pads and the shims, in the positions noted before removal. Ensure that the pad friction material is against the disc.

14 Pivot the caliper body downwards over the brake pads. Ensure that the peg on the back of the inboard pad engages with the nearest notch in the piston. It may be necessary to rotate the piston as described in paragraph 11 to align the nearest notch with the peg.

15 Refit the caliper lower guide pin bolt, and tighten it to the specified torque.

16 Check that the caliper body slides smoothly on the guide pins.

17 Reconnect the end of the handbrake cable to the lever. Refit the cable and mounting bracket to the caliper, or alternatively secure the cable to the bracket with the spring clip. Where applicable, ensure that the locating peg on the mounting bracket engages with the corresponding hole in the caliper body.

18 Refit the brake hydraulic hose to the mounting bracket on the axle beam and secure with the spring clip.

19 Repeat the procedure on the remaining rear caliper.

20 With both sets of rear brake pads refitted, operate the handbrake repeatedly until the correct brake pad-to-disc clearance is established (adjustment occurs automatically when the handbrake is operated).

21 Depress the brake pedal repeatedly to ensure that normal pedal pressure is restored.

22 Refit the roadwheels, and lower the vehicle to the ground.

23 Finally, check the brake hydraulic fluid level as described in *Weekly checks*.

24 Note that new pads will not give full braking efficiency until they have bedded-in. Be prepared for this, and avoid hard braking as far as possible for the first hundred miles or so after pad renewal.

6 Rear brake shoes – renewal

⚠️ *Warning: Renew BOTH sets of rear brake shoes at the same time – NEVER renew the shoes on only one wheel, as uneven braking may result.*

⚠️ *Warning: Before starting work, refer to the warning given at the beginning of Section 4, concerning the dangers of asbestos dust.*

1 Remove the rear brake drum, as described in Section 8.

2 Working carefully, and taking the necessary precautions, remove all traces of brake dust from the brake drum, backplate and shoes.

3 Measure the thickness of the friction material of each brake shoe at several points;

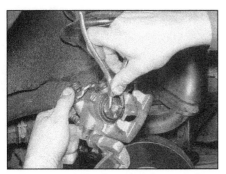

5.11 Use a pair of long-nosed or circlip pliers to retract a rear caliper piston

6.6a Remove the leading shoe hold-down spring cup, then lift off the spring

6.6b Withdraw the retainer pin from the rear of the backplate

6.7a Disengage the leading shoe from the bottom anchor . . .

6.7b . . . then unhook and remove the lower return spring from the leading and trailing shoes

6.12 Unhook the handbrake cable from the operating lever, then remove the trailing shoe

6.17 Apply high-temperature brake grease or anti-seize compound to the brake shoe contact areas on the backplate

if either shoe is worn at any point to the specified minimum thickness or less, all four shoes must be renewed as a set. The shoes should also be renewed if any are fouled with oil or grease; there is no satisfactory way of degreasing friction material, once contaminated.

4 If any of the brake shoes are worn unevenly, or fouled with oil or grease, trace and rectify the cause before reassembly.

5 Note the position of each shoe, and the location of the return and adjuster springs. Also make a note of the adjuster component locations, to aid refitting later.

6 Using pliers, depress the leading shoe hold-down spring cup and turn it through 90°, while holding the retainer pin with your finger from the rear of the backplate. With the cup removed, lift off the spring, then withdraw the retainer pin **(see illustrations)**.

7 Disengage the lower end of the leading shoe from the bottom anchor then, using pliers, unhook and remove the lower return spring from the leading and trailing shoes **(see illustrations)**.

8 Unhook the adjuster spring from the adjuster lever and remove it from the stud on the leading shoe.

9 Free the upper end of the leading shoe from the wheel cylinder piston and disengage the shoe from the adjuster strut. Unhook the upper return spring from the stud on the rear of the leading shoe and lift the leading shoe from the backplate. Unhook the upper return spring from the trailing shoe and remove the spring.

10 Withdraw the adjuster strut and adjuster lever from the trailing shoe.

11 Release the trailing shoe hold-down spring cup, then remove the spring and retainer pin.

12 Unhook the handbrake cable from the handbrake operating lever on the trailing shoe, then remove the trailing shoe **(see illustration)**.

13 Retain the wheel cylinder pistons in the wheel cylinder using a cable tie or a strong elastic band. Do not depress the brake pedal until the brakes are reassembled.

14 Withdraw the forked end from the adjuster strut, and carefully examine the assembly for signs of wear or damage. Pay particular attention to the threads and the toothed adjuster wheel, and renew if necessary.

15 Check the condition of all return springs and renew any that show signs of distortion or other damage.

16 Peel back the rubber protective caps, and check the wheel cylinder for fluid leaks or other damage; check that both cylinder pistons are free to move easily. Refer to Section 11, if necessary, for information on wheel cylinder overhaul.

17 Prior to installation, clean the backplate, and apply a thin smear of high-temperature brake grease or anti-seize compound to all those surfaces of the backplate which bear on the shoes, particularly the wheel cylinder pistons and lower pivot point **(see illustration)**. Do not allow the lubricant to foul the friction material.

18 Prise off the retaining clip, then remove the washer and handbrake operating lever from the trailing brake shoe. Transfer the lever to the new trailing shoe, place the washer in position and secure with a new retaining clip.

19 Offer up the trailing shoe to the backplate, then engage the handbrake cable with the operating lever on the shoe. Ensure that the spring on the handbrake cable is correctly located against the operating lever.

20 Slip the retaining cable tie or elastic band out of the wheel cylinder trailing shoe piston, then manipulate the trailing shoe into position on the backplate. Ensure that the upper edge of the shoe engages with the recess in the end of the wheel cylinder piston, then secure the shoe with the hold-down retainer pin, spring and cup **(see illustrations)**.

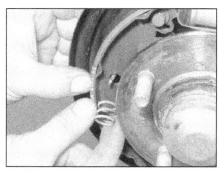

6.20a Refit the trailing shoe and secure the shoe with the hold-down retainer pin and spring . . .

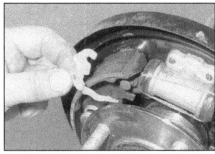

6.20b ... followed by the spring cup

6.21 Fit the adjuster lever to the trailing shoe ensuring that it fully engages with the pivot peg on the shoe

6.23 Place the adjuster strut in position with its slots engaged with the trailing shoe and adjuster lever

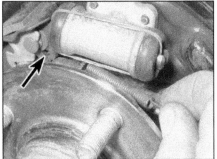

6.24 Attach the adjuster spring to the trailing shoe, with the end of the spring hooked into the hole (arrowed) from the rear

21 Fit the adjuster lever to the trailing shoe ensuring that it fully engages with the pivot peg on the shoe **(see illustration)**.

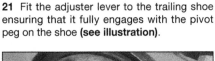
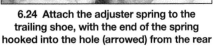

6.25 Hook the other end of the adjuster spring over the stud (arrowed) on the rear of the leading shoe

22 Shorten the adjuster strut to its minimum length by turning the toothed wheel, and apply a smear of brake grease to the contact

faces of the adjuster strut and handbrake operating lever.

23 Place the adjuster strut in position with its large slot engaged with the cut-out on the trailing shoe, and its small slot engaged with the cut-out in the adjuster lever **(see illustration)**.

24 Attach the adjuster spring to the upper edge of the trailing shoe, noting that the end of the spring is hooked into the hole from the rear **(see illustration)**.

25 Offer up the leading shoe to the backplate, then hook the adjuster spring over the stud on the rear surface of the shoe **(see illustration)**.

26 Remove the retaining cable tie or elastic band from the wheel cylinder pistons then manipulate the leading shoe into position, engaging the adjuster strut with the cut-out in the shoe and the upper edge of the shoe with the recess in the wheel cylinder piston **(see illustrations)**.

27 Locate the coiled end of the upper return spring over the stud on the leading shoe, then hook the straight end of the spring into the slot in the adjuster lever **(see illustration)**.

28 Attach the lower return spring to the trailing shoe, noting that the end of the spring is hooked into the hole from the front. Connect the other end of the spring to the leading shoe **(see illustration)**, then engage the leading shoe into position on the lower anchor. Secure the shoe with the hold-down retainer pin, spring and cup.

6.26a Engage the adjuster strut with the cut-out in the leading shoe ...

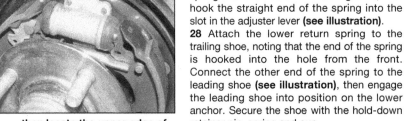
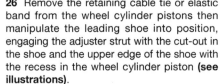
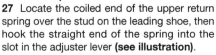

6.26b ... then locate the upper edge of the shoe in the recess in the wheel cylinder piston

29 Check that all components have been correctly refitted, and check that the adjuster mechanism operates correctly.

30 Using a screwdriver, turn the adjuster strut toothed wheel to expand the shoes until the brake drum just slides over the shoes.

31 Refit the brake drum as described in Section 8.

32 Repeat the above procedure on the remaining rear brake.

33 On completion, check the brake hydraulic fluid level in the master cylinder as described in *Weekly checks*.

34 Note that new shoes will not give full braking efficiency until they have bedded-in. Be prepared for this, and avoid hard braking as far as possible for the first hundred miles or so after shoe renewal.

6.27 Locate the coiled end of the upper return spring over the leading shoe stud, then hook the straight end into the adjuster lever slot

6.28 Connect the lower return spring to the slots in the trailing and leading shoes

7 Brake disc – inspection, removal and refitting

 Warning: Before starting work, refer to the warning at the beginning of Section 4 concerning the dangers of asbestos dust.

Note: *If either disc requires renewal, BOTH should be renewed at the same time, to ensure even and consistent braking. New brake pads should also be fitted.*

Front disc

Inspection

1 Firmly apply the handbrake, then jack up the front of the vehicle and support it securely on axle stands (see *Jacking and vehicle support*). Remove the appropriate front roadwheel.

2 Slowly rotate the brake disc so that the full area of both sides can be checked; remove the brake pads (see Section 4) if better access is required to the inboard surface. Light scoring is normal in the area swept by the brake pads, but if heavy scoring or cracks are found, the disc must be renewed.

3 It is normal to find a lip of rust and brake dust around the disc's perimeter; this can be scraped off if required. If, however, a lip has formed due to excessive wear of the brake pad swept area, then the disc's thickness must be measured using a micrometer. Take measurements at several places around the disc, at the inside and outside of the pad swept area; if the disc has worn at any point to the specified minimum thickness or less, the disc must be renewed.

4 If the disc is thought to be warped, it can be checked for run-out. Either use a dial gauge mounted on any convenient fixed point, while the disc is slowly rotated, or use feeler blades to measure (at several points all around the disc) the clearance between the disc and a fixed point, such as the caliper mounting bracket. If the measurements obtained are at the specified maximum or beyond, the disc is excessively warped, and must be renewed; however, it is worth checking first that the hub bearing is in good condition (Chapters 1 and/or 10). Also try the effect of removing the disc and turning it through 180°, to reposition it on the hub; if the run-out is still excessive, the disc must be renewed.

5 Check the disc for cracks, especially around the wheel stud holes, and any other wear or damage, and renew if necessary.

Removal

6 If not already done, firmly apply the handbrake, then jack up the front of the vehicle and support it securely on axle stands (see *Jacking and vehicle support*). Remove the appropriate front roadwheel.

7 Unscrew the two bolts securing the caliper mounting bracket to the hub carrier **(see illustration)**. Withdraw the caliper assembly,

7.7 Unscrew the two bolts securing the caliper mounting bracket to the hub carrier

and suspend it using wire or string. Take care not to strain the brake fluid hose – if necessary release the hose from the securing clip(s).

 HAYNES HINT *Place a clean spacer (of the same thickness as the disc) between the pads, to prevent them from being dislodged.*

8 If the original disc is to be refitted, mark the relationship between the disc and the hub, then pull the disc from the roadwheel studs.

9 If the disc is stuck, it can be pushed off by screwing two M8 bolts into the holes provided in the disc, and evenly tightening the bolts to push the disc from the hub **(see illustration)**.

Refitting

10 Ensure that the mating faces of the disc and the hub are clean and flat. If necessary, wipe the mating surfaces clean.

11 If the original disc is being refitted, align the marks made on the disc and hub before removal, then refit the disc.

12 If a new disc has been fitted, use a suitable solvent to wipe any preservative coating from the disc.

13 Refit the caliper, ensuring that the pads locate correctly over the disc, then tighten the caliper mounting bracket securing bolts to the specified torque. Where applicable, refit the brake fluid hose to the clip(s).

14 Depress the brake pedal repeatedly until the pads are pressed into firm contact with the brake disc, and normal pedal pressure is restored.

15 Refit the roadwheel, and lower the vehicle to the ground.

Rear disc

Inspection

16 Chock the front wheels then jack up the rear of the vehicle and support it securely on axle stands (see *Jacking and vehicle support*). Remove the appropriate rear roadwheel.

17 Fully release the handbrake.

18 Proceed as described for the front disc in paragraphs 2 to 5, but refer to Section 5 if the brake pads are to be removed.

7.9 Using two M8 bolts to remove a front brake disc

Removal

19 If not already done, chock the front wheels then jack up the rear of the vehicle and support it securely on axle stands (see *Jacking and vehicle support*). Remove the appropriate rear roadwheel.

20 Unbolt the handbrake cable mounting bracket from the caliper, then disconnect the end of the cable from the lever on the caliper. It may be necessary to release the spring clip securing the cable to the mounting bracket to enable the cable to be disconnected.

21 Proceed as described for the front brake disc in paragraphs 7 to 9.

Refitting

22 Proceed as described for the front brake disc in paragraphs 10 to 13.

23 Reconnect the handbrake cable to the lever on the caliper, then refit the cable mounting bracket, and where applicable, the spring clip. Where applicable, ensure that the locating peg on the mounting bracket engages with the corresponding hole in the caliper body.

24 Operate the handbrake repeatedly until the correct brake pad-to-disc clearance is established (adjustment occurs automatically when the handbrake is operated).

25 Depress the brake pedal repeatedly to ensure that normal pedal pressure is restored.

26 Refit the roadwheel, and lower the vehicle to the ground.

8 Rear brake drum – removal, inspection and refitting

 Warning: Before starting work, refer to the warning at the beginning of Section 4 concerning the dangers of asbestos dust.

Removal

1 Chock the front wheels then jack up the rear of the vehicle and support it securely on axle stands (see *Jacking and vehicle support*). Remove the appropriate rear roadwheel.

2 Fully release the handbrake.

3 If the drum cannot easily be pulled from the wheel studs, retract the brake shoes as follows.

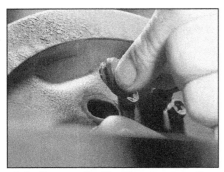

8.4a Remove the blanking plug from the brake backplate . . .

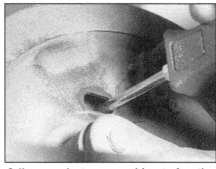

8.4b . . . and use a screwdriver to free the handbrake adjuster mechanism

8.6 Using two M8 bolts to remove a brake drum

4 Remove the blanking plug from the rear of the brake backplate. Insert a small screwdriver through the hole in the backplate and carefully push the adjuster lever on the trailing shoe clear of the toothed wheel on the adjuster strut **(see illustrations)**. Engage the screwdriver with the toothed wheel and pivot the screwdriver downwards to turn the wheel, thus shortening the adjuster strut. Continue doing this until the adjuster is fully retracted.

5 If the original drum is to be refitted, mark the relationship between the drum and the hub, then pull the drum from the roadwheel studs.

6 If the drum is still tight, it can be pushed off by screwing two M8 bolts into the holes provided in the drum, and evenly tightening the bolts to push the drum from the hub **(see illustration)**.

Inspection

Note: *If either drum requires renewal, BOTH should be renewed at the same time, to ensure even and consistent braking. New brake shoes should also be fitted.*

7 Working carefully, remove all traces of brake dust from the drum, but *avoid inhaling the dust, as it is a health hazard.*

8 Clean the outside of the drum, and check it for obvious signs of wear or damage, such as cracks around the roadwheel stud holes; renew the drum if necessary.

9 Carefully examine the inside of the drum. Light scoring of the friction surface is normal, but if heavy scoring is found, the drum must be renewed.

10 It is usual to find a lip on the drum's inboard edge which consists of a mixture of rust and brake dust; this should be scraped away, to leave a smooth surface which can be polished with fine (120- to 150-grade) emery paper. If, however, the lip is due to the friction surface being recessed by excessive wear, then the drum must be renewed.

11 If the drum is thought to be excessively worn, or oval, its internal diameter must be measured at several points using an internal micrometer. Take measurements in pairs, the second at right-angles to the first, and compare the two, to check for signs of ovality. Provided that it does not enlarge the drum to beyond the specified maximum diameter, it

may be possible to have the drum refinished by skimming or grinding; if this is not possible, the drums on both sides must be renewed. Note that if the drum is to be skimmed, BOTH drums must be refinished, to maintain a consistent internal diameter on both sides.

Refitting

12 If a new brake drum is to be installed, use a suitable solvent to remove any preservative coating that may have been applied to its internal friction surfaces. Note that it may also be necessary to shorten the adjuster strut length, by rotating the strut wheel, to allow the drum to pass over the brake shoes.

13 If the original drum is being refitted, align the marks made on the drum and hub before removal, then fit the drum over the wheel studs.

14 Depress the footbrake repeatedly to expand the brake shoes against the drum, and ensure that normal pedal pressure is restored.

15 Check and if necessary adjust the handbrake cable as described in Chapter 1.

16 Refit the roadwheel, and lower the vehicle to the ground.

9 Front brake caliper – removal, overhaul and refitting

⚠ *Warning: Before starting work, refer to the note at the beginning of Section 2 concerning the*

9.10 Remove the piston by applying a jet of compressed air, such as that from a tyre pump, to the fluid inlet port

dangers of hydraulic fluid, and to the warning at the beginning of Section 4 concerning the dangers of asbestos dust.

Removal

1 Firmly apply the handbrake, then jack up the front of the vehicle and support it securely on axle stands (see *Jacking and vehicle support*). Remove the appropriate front roadwheel.

2 To minimise fluid loss during the following operations, remove the master cylinder reservoir cap, then tighten it down onto a piece of polythene to obtain an airtight seal. Alternatively, use a brake hose clamp, a G-clamp or a similar tool to clamp the flexible hose running to the caliper.

3 Clean the area around the fluid hose union on the caliper, then unscrew the hose union banjo bolt. Recover the two sealing washers noting that new washers will be required for refitting. Cover the open ends of the banjo and the caliper, to prevent dirt ingress.

4 Remove the brake pads as described in Section 4.

5 Unscrew the caliper upper guide pin bolt. If necessary, use a slim open-ended spanner to counterhold the head of the guide pin.

6 Withdraw the caliper from the mounting bracket.

7 If desired, the caliper mounting bracket can be unbolted from the hub carrier.

Overhaul

> **HAYNES HINT** *Before commencing work, ensure that the appropriate caliper overhaul kit is obtained.*

8 With the caliper on the bench, wipe away all traces of dust and dirt, but *avoid inhaling the dust, as it is a health hazard.*

9 Extract the caliper guide pins, if necessary by screwing the bolts into the pins, and pulling on the bolts to withdraw the pins. Peel off the rubber dust cover from each guide pin.

10 Place a small block of wood between the caliper body and the piston. Remove the piston, including the dust seal, by applying a jet of low-pressure compressed air, such as that from a tyre pump, to the fluid inlet port **(see illustration)**.

11 Peel the dust seal off the piston, and use a blunt instrument, such as a knitting needle, to extract the piston seal from the caliper cylinder bore **(see illustration)**.

12 Thoroughly clean all components, using only methylated spirit or clean hydraulic fluid. Never use mineral-based solvents such as petrol or paraffin, which will attack the hydraulic system rubber components.

13 The caliper piston seal and the dust seal, the guide pin dust covers, and the bleed nipple dust cap, are only available as part of a seal kit. Since the manufacturers recommend that the piston seal and dust seal are renewed whenever they are disturbed, all of these components should be discarded, and new ones fitted on reassembly as a matter of course.

14 Carefully examine all parts of the caliper assembly, looking for signs of wear or damage. In particular, the cylinder bore and piston must be free from any signs of scratches, corrosion or wear. If there is any doubt about the condition of any part of the caliper, the relevant part should be renewed; note that if the caliper body or the mounting bracket are to be renewed, they are available only as part of the complete assembly.

15 The manufacturers recommend that minor scratches, rust, etc, may be polished away from the cylinder bore using fine emery paper, but the piston must be renewed to cure such defects. The piston surface is plated, and **must not** be polished with emery or similar abrasives.

16 Check that the threads in the caliper body and the mounting bracket are in good condition. Check that both guide pins are undamaged, and (when cleaned) a reasonably tight sliding fit in the mounting bracket bores.

17 Use compressed air to blow clear the fluid passages.

 Warning: Wear eye protection when using compressed air.

18 Before commencing reassembly, ensure that all components are spotlessly-clean and dry.

19 Soak the new piston seal in clean hydraulic fluid, and fit it to the groove in the cylinder bore, using your fingers only (no tools) to manipulate it into place.

20 Fit the new dust seal inner lip to the cylinder groove, smear clean hydraulic fluid over the piston and caliper cylinder bore, and twist the piston into the dust seal. Press the piston squarely into the cylinder, then slide the dust seal outer lip into the groove in the piston.

21 Fit a new rubber dust cover to each guide pin, and apply a smear of brake grease to the guide pins before refitting them to their bores.

Refitting

22 Where applicable, refit the caliper mounting bracket to the hub carrier, and tighten the mounting bolts to the specified torque.

9.11 Use a blunt instrument to extract the piston seal from the caliper cylinder bore

23 Place the caliper in position, refit the upper guide pin bolt, and tighten it to the specified torque.

24 Refit the brake pads as described in Section 4.

25 Check that the caliper slides smoothly on the mounting bracket.

26 Check that the hydraulic fluid hose is correctly routed, without being twisted, then reconnect the union to the caliper, using two new sealing washers. Refit the union banjo bolt, and tighten to the specified torque.

27 Remove the polythene from the master cylinder reservoir cap, or remove the clamp from the fluid hose, as applicable.

28 Bleed the hydraulic fluid circuit as described in Section 2. Note that if no other part of the system has been disturbed, it should only be necessary to bleed the relevant front circuit.

29 Depress the brake pedal repeatedly to bring the pads into contact with the brake disc, and ensure that normal pedal pressure is restored.

30 Refit the roadwheel, and lower the vehicle to the ground.

10 Rear brake caliper – removal, overhaul and refitting

 Warning: Before starting work, refer to the note at the beginning of Section 2 concerning the dangers of hydraulic fluid, and to the warning at the beginning of Section 4 concerning the dangers of asbestos dust.

Removal

1 Chock the front wheels then jack up the rear of the vehicle and support it securely on axle stands (see *Jacking and vehicle support*). Remove the appropriate rear roadwheel.

2 Release the spring clip securing the handbrake cable to the mounting bracket, then disconnect the end of the cable from the lever on the caliper.

3 To minimise fluid loss during the following operations, remove the master cylinder reservoir cap, then tighten it down onto a piece of polythene to obtain an airtight seal. Alternatively, use a brake hose clamp, a G-

clamp or a similar tool to clamp the flexible hose running to the caliper.

4 Clean the area around the fluid hose union on the caliper, then unscrew the hose union banjo bolt. Recover the two sealing washers noting that new washers will be required for refitting. Cover the open ends of the banjo and the caliper, to prevent dirt ingress.

5 Remove the brake pads as described in Section 5.

6 Unscrew the caliper upper guide pin bolt. If necessary, use a slim open-ended spanner to counterhold the head of the guide pin.

7 Withdraw the caliper from the mounting bracket.

8 If desired, the caliper mounting bracket can be unbolted from the hub carrier.

Overhaul

 HAYNES HiNT *Before commencing work, ensure that the appropriate caliper overhaul kit is obtained.*

9 With the caliper on the bench, wipe away all traces of dust and dirt, but *avoid inhaling the dust, as it is a health hazard.*

10 Extract the caliper guide pins by screwing the bolts into the pins, and pulling on the bolts to withdraw the pins. Peel off the rubber dust cover from each guide pin.

11 Engage a suitable pair of long-nosed pliers with the notches in the piston, then turn the piston anti-clockwise (viewed from the outboard end of the caliper), to unscrew it from the caliper. Withdraw the piston from the caliper body, and peel off the dust seal.

12 Working inside the rear of the piston, use circlip pliers to extract the circlip, then remove the first spacer, the wave washer, the second spacer, the ball-bearing, and the adjusting nut. Prise the cup seal off the adjusting nut.

13 Working inside the caliper cylinder bore, use circlip pliers to extract the circlip, whilst using a suitable length of tubing to compress the spring cap against the spring pressure.

14 With the circlip removed from its groove, allow the spring to push out the components until pressure is relaxed, then withdraw the circlip, the spring cap, the spring and the spring seat.

15 Use circlip pliers to extract the remaining circlip, then withdraw the key plate, the pushrod and the plunger. Prise the sealing O-ring off the pushrod.

16 Using a blunt instrument such as a knitting needle, extract the piston seal from the caliper cylinder bore.

17 Ensure that the pushrod and plunger have been removed (paragraph 15), then unhook the return spring from the handbrake lever, and prise the lever and cam assembly from the caliper body.

18 Clamp the cam gently in a soft-jawed vice, then unscrew the retaining nut, and withdraw the spring washer, the return spring, the lever, and the dust cover, from the cam.

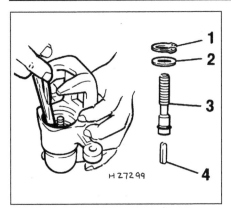
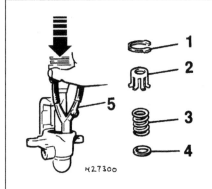

10.30 Rear caliper pushrod components

1	Circlip	3	Pushrod
2	Key plate	4	Plunger

10.31 Rear caliper spring components

1	Circlip	4	Spring seat
2	Spring cap	5	Circlip pliers
3	Spring		

19 Thoroughly clean all components, using only methylated spirit or clean hydraulic fluid. Never use mineral-based solvents such as petrol or paraffin, which will attack the hydraulic system rubber components.

20 Discard all seals, cups, dust covers and other rubber components. These are available as part of a caliper seal kit, and should be renewed as a matter of course whenever they are disturbed.

21 Carefully examine all parts of the caliper assembly, looking for signs of wear or damage. In particular, the cylinder bore and piston must be free from any signs of scratches, corrosion or wear. If there is any doubt about the condition of any part of the caliper, the relevant part should be renewed. Note that if the caliper body or the mounting bracket are to be renewed, they are available only as part of the complete assembly.

22 Minor scratches, rust, etc, may be polished away from the cylinder bore using fine emery paper, but the piston must be

renewed to cure such defects. The piston surface is plated, and **must not** be polished with emery or similar abrasives.

23 Check that the threads in the caliper body and the mounting bracket are in good condition. Check that both guide pins are undamaged, and (when cleaned) a reasonably tight sliding fit in the mounting bracket bores.

24 Use compressed air to blow clear the fluid passages.

> ⚠ **Warning: Wear eye protection when using compressed air.**

25 Before commencing reassembly, ensure that all components are spotlessly-clean and dry.

26 Fit the new dust seal to the handbrake lever cam, and locate the lever on the cam flats. Refit the return spring, spring washer, and nut. Tighten the nut securely, taking care not to overtighten it.

27 Apply a smear of brake grease to the cam, then slide the cam into the caliper body. Check

that the assembly is correctly installed, and that the cam cut-out aligns with the pushrod aperture when the lever is rotated. Hook the return spring over its stop and into the lever.

28 Soak the new piston seal in clean hydraulic fluid, and fit it to the groove in the cylinder bore, using your fingers only (no tools) to manipulate it into position.

29 Fit a new O-ring to the pushrod, and apply a smear of rubber grease to the plunger and the pushrod. Assemble the plunger and pushrod, and fit them to the caliper body.

30 Refit the key plate so that its cut-out fits over the squared section of the pushrod, and its convex locating pip matches the concave depression in the caliper body. Secure the assembly by refitting the circlip **(see illustration)**.

31 Refit the spring seat, the spring and the spring cap, then compress the spring cap (using a suitable length of tubing as during removal) while refitting the securing circlip. Check that the circlip is correctly seated in its groove **(see illustration)**.

32 Fit the new cup seal to the adjusting nut, using only your fingers (no tools) to manipulate it into position. Ensure that the seal is correctly fitted **(see illustration)**.

33 Smear rubber grease over the cup seal lips, and fit the adjusting nut into the piston.

34 Pack the ball-bearing with brake grease and refit it, followed by the spacer, the wave washer, and the remaining spacer. Secure the components with the remaining circlip **(see illustration)**.

35 Apply a smear of rubber grease to the inner and outer lips of the new dust seal, then fit it to the piston.

36 Smear clean hydraulic fluid over the piston and the caliper cylinder bore, then refit the piston assembly, and screw it in clockwise (using a retractor tool or long-nosed pliers if necessary – see Section 5) until it seats.

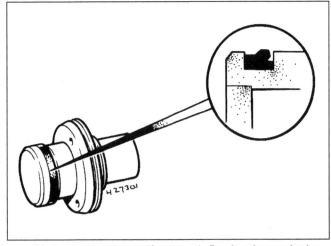

10.32 Ensure that the seal is correctly fitted to the rear brake caliper adjusting nut

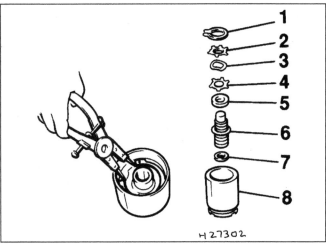

10.34 Rear caliper adjusting nut components

1	Circlip	4	Spacer	7	Cup seal
2	Spacer	5	Ball-bearing	8	Piston
3	Wave washer	6	Adjusting nut		

37 Engage the dust seal with the caliper body.

38 Fit a new rubber dust cover to each guide pin. Apply a smear of brake grease to the guide pins before refitting them to their bores.

Refitting

39 Where applicable, refit the caliper mounting bracket to the hub carrier, and tighten the mounting bolts to the specified torque.

40 Place the caliper in position, refit the caliper upper guide pin bolt and tighten it to the specified torque.

41 Refit the brake pads as described in Section 5.

42 Check that the caliper slides smoothly on the mounting bracket.

43 Check that the brake fluid hose is correctly routed, without being twisted, then reconnect the union to the caliper, using two new sealing washers. Refit the union banjo bolt, and tighten to the specified torque.

44 Reconnect the end of the handbrake cable to the lever. Refit the cable mounting bracket, and where applicable, the spring clip. Where applicable, ensure that the locating peg on the mounting bracket engages with the corresponding hole in the caliper body.

45 Remove the polythene from the master cylinder reservoir cap, or remove the clamp from the fluid hose, as applicable.

46 Bleed the hydraulic fluid circuit as described in Section 2. Note that if no other part of the system has been disturbed, it should only be necessary to bleed the relevant rear circuit.

47 Depress the brake pedal repeatedly to bring the pads into contact with the brake disc, and ensure that normal pedal pressure is restored.

48 Refit the roadwheel, and lower the vehicle to the ground.

11 Rear wheel cylinder – removal, overhaul and refitting

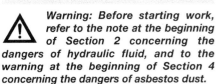

⚠️ **Warning: Before starting work, refer to the note at the beginning of Section 2 concerning the dangers of hydraulic fluid, and to the warning at the beginning of Section 4 concerning the dangers of asbestos dust.**

Removal

1 Remove the brake drum as described in Section 8.

2 Remove the brake shoes as described in Section 6.

3 To minimise fluid loss during the following operations, remove the master cylinder reservoir cap, then tighten it down onto a piece of polythene to obtain an airtight seal.

4 Clean the brake backplate around the wheel cylinder mounting bolts and the hydraulic pipe union, then unscrew the union nut and disconnect the hydraulic pipe. Cover the open ends of the pipe and the master cylinder to prevent dirt ingress.

5 Remove the securing bolts, then withdraw the wheel cylinder from the backplate.

Overhaul

 Before commencing work, ensure that the appropriate wheel cylinder overhaul kit is obtained.

6 Clean the assembly thoroughly, using only methylated spirit or clean brake fluid.

7 Peel off both rubber dust covers, then use paint or similar to mark one of the pistons so that the pistons are not interchanged on reassembly **(see illustration)**.

8 Withdraw both pistons and the spring.

9 Discard the rubber piston cups and the dust covers. These components should be renewed as a matter of course, and are available as part of an overhaul kit, which also includes the bleed screw dust cap.

10 Check the condition of the cylinder bore and the piston – the surfaces must be perfect and free from scratches, scoring and corrosion. It is advisable to renew the complete wheel cylinder if there is any doubt as to the condition of the cylinder bore or piston.

11 Ensure that all components are clean and dry. The pistons, spring and cups should be fitted wet, using hydraulic fluid as a lubricant – soak them in clean fluid before installation.

12 Fit the cups to the pistons, ensuring that they are the correct way round. Use only your fingers (no tools) to manipulate the cups into position.

13 Fit the first piston to the cylinder, taking care not to distort the cup. If the original pistons are being re-used, the marks made on dismantling should be used to ensure that the pistons are refitted to their original bores.

14 Refit the spring and the second piston.

15 Apply a smear of rubber grease to the exposed end of each piston and to the dust cover sealing lips, then fit the dust covers to each end of the wheel cylinder.

Refitting

16 Refitting is a reversal of removal, bearing in mind the following points:
  a) *Tighten the mounting bolts to the specified torque.*
  b) *Refit the brake shoes as described in Section 6, and refit the brake drum as described in Section 8.*
  c) *Before refitting the roadwheel and lowering the vehicle to the ground, remove the polythene from the fluid reservoir, and bleed the hydraulic system as described in Section 2. Note that if no other part of the system has been disturbed, it should only be necessary to bleed the relevant rear circuit.*

12 Master cylinder – removal, overhaul and refitting

Note: *Before starting work, refer to the warning at the beginning of Section 2 concerning the dangers of hydraulic fluid.*

Removal

1 Remove the master cylinder fluid reservoir cap, and syphon the hydraulic fluid from the reservoir. Alternatively, open any convenient bleed screw in the system, and gently pump the brake pedal to expel the fluid through a tube connected to the screw (see Section 2). Disconnect the wiring connector from the brake fluid level sender unit on the side of the reservoir.

⚠️ **Warning: Do not syphon the fluid by mouth, as it is poisonous; use a syringe or an old poultry baster.**

2 Wipe clean the area around the brake pipe unions on the side and top of the master cylinder, and place absorbent rags beneath the pipe unions to catch any surplus fluid. Make a note of the correct fitted positions of the unions, then unscrew the union nuts and carefully withdraw the pipes. Plug or tape over the pipe ends and master cylinder orifices, to minimise the loss of brake fluid, and to

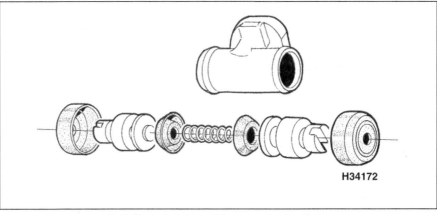

11.7 Exploded view of the rear wheel cylinder

prevent the entry of dirt into the system. Wash off any spilt fluid immediately with cold water.

3 Slacken and remove the two nuts securing the master cylinder to the vacuum servo unit, then withdraw the unit from the engine compartment.

Overhaul

Before attempting overhaul of the master cylinder, ensure that the appropriate overhaul kit is obtained.

4 Tap out the roll pin (later models only) securing the fluid reservoir to the master cylinder body then pull the fluid reservoir from the top of the master cylinder. Prise the reservoir seals from the reservoir or the master cylinder, as applicable.

5 Using a suitable screwdriver, bend back the tangs securing the master cylinder end cap, then withdraw the end cap.

6 On models with ABS, push on the end of the primary piston assembly using a suitable wooden dowel to compress the primary and secondary piston assemblies. On early models unscrew the stop-bolt from the side of the master cylinder. On later models extract the stop-pin from the fluid reservoir inlet port.

7 Noting the order of removal, and the direction of fitting of each component, withdraw the primary and secondary piston assemblies with their springs and seals, tapping the body onto a clean wooden surface to dislodge them. If necessary, clamp the master cylinder body in a vice (fitted with soft jaw covers) and use compressed air (applied through the secondary circuit fluid port) to assist the removal of the secondary piston assembly.

⚠️ *Warning: Wear eye protection when working with compressed air.*

8 Thoroughly clean all components, using only methylated spirit, isopropyl alcohol or clean hydraulic fluid as a cleaning medium. Never use mineral-based solvents such as petrol or paraffin, as they will attack the hydraulic system rubber components. Dry the components immediately, using compressed air or a clean, lint-free cloth.

9 Check all components, and renew any that are worn or damaged (note that individual seal components are not available – the overhaul kit will contain complete primary and secondary piston assemblies, complete with all seals, washers, etc). Check particularly the cylinder bores and pistons; the complete assembly should be renewed if these are scratched, worn or corroded. If there is any doubt about the condition of the assembly or of any of its components, renew it. Check that the cylinder body fluid passages are clear.

10 Before reassembly, soak the pistons and the new seals in clean hydraulic fluid. Smear clean fluid into the cylinder bore.

11 Insert the piston assemblies into the cylinder bore (make sure that the assemblies are inserted squarely), using a twisting motion to avoid trapping the seal lips. Ensure that all components are refitted in the correct order and the right way round. Where applicable, follow the assembly instructions supplied with the repair kit. On models with ABS, ensure that the slot in the secondary piston assembly aligns with the stop-bolt hole, or stop-pin hole in the master cylinder body.

12 On models with ABS, compress the piston assemblies into the cylinder bore, using a clean wooden dowel as during removal, then refit the stop-bolt or stop-pin, as applicable, ensuring that it engages with the slot in the secondary piston assembly.

13 Press the piston assemblies fully into the bore using a clean wooden dowel, and secure them in position with the new end cap (supplied in the overhaul kit). Bend the tangs into position to secure the end cap.

14 Examine the fluid reservoir seals, and if necessary renew them. Fit the reservoir seals to the master cylinder body, then refit the reservoir. On later models, secure the reservoir by tapping in the retaining roll pin.

Refitting

15 Remove all traces of dirt from the master cylinder and servo unit mating surfaces.

16 Fit the master cylinder to the servo unit, ensuring that the servo unit pushrod enters the master cylinder bore centrally. Refit the

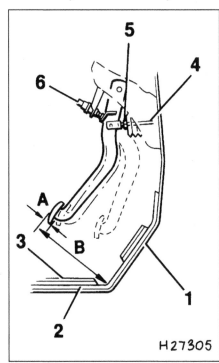

13.4 Brake pedal free height adjustment

A *Pedal free play*	2 *Sound insulation*
B *Pedal free height*	3 *Carpet panel*
1 *Floor*	4 *Servo pushrod*
reinforcement	5 *Locknut*
panel	6 *Stop-light switch*

master cylinder mounting nuts, and tighten them to the specified torque.

17 Place absorbent rags around and beneath the master cylinder, then fill the reservoir with fresh hydraulic fluid.

18 Have an assistant slowly depress the brake pedal fully, then hold it in the fully depressed position. Cover the outlet ports on the master cylinder body with your fingers then have the assistant slowly release the brake pedal. Continue this procedure until the fluid emerging from the master cylinder is free from air bubbles. Take care to collect the expelled fluid in the rags and wash off any spilt fluid immediately with cold water.

19 When all air has been bled from the master cylinder, wipe clean the brake pipe unions, then refit them to the correct master cylinder ports, as noted before removal, and tighten the union nuts securely.

20 On completion, bleed the complete hydraulic system as described in Section 2.

13 Brake pedal – adjustment, removal and refitting

Adjustment

1 The pedal free height should be measured from the top face of the pedal to the floor reinforcement panel.

2 If desired, to improve access, remove the driver's side lower facia panel, as described in Chapter 11, Section 29.

3 To take the measurement, remove the driver's side footwell trim panel, with reference to Chapter 11, Section 27, if necessary, then release the securing clip(s) and lift the carpet trim panel to expose the flap in the floor insulation.

4 Lift the flap in the floor insulation, and measure the pedal free height **(see illustration)**. Check the measured height against the value given in the Specifications.

5 If the height of the pedal requires adjustment, proceed as follows.

6 Loosen the locknut on the servo pushrod, and turn the pushrod as required until the specified height is achieved. Retighten the locknut on completion.

7 Check the free play of the pedal by pressing the pedal slowly until resistance is felt. The free play should be as specified.

8 Check that the stop-lights go out when the pedal is released. If not, adjust the switch as described in Section 18.

9 On completion, refit the carpet and the trim panel(s).

Removal

10 To improve access, if not already done, remove the driver's side lower facia panel as described in Chapter 11, Section 29.

11 Working in the driver's footwell, remove the split-pin from the end of the servo pushrod clevis pin, then withdraw the clevis pin.

14.2a Unscrew the securing nuts . . .

14.2b . . . and pull the master cylinder from the servo

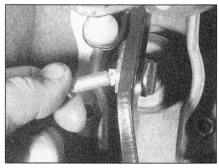

14.5 Removing the servo pushrod clevis pin

12 Disconnect the wiring plug from the stop-light switch.

13 Unscrew the four nuts securing the pedal bracket to the bulkhead (note that these nuts also secure the vacuum servo).

14 Unscrew the pedal bracket upper securing bolt, then withdraw the pedal/bracket assembly from the footwell.

15 The brake pedal is integral with the bracket assembly, and cannot be renewed individually.

Refitting

16 Refitting is a reversal of removal but, on completion, check the pedal height as described previously in this Section.

14 Vacuum servo unit – removal and refitting

Removal

1 Disconnect the battery negative terminal (refer to *Disconnecting the battery* in the Reference Chapter).

2 Disconnect the wiring plug from the brake fluid level sensor, then remove the two nuts securing the master cylinder to the servo. Release the master cylinder fluid pipes from any securing clips, then pull the master cylinder forwards from the servo **(see illustrations)**.

3 If this does not provide sufficient clearance to remove the servo, the master cylinder must be removed completely, with reference to Section 12.

4 Disconnect the vacuum hose from the servo.

5 Working in the driver's footwell, remove the spring clip from the end of the servo pushrod clevis pin, then withdraw the clevis pin **(see illustration)**. If desired, remove the driver's side lower facia panel, as described in Chapter 11, Section 29, to improve access.

6 Again working in the driver's footwell, unscrew the four nuts securing the brake pedal mounting bracket to the servo studs.

7 Working in the engine compartment, withdraw the servo.

Refitting

8 Refitting is a reversal of removal, bearing in mind the following points:

a) *Before refitting the servo on models without ABS, check that the length of the pushrod is as specified (see Specifications), and adjust if necessary by loosening the locknut and turning the pushrod* **(see illustration)**.

b) *Where applicable, tighten all fixings to the specified torque.*

c) *Where applicable, refit the master cylinder as described in Section 12.*

d) *On completion, check the brake pedal height as described in Section 13.*

15 Vacuum servo unit check valve – removal, testing and refitting

Removal

1 The valve is located in the vacuum hose leading to the servo, and is secured to the body panel by a clip.

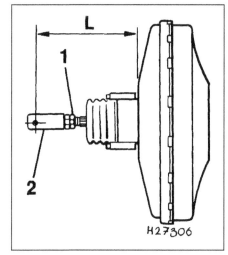

14.8 Ensure that the vacuum servo pushrod length is as specified

L = Approx   1 Locknut
     125.0 mm   2 Pushrod clevis

2 Release the valve from the securing clip. Take note of the direction of the arrow on the valve body, which should point in the direction of the hose connected to the engine.

3 Release the retaining clips (where fitted), and disconnect the vacuum hoses from the valve, then withdraw the valve.

Testing

4 Examine the check valve for signs of damage, and renew if necessary. The valve may be tested by blowing through it in both directions. Air should flow through the valve in one direction only – when blown through from the servo unit end of the valve. Renew the valve if this is not the case.

Refitting

5 Refitting is a reversal of removal, ensuring that the arrow on the valve body points towards the engine **(see illustration)**.

6 On completion, start the engine and check the hose connections to the valve for air leaks.

16 Handbrake lever – removal and refitting

Removal

1 Disconnect the battery negative terminal (refer to *Disconnecting the battery* in the Reference Chapter).

2 Remove the centre console as described in Chapter 11.

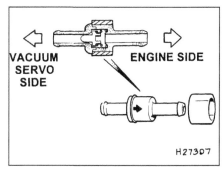

15.5 The arrow on the vacuum servo unit check valve must point towards the engine

3 Chock the wheels, and fully release the handbrake.

4 Disconnect the wiring plug from the handbrake 'on' warning light switch.

5 Count the number of exposed threads on the cable adjuster rod (to ease adjustment on refitting), then unscrew the adjuster nut, and disconnect adjuster rod from the handbrake lever.

6 Remove the two securing bolts, and withdraw the handbrake lever assembly **(see illustration)**.

Refitting

7 Refitting is a reversal of removal, bearing in mind the following point:

a) Screw the adjuster nut onto the cable adjuster rod to give the number of exposed threads noted before removal, then check the handbrake operation, and adjust if necessary, as described in Chapter 1.

b) Before refitting the centre console, check the operation of the handbrake 'on' warning light.

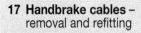

17 Handbrake cables –
removal and refitting

Rear cables

Removal

1 There are two rear handbrake cables, one on each side of the vehicle. To renew either rear cable, proceed as follows.

2 Chock the front wheels then jack up the rear of the vehicle and support it securely on axle stands (see *Jacking and vehicle support*). Release the handbrake fully.

3 On models with rear drum brakes, remove the brake shoes, and disconnect the end of the handbrake cable from the lever on the trailing shoe, as described in Section 6. Unbolt the cable bracket from the rear of the brake backplate.

4 On models with rear disc brakes, release the spring clip securing the cable to the mounting bracket, then disconnect the end of the cable from the lever on the caliper.

5 Unscrew the nuts and bolts securing the

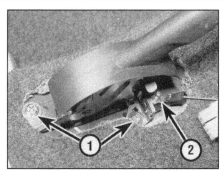

16.6 Handbrake lever securing bolts (1) and handbrake 'on' warning light switch (2)

handbrake cable brackets to the underbody, and to the rear suspension trailing arms.

6 Disconnect the front end of the cable from the cable equaliser, then withdraw the cable **(see illustrations)**. Note that on certain models, it will be necessary to remove the exhaust heat shield for access to the handbrake cable equaliser.

Refitting

7 Refitting is a reversal of removal, bearing in mind the following points:

a) On models with rear drum brakes, reconnect the cable end to the lever on the trailing shoe, and refit the brake shoes, as described in Section 6.

b) On completion, check the handbrake adjustment, as described in Chapter 1.

Front cable

Removal

8 Disconnect the battery negative terminal (refer to *Disconnecting the battery* in the Reference Chapter).

9 Chock the rear wheels and fully release the handbrake.

10 Working inside the vehicle, remove the centre console as described in Chapter 11.

11 Disconnect the wiring plug from the handbrake 'on' warning light switch.

12 Count the number of exposed threads on the cable adjuster rod (to ease adjustment on refitting), then unscrew the adjuster nut, and disconnect adjuster rod from the handbrake lever **(see illustration)**.

13 For improved access, firmly apply the

---

handbrake, then jack up the front of the vehicle and support it securely on axle stands (see *Jacking and vehicle support*).

14 Working under the vehicle, disconnect the two rear cables from the cable equaliser, then withdraw the front section of the cable (which includes the cable equaliser and the adjuster rod) down through the floor panel. Note that on certain models, it will be necessary to remove the exhaust heat shield for access to the handbrake cable equaliser.

Refitting

15 Refitting is a reversal of removal, bearing in mind the following points:

a) Screw the adjuster nut onto the cable adjuster rod to give the number of exposed threads noted before removal, then check the handbrake adjustment as described in Chapter 1.

b) Before refitting the centre console, check the operation of the handbrake 'on' warning light.

---

18 Stop-light switch –
adjustment, removal and refitting

Adjustment

1 Check that the stop-lights are extinguished when the brake pedal is released, and illuminated within the first few millimetres of brake pedal travel.

2 If adjustment is required, remove the driver's side lower facia panel, as described in Chapter 11, Section 29, for access to the switch.

3 Disconnect the switch wiring connector, then slacken the locknut securing the switch to the brake pedal bracket.

4 Screw the switch body in or out as necessary, then reconnect the wiring connector and test the switch operation. When the adjustment is correct, tighten the locknut and refit the lower facia panel.

Removal

5 Remove the driver's side lower facia panel, as described in Chapter 11, Section 29.

6 Disconnect the wiring connector from the switch.

17.6a Exhaust heat shield removed to expose handbrake cable equaliser (arrowed)

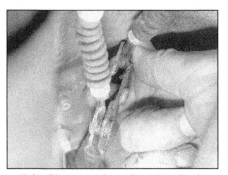

17.6b Disconnecting a front handbrake cable from the equaliser

17.12 Unscrewing the handbrake cable adjuster nut

7 Slacken the locknut and unscrew the switch body from the brake pedal bracket.

Refitting

8 Screw the switch into position and carry out the adjustment procedure as described previously.

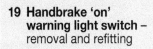

19 Handbrake 'on' warning light switch – removal and refitting

Removal

1 Disconnect the battery negative terminal (refer to *Disconnecting the battery* in the Reference Chapter).
2 Remove the centre console as described in Chapter 11.
3 Disconnect the wiring plug from the switch, located at the front of the handbrake lever assembly **(see illustration)**.
4 Remove the securing screw, and withdraw the switch.

Refitting

5 Refitting is a reversal of removal, but before refitting the centre console, check that the warning light comes on after the specified number of handbrake clicks (see

19.3 Disconnecting the handbrake 'on' warning light switch wiring

Specifications). If necessary, bend the switch bracket to give the correct adjustment.

20 Anti-lock braking system (ABS) – general information and component renewal

General information

Anti-lock Braking is available as an option or as standard equipment to the majority of models covered by this manual. Two different systems may be fitted, however the operation of each is virtually identical.

The system is fail-safe, and is fitted in addition to the conventional braking system, meaning that the vehicle retains conventional braking in the event of an ABS failure.

To prevent wheel locking, the system provides a means of modulating (varying) the hydraulic pressure in the braking circuits, to control the amount of braking effort at each wheel. To achieve this, sensors mounted at all four wheels monitor the rotational speeds of the wheels, and are thus able to detect when there is a risk of wheel locking (low rotational speed, relative to vehicle speed). Solenoid

valves are positioned in the brake circuits to each wheel, and the solenoid valves are incorporated in a modulator assembly, which is controlled by an electronic control unit. The electronic control unit controls the braking effort applied to each wheel, according to the information supplied by the wheel sensors.

Should an ABS fault develop, the system can only be satisfactorily tested using specialist diagnostic equipment available to a Nissan dealer. For safety reasons, owners are strongly advised against attempting to diagnose complex problems with the ABS using standard workshop equipment.

Component renewal

Wheel speed sensors

1 Jack up the front or rear of the vehicle (as applicable), and support it securely using axle stands (see *Vehicle jacking and support*). Remove the relevant roadwheel.
2 The front speed sensors are located on the rear side of each wheel hub carrier, whilst the rear sensors are located on the inner face of the stub axle assembly on each side, Trace the sensor wiring back to the connector, and separate the two halves of the wiring plug. Release the wiring from any retaining brackets/clips.
3 Undo the retaining bolt and pull the sensor from the hub carrier/stub axle assembly **(see illustrations)**. If the sensor is reluctant to move, apply releasing/penetrating fluid to the assembly, and leave it to soak for a few minutes before trying again. If the sensor still will not move, the hub carrier (see Chapter 10) or rear brake drum/disc must be removed, and the sensor driven from place.
4 Refitting is the reversal of removal, noting the following points:
a) Ensure the mating faces of the hub carrier/stub axle assembly and sensor are clean and free from corrosion.

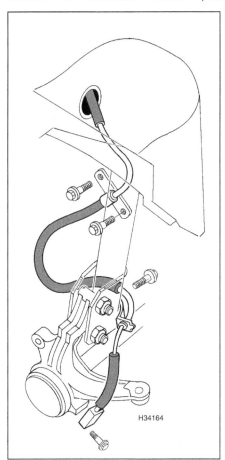

20.3a Front wheel speed sensor

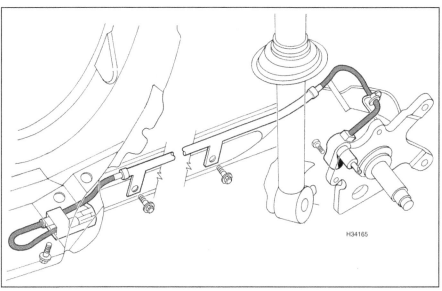

20.3b Rear wheel speed sensor

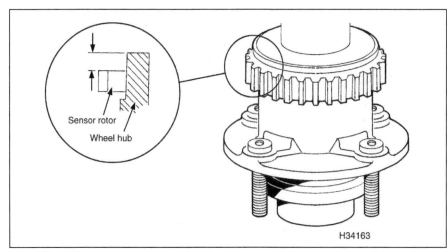

20.11 The gap between the edge of the rotor and the edge of the hub should be 4.5 to 5.5 mm

b) *Apply a thin smear of anti-seize compound to mounting surfaces of the sensor and hub carrier/stub axle assembly.*

c) *Tighten the sensor retaining bolt to the specified torque.*

Electronic Control Module (ECM)

5 The ECM is located behind the kick panel trim in the passenger side front footwell. Pull the weatherstrip from the door aperture adjacent to the kick panel, and remove the kick panel trim as described in Chapter 11, Section 27.

6 Remove the retaining bolts/nut and withdraw the ECM. Open the locking lever and disconnect the wiring plug as the unit is removed.

7 Refitting is a reversal of removal, ensuring that the earth shield cable is secured by the ECM mounting bolt.

Wheel speed sensor rotors

8 The rotors should be inspected for damage and/or corrosion. Any damage or deformation of the rotor teeth will cause an erroneous signal from the sensor, and may illuminate the ABS warning light as well as set a fault code in the ECM. The front rotor is pressed onto the driveshaft outer CV joint, whilst the rear rotor is pressed onto the inner edge of the rear wheel hub.

9 Remove the relevant driveshaft or wheel hub, as described in Chapters 8 or 10, as applicable.

10 Use a bearing puller to remove the rotor from the CV joint/hub.

11 Position the new rotor over then end of the CV joint/hub and press it into place, or use a block of wood and a hammer to install the rotor. Ensure that the rotor is fitted squarely into place. Note that the rear rotor must be fitted with a gap of 4.5 to 5.5 mm between the inner edge of the rotor and the inner edge of the hub **(see illustration)**.

12 The remainder of refitting is a reversal of removal.

Actuator

13 The actuator is located in the left-hand rear corner of the engine compartment. Disconnect the battery negative lead (refer to *Disconnecting the battery* in the Reference Chapter).

14 Open any convenient bleed screw in the system, and gently pump the brake pedal to expel the fluid through a tube connected to the screw (see Section 2). Alternatively, have some caps handy to plug the open ends of the brake pipes once they are disconnected from the actuator.

15 On models equipped with air conditioning, have the refrigerant discharged by a suitably-equipped specialist. This is necessary as the air conditioning low pressure pipes adjacent to the actuator must be moved.

16 On all models, note their fitted locations, then disconnect the wiring plugs from the ABS relay bracket alongside the actuator.

17 Remove the mounting bolt, and remove the ABS relay box complete with bracket.

18 On models with air conditioning, undo the union nut and disconnect the low pressure refrigerant pipe from the engine compartment bulkhead. Remove the pipe retaining bracket bolts and place the pipe to one side.

19 Note their fitted locations, then undo the union nuts and disconnect the brake pipes from the ABS actuator. If the system has not been drained, be prepared for fluid spillage. Plug the end of the pipes to prevent dirt ingress.

20 Undo the actuator mounting nuts, and manoeuvre the actuator from the mounting bracket.

21 To refit the actuator, engine the locating lug at the rear of the unit with the corresponding hole in the mounting bracket, then tighten the mounting nuts securely.

22 The remainder of refitting is a reversal of removal, noting the following points:

a) *Bleed the brake system as described in Section 2.*

b) *Renew the refrigerant pipe O-ring seal, and have the air conditioning system recharged by a suitably-equipped specialist.*

Chapter 10
Suspension and steering

Contents

Degrees of difficulty

Easy, suitable for novice with little experience	Fairly easy, suitable for beginner with some experience	Fairly difficult, suitable for competent DIY mechanic 	Difficult, suitable for experienced DIY mechanic	Very difficult, suitable for expert DIY or professional

Specifications

Front suspension
Type . . . .   Independent by MacPherson struts, with coil springs and integral shock absorbers, and lower arms. Anti-roll bar fitted to certain models

Rear suspension
Type . . . .   Semi-independent multi-link beam axle with coil springs and integral shock absorbers

Wheel bearings
Maximum endfloat at hub (front and rear) . . . .   0.05 mm

Steering
Type . . . .   Rack-and-pinion, manual or power-assisted, depending on model
Maximum steering wheel free play . . . .   35.0 mm
Steering column length (see Section 11) . . . .   533.0 to 535.0 mm

Roadwheels and tyres
See *Weekly checks* and Chapter 1

Front wheel alignment
Front wheel toe setting . . . .   2.0 mm ± 2.0 mm toe-in

Torque wrench settings

	Nm	lbf ft
Front suspension		
Anti-roll bar drop link-to-anti-roll bar nuts*	50	37
Anti-roll bar drop link-to-lower arm nuts*	20	15
Anti-roll bar mounting bracket bolts	35	26
Hub carrier-to-suspension strut retaining bolt nuts*	124	92
Lower arm balljoint-to-hub carrier nut	66	49
Lower arm front mounting bolt nut*	110	81
Lower arm rear mounting clamp bolts	88	65
Suspension strut damper rod top nut*	70	52
Suspension strut upper mounting nuts*	27	20

Torque wrench settings (continued)

	Nm	lbf ft
Rear suspension		
Control rod-to-lateral link mounting bolt nut*	69	51
Control rod-to-axle beam mounting nut*	108	80
Lateral link-to-axle beam mounting nut*	108	80
Lateral link-to-body mounting bolt nut*	108	80
Rear hub nut	230	170
Suspension strut damper rod top nut*	20	15
Suspension strut lower mounting bolt nut*	108	80
Suspension strut upper mounting-to-body nuts*	20	15
Trailing arm front mounting bolt nuts*	108	80
Steering		
Power steering pump fluid pipe banjo union bolt	59	44
Power steering pump-to-mounting bracket through-bolt nut	35	26
Steering column securing nuts	17	13
Steering column universal joint clamp bolts	27	20
Steering gear mounting bracket bolts	85	63
Steering wheel securing nut	35	26
Track rod end locknut	45	33
Track rod end-to-steering arm/hub carrier nut:		
Recommended torque	35	26
Maximum permissible torque	49	36
Roadwheels		
Roadwheel nuts	110	81

Use new nut(s)

1 General information

Front suspension

The independent front suspension is of the MacPherson strut type, incorporating coil springs and integral telescopic shock absorbers. The upper ends of the MacPherson struts are connected to the bodyshell front suspension turrets; the lower ends are bolted to the hub carriers, which carry the wheel bearings, brake calipers and hub/disc assemblies. The hub carriers are located at their lower ends by transverse lower arms. A front anti-roll bar is fitted to certain models.

Rear suspension

The rear suspension is of the semi-independent multi-link beam axle type, with two trailing arms, linked by a torsion beam. Suspension strut units comprising coil springs and integral telescopic shock absorbers are fitted between the axle beam and the vehicle body. Lateral location of the axle is by a unique arrangement consisting of a lateral link and control rod connected between the axle beam and the body.

Steering

The one-piece steering shaft has a universal joint fitted at its lower end, which is clamped to both the steering shaft and the steering gear pinion by means of clamp bolts.

The steering gear is mounted on the engine compartment bulkhead, and is connected to the steering arms projecting rearwards from the hub carriers. The track rods are fitted with balljoints at their inner and outer ends, to allow for suspension movement, and are threaded to facilitate adjustment.

Power steering is fitted as standard on some models. The hydraulic system is powered by a belt-driven pump, which is driven from the crankshaft pulley.

2 Front hub bearings – renewal

Note: *A balljoint separator tool, and suitable metal tubes, or large sockets, together with threaded bar, spacers and nuts (see text) will be required for this operation. The bearing will be destroyed during the removal procedure.*

Removal

1 Firmly apply the handbrake, then jack up the front of the vehicle and support it securely on axle stands (see *Jacking and vehicle support*). Remove the appropriate roadwheel.
2 Disconnect the outboard end of the driveshaft from the hub carrier, as described in Chapter 8, Section 2. Note that there is no need to drain the transmission oil/fluid, or disconnect the inboard end of the driveshaft from the transmission. **Do not** allow the end of the driveshaft to hang down under its own weight – support the end of the driveshaft using wire or string.
3 Remove the brake disc, as described in Chapter 9.
4 On models with ABS, if not already done, unbolt the ABS wheel sensor, and remove the screw(s) securing the ABS sensor wiring to the hub carrier. Suspend the sensor away from the working area, to avoid the possibility of damage.
5 Remove the split-pin, then partially unscrew the castellated nut securing the track rod end to the steering arm. Using a balljoint separator tool, separate the track rod end from the steering arm. Remove the nut. Discard the split-pin – a new one must be used on refitting.
6 Remove the split-pin, then partially unscrew the nut securing the lower end of the hub carrier to the lower arm balljoint. Discard the split-pin – a new one must be used on refitting.
7 Separate the hub carrier from the lower arm using a balljoint separator tool. Remove the nut, and lift the hub carrier off the balljoint.
8 Clamp the hub carrier in a vice, then using a metal bar or tube of suitable diameter, drive the wheel hub from the bearing. Note that one half of the bearing inner race will remain on the hub **(see illustrations)**.

2.8a Using a metal bar or tube of suitable diameter, drive the wheel hub from the bearing . . .

2.8b . . . then withdraw the hub, noting that one half of the bearing inner race will remain on the hub

2.9 Draw off the bearing inner race which is still attached to the hub, using a suitable puller

2.11 Extract the outer oil seal from the hub carrier . . .

2.12 . . . then extract the bearing retaining circlip

2.13 Using a suitable metal tube in contact with the bearing outer race, drive the bearing from the hub carrier

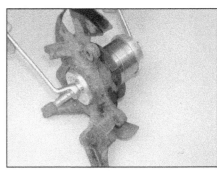

2.15 Draw the new bearing into the hub carrier using an arrangement of suitable tubing, threaded bar, spacers and nuts

9 Support the wheel hub in a vice then draw off the bearing inner race which is still attached to the hub, using a suitable puller **(see illustration)**.

10 Working at the rear of the hub carrier, prise out the inner oil seal.

11 Using a screwdriver, extract the outer oil seal from the hub carrier **(see illustration)**.

12 Using a screwdriver, extract the bearing retaining circlip from the hub carrier **(see illustration)**.

13 Using a suitable metal tube which bears on the bearing outer race, drive the bearing from the hub carrier **(see illustration)**.

Refitting

14 Before installing the new bearing, thoroughly clean the wheel hub and the bearing location in the hub carrier.

15 Fit the new bearing from the front of the hub carrier, and draw the bearing into position until it contacts the shoulder in the hub carrier. This can be achieved using metal tubing, or large sockets of suitable diameter, which bear on the outer race of the bearing only, together with threaded bar, spacers and nuts **(see illustration)**.

16 With the bearing fully home, fit the bearing retaining circlip to its groove in the hub carrier **(see illustration)**.

17 Pack the lips of the new outer oil seal with grease then, using a tube of suitable diameter, carefully tap the seal into position in the outer face of the hub carrier **(see illustrations)**.

18 Pack the lips of the new inner oil seal with

grease, then fit the seal using the same procedure as for the outer seal. It may be necessary to finish the installation by tapping

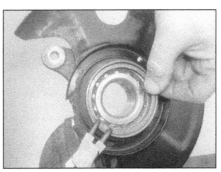

2.16 Fit the bearing retaining circlip to its groove in the hub carrier

2.17b . . . then tap the seal into position using a tube of suitable diameter

around the periphery of the seal with a suitable punch to ensure it is fully seated **(see illustrations)**.

2.17a Locate the new outer oil seal on the hub carrier . . .

2.18a Locate the new inner oil seal on the hub carrier . . .

2.18b . . . initially fit the seal using the same procedure as for the outer oil seal . . .

2.18c . . . then finish the installation by tapping around the periphery of the seal with a suitable punch

support). Remove the appropriate roadwheel.

2 Extract the retaining clip and release the brake hydraulic hose from the support bracket on the suspension strut.

3 Undo the two nuts and withdraw the bolts securing the suspension strut to the hub carrier **(see illustration)**. Discard the nuts – new nuts must be used on refitting.

4 Have an assistant support the strut from underneath the wheel arch then, working in the engine compartment, unscrew the three nuts securing the top of the strut to the suspension turret **(see illustration)**. Release the lower end of the strut from the hub carrier, then withdraw the assembly from under the wheel arch.

> ⚠ **Warning: Do not unscrew the centre damper rod nut at this stage.**

Overhaul

Note: *Suitable coil spring compressor tools will be required for this operation, and a new damper rod top nut must be used on reassembly.*

5 Temporarily refit two of the strut upper securing nuts to their studs. Using a suitable bar inserted between the upper mounting studs, counterhold the strut upper mounting whilst loosening the damper rod top nut. **Do not** remove the nut.

6 Fit suitable spring compressors to the spring, and compress the spring sufficiently to enable the upper spring seat to be turned by hand **(see illustration)**.

> ⚠ **Warning: Do not use makeshift or improvised tools to compress the spring, as there is a danger of serious injury if the spring is not retained properly and released slowly.**

7 Fully unscrew and remove the damper rod top nut. Note that it will be necessary to counterhold the damper rod, using a suitable spanner, as the nut is unscrewed **(see illustration)**. Discard the nut – a new one must be used on reassembly.

8 Withdraw the gasket, upper mounting plate, upper mounting insulator, bearing, upper spring seat, and upper spring seat rubber. If required, separate the dust cover and bump rubber from the upper spring seat.

9 Withdraw the spring, complete with the

2.19a Locate the wheel hub on the bearing inner race . . .

2.19b . . . and draw the hub into the bearing using an arrangement of suitable tubing, threaded bar, spacers and nuts

19 Locate the wheel hub on the bearing inner race and draw the hub into the bearing using the same method as described previously to fit the bearing itself **(see illustrations)**. Ensure that the tube or socket which supports the bearing is in contact with the bearing inner race only. With the wheel hub installed, check that it rotates freely in the bearing.

20 Ensure that the balljoint taper is clean then engage the hub carrier with the lower arm balljoint, refit the castellated nut and tighten the nut to the specified torque. If necessary, tighten the nut further slightly to align the grooves in the nut with the split-pin hole in the balljoint and secure the nut using a new split-pin.

21 Similarly, reconnect the track rod end to the hub carrier, secure with the castellated nut tightened to the specified torque, then fit a new split-pin.

22 Refit the brake disc, as described in Chapter 9.

23 Reconnect the outboard end of the driveshaft to the hub as described in Chapter 8, Section 2.

3 Front suspension strut – removal, overhaul and refitting

Note: *New hub carrier-to-suspension strut retaining bolt nuts, and suspension strut top mounting nuts and gasket, must be used on refitting.*

Removal

1 Firmly apply the handbrake, then jack up the front of the vehicle and support it securely on axle stands (see *Jacking and vehicle*

3.3 Front suspension strut-to-hub carrier retaining bolt nuts (arrowed)

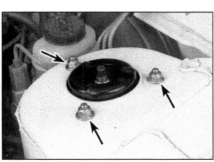

3.4 Front suspension strut upper mounting nuts (arrowed). DO NOT unscrew the centre damper rod nut at this stage

3.6 Fit suitable spring compressors and compress the coil spring

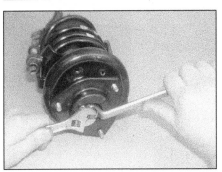

3.7 Unscrew the damper rod top nut while counterholding the damper rod

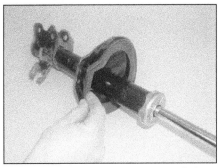

3.13 Commence reassembly by refitting the lower spring seat rubber

3.14 Locate the spring on the strut, ensuring that the lower end of the spring is correctly located

3.16a Position the upper spring seat rubber on the upper spring seat . . .

3.16b . . . fit this assembly to the strut . . .

3.16c . . . ensuring that the arrow on the upper spring seat is positioned nearest the outer edge of the suspension turret

compressors, then withdraw the lower spring seat rubber.

10 With the strut assembly now dismantled, examine all the components for wear, damage or deformation. Check the rubber components for deterioration. Renew any of the components as necessary.

11 Examine the damper for signs of fluid leakage. Check the damper rod for signs of pitting along its entire length, and check the strut body for signs of damage. While holding it in an upright position, test the operation of the strut by moving the damper rod through a full stroke, and then through short strokes of 50 to 100 mm. In both cases, the resistance felt should be smooth and continuous. If the resistance is jerky, or uneven, or if there is any visible sign of wear or damage to the strut, renewal is necessary. Note that the damper

cannot be renewed independently, and if leakage or damage is evident, the complete strut/damper assembly must be renewed (in which case, the spring, upper mounting components, bushes, and associated components can be transferred to the new strut).

12 If any doubt exists about the condition of the coil spring, carefully remove the spring compressors, and check the spring for distortion and signs of cracking. Renew the spring if it is damaged or distorted, or if there is any doubt as to its condition.

13 Commence reassembly by refitting the lower spring seat rubber, ensuring that it is correctly located in the recess in the lower spring seat **(see illustration)**.

14 Ensure that the coil spring is compressed sufficiently to enable the upper mounting components to be fitted, then locate the

spring on the strut, ensuring that the lower end of the spring is correctly located on the lower spring seat rubber **(see illustration)**.

15 If previously separated, refit the dust cover and bump rubber to the upper spring seat, ensuring that the lip on the end of the dust cover is fully engaged in the upper spring seat.

16 Refit the upper spring seat rubber, and the upper spring seat, ensuring that the arrow on the upper spring seat is positioned nearest the outer edge of the suspension turret (ie, the side nearest the roadwheel) **(see illustrations)**.

17 Refit the bearing and the upper mounting insulator, with the arrow on the insulator aligned with the arrow on the upper spring seat **(see illustrations)**.

18 Refit the upper mounting plate to the

3.17a Refit the bearing . . .

3.17b . . . and the upper mounting insulator . . .

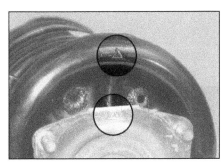

3.17c . . . with the arrow on the insulator aligned with the arrow on the upper spring seat

upper mounting insulator, with the arrow on the plate aligned with the arrow on the insulator **(see illustration)**.

19 Fit a new damper rod top nut, then tighten the top nut to the specified torque, counterholding the damper rod as during dismantling. Note that a suitable crows-foot adapter will be required to tighten the top nut to the specified torque.

20 Check that the spring ends are correctly located in the upper and lower spring seats, then remove the spring compressors.

Refitting

21 Fit a new gasket to the top mounting, then manoeuvre the strut assembly into position under the wheel arch, passing the mounting studs through the holes in the body turret. Note that the arrow on the upper spring seat must be positioned nearest the outer edge of the suspension turret (ie, the side nearest the roadwheel). Fit the new upper mounting nuts, and tighten them to the specified torque.

22 Engage the lower end of the strut with the hub carrier and refit the securing bolts with the bolt heads toward the rear of the vehicle. Fit the new nuts to the bolts, and tighten to the specified torque.

23 Locate the brake hydraulic hose in the support bracket on the suspension strut and secure with the retaining clip.

24 Refit the roadwheel, and lower the vehicle to the ground.

---

4 Front suspension lower arm – removal, inspection and refitting

Note: A balljoint separator tool will be required for this operation. The lower arm balljoint split-pin, lower arm front mounting nut and, on models so equipped, the anti-roll bar drop link nut, must be renewed on refitting.

Removal

1 Firmly apply the handbrake, then jack up the front of the vehicle and support it securely on axle stands (see *Jacking and vehicle support*). Remove the appropriate roadwheel.

2 Disconnect the outboard end of the

3.18 Refit the upper mounting plate to the upper mounting insulator, again aligning the arrows

driveshaft from the hub carrier, as described in Chapter 8, Section 2. Note that there is no need to drain the transmission oil/fluid, or disconnect the inboard end of the driveshaft from the transmission. **Do not** allow the end of the driveshaft to hang down under its own weight – support the end of the driveshaft using wire or string.

3 Remove the split-pin, then partially unscrew the nut securing the lower arm balljoint to the lower end of the hub carrier. Discard the split-pin – a new one must be used on refitting.

4 Separate the balljoint from the hub carrier using a balljoint separator tool. Remove the nut, and lift the hub carrier off the balljoint. Locate the hub carrier back on the suspension strut and temporarily insert one of the retaining bolts to retain it while the lower arm is removed.

5 On models with a front anti-roll bar, unscrew the nut securing the anti-roll bar drop link to the lower arm, and recover the washer and bush **(see illustration)**. Note that it may be necessary to counterhold the drop link pin in order to unscrew the nut. Discard the nut – a new one should be used on refitting.

6 Unscrew the two bolts securing the lower arm rear mounting clamp to the subframe, and remove the clamp **(see illustration)**.

7 Unscrew the nut and remove the lower arm front mounting bolt **(see illustration)**.

8 Disengage the lower arm from the anti-roll bar drop link pin (where applicable) and recover the remaining bush and washer from the drop link. Withdraw the front of the lower

arm from the subframe mounting bracket, and remove the lower arm from under the vehicle.

Inspection

9 With the lower arm removed, examine the lower arm itself, and the mounting bushes, for wear, cracks or damage.

10 Check the balljoint for wear, excessive play, or stiffness. Also check the balljoint dust boot for cracks or damage.

11 The mounting bushes and balljoint assembly are integral with the lower arm, and cannot be renewed independently. If either the bushes or the balljoint are worn or damaged, the complete lower arm assembly must be renewed.

Refitting

Note: Final tightening of all lower arm mountings should be carried out with the vehicle resting on its wheels.

12 Slide the front of the lower arm into position on the subframe and, where applicable, engage the lower arm with the anti-roll bar drop link (ensure that the upper washer and bush are fitted to the drop link – the concave side of the washer should be against the bush). Refit the lower arm front mounting bolt (with the bolt head toward the front of the vehicle) and screw on the new retaining nut. Do not fully tighten the nut at this stage.

13 Refit the lower arm rear mounting clamp, and refit the securing bolts. Do not fully tighten the bolts at this stage.

14 Ensure that the balljoint taper is clean then engage the hub carrier with the lower arm balljoint, refit the castellated nut and tighten the nut to the specified torque. If necessary, tighten the nut further slightly to align the grooves in the nut with the split-pin hole in the balljoint and secure the nut using a new split-pin.

15 Where applicable, refit the lower bush and washer to the anti-roll bar drop link (the concave side of the washer should be against the bush), then fit a new securing nut. Again, do not fully tighten the nut at this stage.

16 Reconnect the outboard end of the driveshaft to the hub as described in Chapter 8, Section 2.

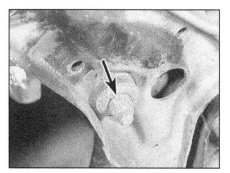

4.5 Unscrew the nut (arrowed) securing the anti-roll bar drop link to the lower arm, and recover the washer and bush

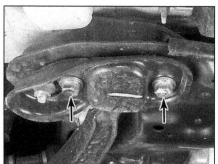

4.6 Unscrew the two bolts (arrowed) securing the lower arm rear mounting clamp to the subframe

4.7 Unscrew the nut and remove the lower arm front mounting bolt (arrowed)

6.2 Unscrew the nut (A) securing the end of the anti-roll bar to the drop link (B)

6.4 Unscrew the bolts (arrowed), and withdraw the clamps securing the anti-roll bar to the subframe

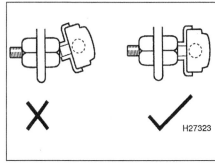

6.7 Correct positioning of anti-roll bar drop link balljoint

17 With the roadwheel refitted and the vehicle lowered to the ground, roll the vehicle backwards and forwards, and bounce the front of the vehicle to settle the suspension components.

18 Chock the wheels, then tighten all the lower arm mounting nuts and bolts and, where applicable, the anti-roll bar drop link nut to the specified torque.

19 On completion the front wheel alignment should be checked, with reference to Chapter 1.

5 Front suspension lower arm balljoint – renewal

The balljoint is integral with the suspension lower arm. If the balljoint is worn or damaged, the complete lower arm must be renewed as described in Section 4.

6 Front suspension anti-roll bar – removal and refitting

Anti-roll bar

Note: *New anti-roll bar-to-drop link nuts should be used on refitting.*

Removal

1 Firmly apply the handbrake, then jack up the front of the vehicle and support it securely on axle stands (see *Jacking and vehicle support*). Remove the front roadwheels.

2 Working at one side of the anti-roll bar, where necessary, counterhold the drop link pin, then unscrew the nut securing the end of the anti-roll bar to the drop link **(see illustration)**. Recover the washer. Discard the nut – a new one should be used on refitting.

3 Repeat the operation on the other side of the anti-roll bar.

4 Unscrew the bolts, and withdraw the clamps securing the anti-roll bar to the front subframe **(see illustration)**.

5 Manipulate the anti-roll bar out from under the vehicle.

Refitting

6 Inspect the mounting clamp rubbers for cracks or deterioration. If renewal is necessary, slide the old rubbers from the bar, and fit the new rubbers. Note that the rubbers should be positioned with the paint marks on the bar against their inner edges.

7 Refitting is a reversal of removal, bearing in mind the following points:

a) *When reconnecting the drop links to the anti-roll bar, make sure that the drop link balljoint is positioned centrally in its arc of movement, and is not under strain* **(see illustration)***. Use new nuts to secure the drop link balljoints.*

b) *Do not fully tighten the clamp bolts until the vehicle is resting on its roadwheels, and the suspension has been settled.*

Drop link

Note: *New securing nuts should be used on refitting.*

Removal

8 To improve access, firmly apply the handbrake, then jack up the front of the vehicle and support it securely on axle stands (see *Jacking and vehicle support*). Remove the relevant front roadwheel.

9 Counterhold the drop link pin, then unscrew the nut securing the end of the anti-roll bar to the drop link. Recover the washer. Discard the nut – a new one should be used on refitting.

10 Again, counterhold the drop link pin, then unscrew the nut securing the anti-roll bar drop

7.5 Prise the dust cap from the rear hub, then remove the split-pin from the end of the stub axle

link to the lower arm. Recover the washer and bush. Again, discard the nut – a new one should be used on refitting.

11 Lift out the drop link, and recover the remaining bush and washer.

Refitting

12 Check the condition of the drop link bushes, and renew if necessary.

13 Refitting is a reversal of removal, bearing in mind the following points:

a) *Note that the concave sides of the washers should be positioned against the bushes.*

b) *When reconnecting the drop links to the anti-roll bar, make sure that the drop link balljoint is positioned centrally in its arc of movement, and is not under strain* **(see illustration 6.7)***.*

c) *Use new drop link securing nuts, and tighten them to the specified torque.*

7 Rear hub bearings – renewal

1 The rear hub bearings are integral with the rear hubs, and cannot be renewed independently. If the bearings require renewal, the complete hub assembly must be renewed as follows.

2 Chock the front wheels then jack up the rear of the vehicle and support it securely on axle stands (see *Jacking and vehicle support*). Remove the appropriate rear roadwheel.

3 Referring to Chapter 9, remove the brake drum (drum brake models), or brake caliper and brake disc (disc brake models). Note that there is no need to disconnect the brake hydraulic fluid hose from the caliper – suspend the caliper with wire or string, to avoid straining the hose. **Do not** depress the brake pedal whilst the drum/caliper is removed.

4 On models with ABS, undo the mounting bolt and withdraw the wheel sensor from the hub carrier. Move the sensor away from the hub carrier to avoid damage to the sensor tip.

5 Prise the dust cap from the hub, then remove the split-pin from the end of the stub axle **(see illustration)**. Discard the split-pin – a new one should be used on refitting.

7.6 Slacken and remove the hub nut and, on early models, collect the washer

6 Using a suitable socket and extension bar, slacken and remove the hub nut and, on early models, the washer **(see illustration)**.

 Warning: Take care, the nut is very tight.

7 Withdraw the hub assembly from the stub axle **(see illustration)**.
8 Thoroughly clean the stub axle, then slide the new hub assembly into position.
9 Fit the washer (early models only), followed by the hub nut and tighten the nut to the specified torque.
10 Check that the hub spins freely, then fit a new split-pin, and tap the dust cap into position.
11 Refit the brake drum, or brake disc and caliper, as described in Chapter 9.
12 On models with ABS, refit the wheel

8.4 Unscrew the nut then withdraw the bolt securing the lower end of the rear suspension strut to the axle beam

8.6 Unscrew the two upper mounting securing nuts, then remove the rear strut from under the wheel arch

7.7 Withdraw the hub assembly from the stub axle

sensor and tighten its retaining bolt to the specified torque (see Chapter 9).
13 Refit the roadwheel, then lower the vehicle to the ground.

---

8 Rear suspension components – removal, overhaul and refitting

Suspension strut

Note: *New suspension strut upper and lower securing nuts, must be used on refitting.*

Removal

1 On Saloon models, remove the luggage compartment trim panel on the appropriate side, as described in Chapter 11, Section 27, for access to the suspension strut upper mounting.

8.5 On Hatchback models, undo the retaining bolts and remove the appropriate rear floor stiffener panel

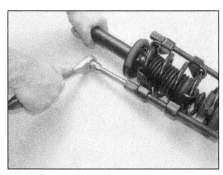

8.9 Fit suitable spring compressors and compress the coil spring

2 On Hatchback models, if working on the left-hand suspension strut, remove the luggage compartment trim panel on the left-hand side, as described in Chapter 11, Section 27. if working on the right-hand suspension strut, remove the luggage compartment trim panels on both sides, as described in Chapter 11, Section 27.
3 On all models, chock the front wheels then jack up the rear of the vehicle and support it securely on axle stands (see *Jacking and vehicle support*). Remove the appropriate rear roadwheel.
4 Unscrew the nut from the bolt securing the lower end of the suspension strut to the axle beam, then withdraw the bolt **(see illustration)**. Discard the nut – a new one must be used on refitting.
5 On Hatchback models, undo the retaining bolts and remove the appropriate rear floor stiffener panel for access to the strut upper mounting **(see illustration)**. If the right-hand side suspension strut is being removed, it will be necessary to remove both stiffener panels as they overlap each other at the centre.
6 Have an assistant support the suspension strut from under the wheel arch then, from inside the car, unscrew the two upper mounting securing nuts **(see illustration)**. Discard the nuts – new ones must be used on refitting. Lower the strut assembly and remove it from under the wheel arch. Recover the gasket from the top of the upper mounting.

Overhaul

Note: *Suitable coil spring compressor tools and a new damper rod top nut will be required for this operation.*
7 Clamp the lower end of the strut in a vice fitted with jaw protectors.
8 Temporarily fit two old bolts and nuts to the upper mounting stud holes. Using a suitable bar inserted between the bolts, counterhold the strut upper mounting whilst loosening the damper rod top nut. **Do not** remove the nut.
9 Fit suitable spring compressors to the spring, and compress the spring sufficiently to enable the upper mounting to be turned by hand **(see illustration)**.

 Warning: Do not use makeshift or improvised tools to compress the spring, as there is a danger of serious injury if the spring is not retained properly and released slowly.

10 Fully unscrew and remove the damper rod top nut. Note that it will be necessary to counterhold the damper rod, using a suitable spanner, as the nut is unscrewed **(see illustration)**. Discard the nut – a new one must be used on refitting.
11 Withdraw the washer, upper bushing, upper mounting plate and upper mounting insulator.
12 Withdraw the spring, complete with compressors, then lift off the lower bushing, and bump rubber cover.
13 Slide the rubber gaiter and bump rubber from the damper rod.

8.10 Unscrew the damper rod top nut while counterholding the damper rod

8.17 Fit the spring over the damper rod ensuring that the lower end of the spring is correctly located on the lower spring seat

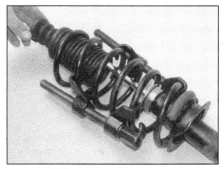

8.18a Refit the rubber gaiter and bump rubber . . .

8.18b . . . followed by the bump rubber cover . . .

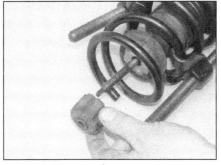

8.18c . . . and lower bushing

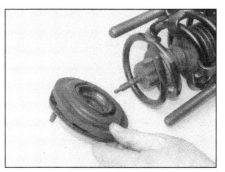

8.19a Refit the upper mounting plate and upper mounting insulator assembly

14 With the strut assembly now dismantled, examine all the components for wear, damage or deformation. Renew any of the components as necessary.

15 Examine the damper for signs of fluid leakage. Check the damper rod for signs of pitting along its entire length, and check the strut body for signs of damage. While holding it in an upright position, test the operation of the strut by moving the damper rod through a full stroke, and then through short strokes of 50 to 100 mm. In both cases, the resistance felt should be smooth and continuous. If the resistance is jerky, or uneven, or if there is any visible sign of wear or damage to the strut, renewal is necessary. Note that the damper cannot be renewed independently, and if leakage or damage is evident, the complete strut/damper assembly must be renewed (in which case, the spring, upper mounting components, bushes, and associated components can be transferred to the new strut).

16 If any doubt exists about the condition of the coil spring, carefully remove the spring compressors, and check the spring for distortion and signs of cracking. Renew the spring if it is damaged or distorted, or if there is any doubt as to its condition.

17 Ensure that the coil spring is compressed sufficiently to enable the upper mounting components to be fitted, then fit the spring over the damper rod, noting that flat end of the spring coils must be uppermost. Make sure that the lower end of the spring is correctly located on the lower spring seat (see illustration).

18 Refit the rubber gaiter and bump rubber, followed by the bump rubber cover and lower bushing (see illustrations).
19 Refit the upper mounting plate and upper

mounting insulator assembly, noting that the mounting must be correctly positioned (see illustrations).
20 Refit the upper bushing and washer then

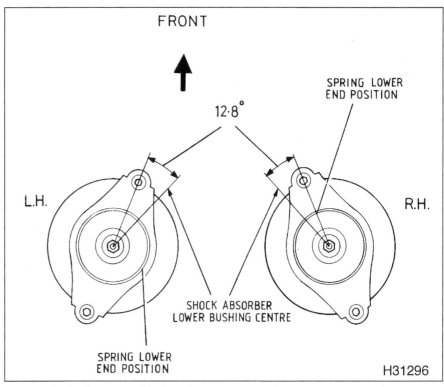

FRONT

12·8°

SPRING LOWER
END POSITION

L.H.

R.H.

SHOCK ABSORBER
LOWER BUSHING CENTRE

SPRING LOWER
END POSITION

H31296

8.19b Correct positioning of the rear suspension strut upper mounting

8.20a Refit the upper bushing . . .

8.20b . . . and washer . . .

8.20c . . . followed by a new damper rod top nut

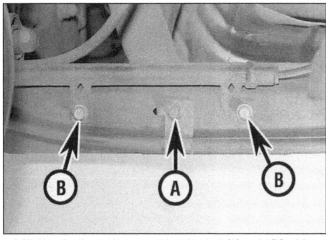

8.27 Handbrake cable support bracket bolt (A) and ABS wiring harness guide retaining bolts (B)

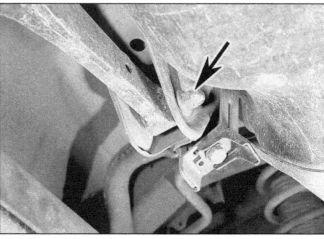

8.33 Unscrew the nut (arrowed) from the bolt securing the lateral link to the body mounting bracket

screw on a new damper rod top nut **(see illustrations)**. Tighten the top nut to the specified torque, counterholding the damper rod as during dismantling. Note that a suitable crows-foot adapter will be required to tighten the top nut to the specified torque.

21 Remove the spring compressors.

Refitting

22 Fit a new gasket to the upper mounting, then manoeuvre the strut assembly into position under the wheel arch, passing the mounting studs through the holes in the body turret. Fit the new upper mounting nuts, and tighten them to the specified torque.

23 On Hatchback models, refit the rear floor

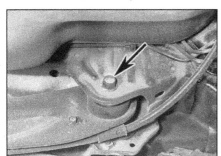

8.34 Unscrew the nuts from the trailing arm front mounting bolts (arrowed) each side

stiffener panel(s) and securely tighten the retaining bolts.

24 Refit the interior trim panels and associated components, as necessary, with reference to Chapter 11, Section 27.

25 Refit the roadwheel, and lower the vehicle to the ground.

Rear axle assembly

Note: *New shock absorber lower mounting bolt nut, new lateral link mounting nut, and trailing arm mounting nuts, must be used on refitting.*

Removal

26 Chock the front wheels then jack up the rear of the vehicle and support it securely on axle stands (see *Jacking and vehicle support*). Remove the rear roadwheels.

27 Disconnect the handbrake cables from the rear brake shoes or calipers, as applicable, as described in Chapter 9, then unbolt the handbrake cable support brackets from the trailing arms **(see illustration)**.

28 On models with rear disc brakes, extract the retaining clips and release the brake fluid hoses from the brackets on the axle beam. Referring to Chapter 9, remove the brake calipers from the hub carriers. Note that there is no need to disconnect the brake fluid hoses from the calipers – suspend the calipers with

wire or string, to avoid straining the hoses. **Do not** depress the brake pedal whilst the calipers are removed.

29 On models with rear drum brakes, disconnect the brake fluid hoses from the pipes at the brackets on the axle beam, with reference to Chapter 9. Be prepared for fluid spillage, and plug the open ends of the pipes and hoses, to reduce fluid spillage and to prevent dirt ingress.

30 Where applicable, undo the mounting bolts and withdraw the ABS wheel sensors from the hub carriers. Unbolt the wiring harness guides from the trailing arms and move the sensors clear of the axle assembly.

31 Position a trolley jack under the centre of the axle beam and raise the jack slightly to just take the weight of the axle.

32 Unscrew the nut from the bolt securing the lower end of each suspension strut to the axle beam, then withdraw the bolt. Discard the nuts – new ones must be used on refitting.

33 Unscrew the nut from the bolt securing the lateral link to the body mounting bracket then withdraw the bolt **(see illustration)**. Discard the nut – a new one must be used on refitting.

34 Unscrew the nuts from the trailing arm front mounting bolts each side **(see illustration)**. Discard the nuts – new ones must be used on refitting.

35 Ensure that the axle assembly is adequately supported, then withdraw the bolts securing the trailing arms to the body brackets. Lower the assembly sufficiently to withdraw it from under the vehicle. Take care not to drop the assembly off the jack – the help of an assistant to support the assembly on the jack will ease the operation greatly.

36 If required, remove the lateral link and control rod from the axle beam as described later in this Section.

Overhaul

37 Examine the metal components of the assembly for cracks or signs of damage. Similarly, examine the trailing arm front mounting bushes.

38 The mounting bushes are integral with the rear axle assembly and cannot be renewed independently. If the bushes are worn or damaged, the complete rear axle assembly must be renewed.

Refitting

39 If removed, refit the lateral link and control rod to the axle beam as described later in this Section.

40 Support the axle assembly on a jack, as during removal, then manoeuvre the assembly into position under the vehicle.

41 Refit the bolts securing the trailing arm front mountings to the body brackets, then fit the new nuts, but do not tighten them at this stage.

42 Locate the lateral link in the body mounting bracket then fit the new nut, but do not tighten the nut at this stage.

43 Raise the axle assembly by means of the jack until the lateral link and control rod are in a horizontal position in relation to the axle beam. With the axle in this position, tighten the lateral link-to-body mounting bracket nut to the specified torque.

44 Lower the jack slightly, if necessary, and engage the lower ends of the suspension struts with the axle beam. Fit the bolts and new nuts and tighten to the specified torque.

45 Lower the jack fully and remove it from under the axle assembly. With the axle assembly in the fully extended position, tighten both trailing arm front mounting nuts to the specified torque.

46 Where applicable, refit the ABS wheel sensors to the hub carriers and secure with the mounting bolts tightened to the specified torque (see Chapter 9). Secure the wiring harness guides to the trailing arms.

47 On models with rear drum brakes, reconnect the brake fluid hoses to the pipes at the brackets on the axle beam, with reference to Chapter 9 if necessary.

48 On models with rear disc brakes, refit the brake calipers to the hub carriers as described in Chapter 9. Refit the brake fluid hoses to the brackets on the axle beam and secure with the retaining clips.

49 Reconnect the handbrake cables to the rear brake shoes or calipers, as applicable, as described in Chapter 9, then bolt the handbrake cable support brackets to the trailing arms.

50 On drum brake models, bleed the brake hydraulic system as described in Chapter 9.

51 Check the handbrake adjustment as described in Chapter 1.

52 Refit the roadwheels, and lower the vehicle to the ground.

Lateral link and control rod

Note: *New lateral link and control rod mounting nuts should be used on refitting.*

Removal

53 Remove the rear axle assembly as described previously in this Section.

54 Undo the mounting nuts and remove the lateral link and control rod from the studs on the axle beam **(see illustrations)**. Discard the nuts – new ones must be used on refitting.

55 Undo the nut from the bolt securing the control rod to the lateral link. Withdraw the bolt and separate the two components. Discard the nut – a new one must be used on refitting.

Overhaul

56 Examine the control rod and lateral link for distortion, cracks or signs of damage. Similarly, examine the mounting bushes for signs of deterioration. If any signs of wear or damage are apparent, renew the component(s).

Refitting

57 Locate the control rod in position on the lateral link and insert the mounting bolt with its bolt head toward the rear of the vehicle. Ensure that the control rod is fitted correctly – the bush with the smaller internal diameter is fitted to the lateral link, and the bush with the larger internal diameter locates on the axle beam stud.

58 Fit the new nut to the control rod mounting bolt, but do not tighten the nut at this stage.

59 Engage the lateral link and control rod ends with the studs on the axle beam, ensuring that the lateral link is the correct way up – the arrow on the link adjacent to the axle beam mounting must point upward.

60 Fit the new nuts to the control rod and lateral link mounting studs on the axle beam, but do not tighten the nuts at this stage.

61 Position the lateral link and control rod horizontally in relation to the axle beam, hold the components in this position and tighten the two mounting nuts to the specified torque.

62 Refit the rear axle assembly as described previously in this Section.

---

9 Steering wheel –
removal and refitting

Models without airbag

Removal

1 Disconnect the battery negative terminal (refer to *Disconnecting the battery* in the Reference Chapter).

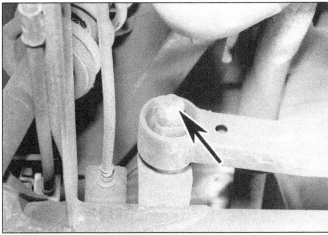

8.54a Lateral link-to-axle beam mounting nut (arrowed)

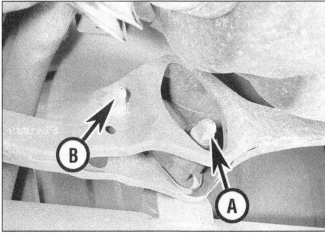

8.54b Control rod-to-axle beam mounting nut (A) and control rod-to-lateral link mounting bolt nut (B)

9.15 Undo and remove the steering wheel retaining nut

2 Set the front wheels in the straight-ahead position, then release the steering lock by inserting the ignition key.

3 Using a crosshead screwdriver inserted through the recess either in the base of the steering wheel hub, or on the reverse of the lower spoke (according to steering wheel type), undo the horn pad clamp screw. Release the horn pad, disconnect the wiring connector and withdraw the pad from the steering wheel.

4 Slacken the steering wheel retaining nut but do not remove the nut at this stage.

5 Make alignment marks between the steering wheel and the end of the steering column shaft.

6 Release the wheel from the column shaft taper by tapping it upward near the centre, using the palm of your hand, or twist it from side-to-side, whilst pulling upwards. If the wheel is particularly tight, a suitable puller should be used.

7 Once the wheel is free, unscrew the retaining nut and lift the wheel off the column shaft.

Refitting

8 Before commencing refitting, lightly coat the surfaces of the direction indicator cancelling mechanism components and the horn contact slip ring components with grease.

9 Refitting is a reversal of removal, bearing in mind the following points:
 a) Ensure that the direction indicator switch is in the central (cancelled/off) position, otherwise the switch may be damaged as the wheel is refitted.

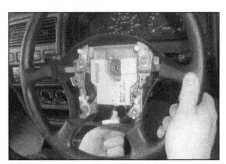

9.16 Unclip the wiring harnesses, and feed them through the steering wheel as the wheel is withdrawn

 b) Align the marks made on the wheel and the steering column shaft before removal.
 c) Tighten the securing nut to the specified torque.

10 Note that, if necessary, the position of the steering wheel on the column shaft can be altered in order to centralise the wheel (ensure that the front roadwheels are pointing in the straight-ahead position), by moving the wheel the required number of splines on the shaft.

Models with airbag

⚠️ **Warning: Later models are equipped with an airbag system. The airbag is mounted in the steering wheel centre pad. Make sure that the safety recommendations given in Chapter 12 are followed, to prevent personal injury.**

Removal

11 Ensure that the ignition is switched off, then disconnect the battery negative terminal (refer to *Disconnecting the battery* in the Reference Chapter). *Wait for at least ten minutes before carrying out any further work.*

12 Remove the airbag unit from the steering wheel as described in Chapter 12.

13 Set the front wheels in the straight-ahead position, then release the steering lock by inserting the ignition key.

14 Disconnect the horn wiring connector.

15 Undo and remove the steering wheel retaining nut **(see illustration)**, then make alignment marks between the steering wheel and the end of the steering column shaft.

16 Release the wheel from the column shaft taper using a suitable puller. Do not tap or strike the steering wheel. Unclip the wiring harnesses, and feed them through the steering wheel as the wheel is withdrawn **(see illustration)**.

Refitting

17 Refitting is a reversal of removal, bearing in mind the following points:
 a) Ensure that the front wheels are in the straight-ahead position.
 b) Centralise the airbag rotary connector by first turning the rotating centre part fully clockwise. From this position, turn it anti-clockwise 2.5 turns, and align the arrow on the rotating centre part with the corresponding arrow on the connector body.
 c) Ensure that the direction indicator switch is in the central (cancelled/off) position, otherwise the switch may be damaged as the wheel is refitted.
 d) Feed the wiring harnesses through the steering wheel, and clip the connector into position as the wheel is refitted.
 e) Align the marks made on the wheel and the steering column shaft before removal, and align the steering wheel with the guide pins on the airbag rotary connector.
 f) Tighten the steering wheel securing nut to the specified torque.
 g) Refit the airbag unit to the steering wheel as described in Chapter 12.

10 Ignition switch/ steering column lock – removal and refitting

Note: *New shear-screws must be used when refitting the lock assembly.*

Removal

1 Disconnect the battery negative terminal (refer to *Disconnecting the battery* in the Reference Chapter).

2 Remove the steering column shrouds as described in Chapter 11, Section 29.

3 To remove the ignition switch, disconnect the wiring plug from the switch, then remove the grub screw from the rear of the lock (using a suitable cranked key), and withdraw the switch.

4 To remove the lock assembly, drill out and remove the two shear-screws, then withdraw the two sections of the lock casting from the steering column. Note that the lock assembly cannot be removed from the casting.

Refitting

5 Refitting is a reversal of removal, but when refitting the lock assembly, use new shear-screws, and tighten the screws until the heads break off.

11 Steering column – removal, inspection and refitting

⚠️ **Warning: Later models are equipped with an airbag system. Ensure that the safety recommendations given in Chapter 12 are followed, to prevent personal injury.**

Removal

1 Disconnect the battery negative terminal (refer to *Disconnecting the battery* in the Reference Chapter).

2 Remove the steering wheel as described in Section 9.

3 Remove the steering column stalk switches with reference to Chapter 12.

4 Disconnect the wiring plug from the ignition switch.

5 If desired, to improve access, remove the driver's side lower facia panel as described in Chapter 11, Section 29.

6 Working in the driver's footwell, release the steering column lower gaiter from the bulkhead, then peel it back to expose the lower end of the shaft.

7 Temporarily refit the steering wheel, and turn the steering column shaft as necessary for access to the universal joint clamp bolt.

8 Unscrew and remove the universal joint clamp bolt.

9 Unscrew the two steering column upper securing nuts, and the two lower nuts, and withdraw the steering column **(see illustrations)**.

Inspection

10 The steering column incorporates a telescopic safety feature. In the event of a front-end crash, the shaft collapses and prevents the steering wheel injuring the driver. Before refitting the steering column, examine the column and mountings for signs of damage and deformation, and renew as necessary.

11 Check the steering column shaft for signs of free play in the column bushes, and check the universal joints for signs of damage or roughness in the joint bearings. If any damage or wear is found in the steering column universal joints or shaft bushes, the column must be renewed as an assembly.

12 Measure the length of the steering column from the centre of the universal joint pin to the top of the column shaft (see illustration). If the measurement is outside the specified limits, this is probably due to accident damage, and the complete column assembly should be renewed.

Refitting

13 Offer the steering column into position, and engage the lower end of the column shaft with the universal joint. Note the cut-out in the splined section at the lower end of the column shaft – the joint must be reassembled such that the clamp bolt bears on this cut-out when it is tightened.

14 Refit the steering column upper and lower securing nuts, then tighten them to the specified torque.

15 Refit and tighten the universal joint clamp bolt.

16 Refit and secure the steering column lower gaiter to the bulkhead.

17 Where applicable, refit the driver's side lower facia panel.

18 Reconnect the ignition switch wiring plug.

19 Refit the steering column stalk switches.

20 Refit the steering wheel, as described in Section 9.

21 Reconnect the battery negative terminal.

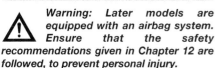

12 Steering gear assembly – removal, inspection and refitting

> ⚠️ **Warning: Later models are equipped with an airbag system. Ensure that the safety recommendations given in Chapter 12 are followed, to prevent personal injury.**
> Note: A balljoint separator tool will be required for this operation. New track rod end retaining nut split-pins must be used on refitting.

Manual steering gear

Removal

1 Firmly apply the handbrake, then jack up the front of the vehicle and support it securely on axle stands (see Jacking and vehicle support). Remove the front roadwheels.

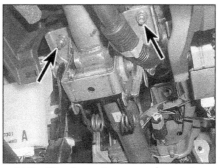

11.9a Steering column upper securing nuts (arrowed) . . .

11.9b . . . and lower securing nuts (arrowed)

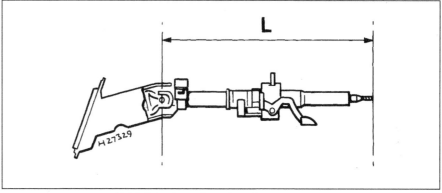

11.12 Steering column length measurement

L See Specifications

2 Move the steering wheel so that the front wheels are pointing in the straight-ahead position.

3 Working on one side of the vehicle, remove the split-pin, then partially unscrew the castellated nut securing the track rod end to the steering arm. Using a balljoint separator tool, separate the track rod end from the steering arm. Remove the nut. Discard the split-pin – a new one must be used on refitting.

4 Repeat the procedure to disconnect the track rod end on the other side of the vehicle.

5 Working in the engine compartment at the bulkhead, undo the retaining nuts, remove the retaining plates (where applicable) and peel back the gaiter covering the steering column lower universal joint (see illustration).

6 Mark the relationship between the steering column lower universal joint and the steering gear pinion. Unscrew the clamp bolt and separate the joint from the steering gear pinion (see illustration). Do not rotate the steering column shaft while it is separated from the steering gear pinion.

7 Unscrew the bolts securing the steering gear mounting brackets to the bulkhead (see illustrations). Support the steering gear, then withdraw the brackets.

8 Rotate the steering gear, and manipulate it out through the right-hand wheel arch. Note that on left-hand-drive models, it will be necessary to lift the steering gear over the mounting brackets before it can be withdrawn.

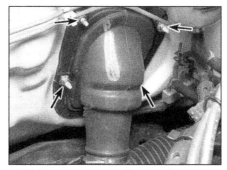

12.5 Remove the nuts (arrowed) and peel back the gaiter covering the steering column lower universal joint

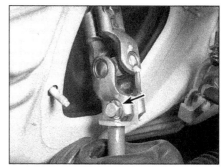

12.6 Unscrew the clamp bolt (arrowed) and separate the universal joint from the steering gear pinion

12.7a Unscrew the bolts from the steering gear mounting brackets on the right-hand . . .

12.7b . . . and left-hand sides

Inspection

9 Examine the assembly for obvious signs of wear or damage.

10 Check the rack for smooth operation through its full stroke of movement, and check that there is no binding or free play.

11 Check the track rods for deformation and cracks.

12 Check the condition of the steering gear rubber gaiters, and renew if necessary with reference to Section 13.

13 Examine the track rod ends for wear or damage, and renew if necessary with reference to Section 16.

14 Apart from renewal of the track rod ends and steering gear rubber gaiters, any further overhaul necessary should be entrusted to a Nissan dealer.

Refitting

15 Ensure that the steering gear is centralised as follows:
a) *Turn the pinion to rotate the steering from lock-to-lock, and count the number of turns of the pinion.*
b) *With the steering on full lock, turn the pinion back through half the number of turns noted for lock-to-lock, to achieve the centralised position.*

16 Manoeuvre the steering gear into position through the right-hand wheel arch, then refit the mounting brackets, and refit and tighten the securing bolts.

17 Engage the steering column universal joint with the steering gear pinion, then refit and tighten the clamp bolt. **Note:** *If the original steering gear is being refitted, the marks made during removal can be used to align the universal joint and pinion.*

18 Position the gaiter over the universal joint, fit the retaining plates (where applicable) then refit and tighten the securing nuts.

19 Working on one side of the vehicle, reconnect the track rod end to the steering arm, then refit the castellated nut, and tighten to the specified torque.

20 If necessary, tighten the nut further (ensure that the maximum torque for the nut is not exceeded) until the nearest grooves in the nut are aligned with the split-pin hole in the track rod end, then fit a new split-pin.

21 Repeat the procedure to reconnect the track rod end on the other side of the vehicle.

22 Refit the roadwheels, and lower the vehicle to the ground.

Power steering gear

Removal

23 Firmly apply the handbrake, then jack up the front of the vehicle and support it securely on axle stands (see *Jacking and vehicle support*). Remove the front roadwheels.

24 Position a suitable container beneath the fluid feed pipe union on the steering gear, then unscrew the union nut, and disconnect the pipe from the steering gear. Similarly, disconnect the fluid return hose from the steering gear. Drain the fluid into the container, then plug or cover the open ends of the pipe/hose and steering gear to reduce further fluid loss, and to prevent dirt ingress.

25 To improve access, remove the exhaust system front pipe as described in Chapter 4A.

26 Disconnect the track rod ends from the steering arms as described previously for the manual steering gear in paragraphs 2 to 4.

27 Disconnect the steering column universal joint from the steering gear pinion as

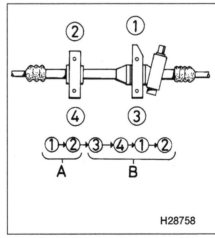

H28758

12.32 Power steering gear mounting bracket retaining bolt tightening sequence

A *Initial tightening*
B *Final tightening*

described previously for the manual steering gear in paragraphs 5 and 6.

28 Unscrew the bolts securing the steering gear mounting brackets to the bulkhead. Support the steering gear, then withdraw the brackets.

29 Manipulate the steering gear around the wiring harness (where applicable), then slide the assembly to the right-hand side of the vehicle, and withdraw the assembly from under the left-hand side of the vehicle.

Inspection

30 Refer to paragraphs 9 to 14.

Refitting

31 Ensure that the steering gear is centralised as follows:
a) *Turn the pinion to rotate the steering from lock-to-lock, and count the number of turns of the pinion.*
b) *With the steering on full lock, turn the pinion back through half the number of turns noted for lock-to-lock, to achieve the centralised position.*

32 Manoeuvre the steering gear into position from under the left-hand side of the vehicle, then refit the mounting brackets and the securing bolts. **Note:** *Lightly tighten the upper mounting bracket bolts (do not fully tighten the bolts at this stage), then tighten all the bolts to the specified torque, starting with the lower bolts, in the order shown (see illustration).*

33 Where applicable, refit the exhaust system front pipe with reference to Chapter 4A.

34 Proceed as described in paragraphs 17 to 21.

35 Reconnect the fluid feed pipe union and the return hose to the steering gear.

36 Refit the roadwheels, then bleed the power steering hydraulic system as described in Section 14.

37 On completion, lower the vehicle to the ground.

---

13 Steering gear rubber gaiters – renewal

Note: *New gaiter retaining clips should be used on refitting.*

1 Remove the relevant track rod end as described in Section 16.

2 If not already done, unscrew the track rod end locknut from the end of the track rod.

3 Mark the correct fitted position of the gaiter on the track rod, then release the gaiter securing clips. Slide the gaiter from the steering gear, and off the end of the track rod.

4 Thoroughly clean the track rod and the steering gear housing, using fine abrasive paper to polish off any corrosion, burrs or sharp edges, which might damage the new gaiter sealing lips on installation. Scrape off all the grease from the old gaiter, and apply it to the track rod inner balljoint. (This assumes that grease has not been lost or contaminated

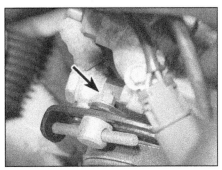

15.5a Unscrew the nut (arrowed) from the power steering pump adjuster stud

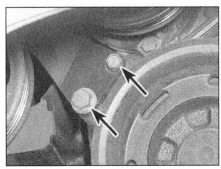

15.5b Power steering pump adjuster bracket lower securing bolts (arrowed)

as a result of damage to the old gaiter. Use fresh grease if in doubt.)

5 Carefully slide the new gaiter onto the track rod, and locate it on the steering gear housing. Align the outer edge of the gaiter with the mark made on the track rod prior to removal, then secure it in position with new retaining clips.

6 Screw the track rod end locknut onto the end of the track rod.

7 Refit the track rod end as described in Section 16.

---

14 Power steering hydraulic system – bleeding

General

1 The following symptoms indicate that there is air present in the power steering hydraulic system:
a) *Generation of air bubbles in fluid reservoir.*
b) *'Clicking' noises from power steering pump.*
c) *Excessive 'buzzing' from power steering pump.*

2 Note that when the vehicle is stationary, or while moving the steering wheel slowly, a 'hissing' noise may be produced in the steering gear or the fluid pump. This noise is inherent in the system, and does not indicate any cause for concern.

Bleeding

3 Firmly apply the handbrake, then jack up the front of the vehicle and support it securely on axle stands (see *Jacking and vehicle support*).

4 Check the fluid level in the power steering fluid reservoir as described in *Weekly checks*. Bear in mind that the vehicle will be slightly tilted, so the level cannot be read accurately. If necessary top-up to just above the relevant level mark.

5 Have an assistant turn the steering quickly from lock-to-lock, and observe the fluid level. If the fluid level drops, add more fluid, and repeat the operation until the fluid level no longer drops. Failure to achieve this within a

reasonable period may indicate a leak in the system.

6 Start the engine and repeat the procedure described in the previous paragraph.

7 Once the fluid level has stabilised, and all air has been bled from the system, lower the vehicle to the ground.

8 Check and if necessary top-up the fluid to the relevant mark as described in *Weekly checks*.

---

15 Power steering pump – removal and refitting

Note: *New sealing rings must be used when reconnecting the high-pressure fluid hose to the pump.*

Removal

1 Firmly apply the handbrake, then jack up the front of the vehicle and support it securely on axle stands (see *Jacking and vehicle support*).

2 Remove the power steering pump (auxiliary) drivebelt as described in Chapter 1.

3 Place a suitable container beneath the power steering pump, then unscrew the banjo bolt from the high-pressure fluid hose union, and disconnect the hose from the pump. Recover the sealing rings and discard them – new ones should be used on refitting. Drain the escaping fluid into the container. Plug or cover the open ends of the hose and pump, to reduce further fluid loss, and to prevent dirt ingress.

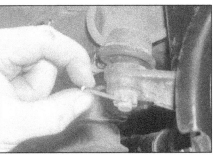

16.2a Remove the split-pin, then partially unscrew the castellated nut securing the track rod end to the steering arm

4 Similarly, disconnect the fluid return hose from the pump, noting that the hose is secured by a hose clip rather than a banjo union.

5 Unscrew the nut from the adjuster stud, then unscrew the two lower bolts securing the adjuster bracket to the engine, and withdraw the adjuster bracket assembly **(see illustrations).**

6 Turn the pump pulley until one of the holes in the pulley lines up with the through-bolt, then unscrew the nut from the rear of the through-bolt, and recover the washers.

7 Support the pump, then withdraw the through-bolt, and manipulate the pump out from above the engine. Note that on certain models surrounding components may prevent the pump from being withdrawn from above – in this case, it will be necessary to unbolt the pump mounting bracket from the engine, and withdraw the pump from underneath the vehicle.

Refitting

8 Refitting is a reversal of removal, but use new sealing rings when reconnecting the high-pressure fluid hose, and on completion, refit and tension the drivebelt as described in Chapter 1.

---

16 Track rod end – removal and refitting

Note: *A balljoint separator tool will be required for this operation. A new track rod end retaining nut split-pin must be used on refitting.*

Removal

1 Firmly apply the handbrake, then jack up the front of the vehicle and support it securely on axle stands (see *Jacking and vehicle support*). Remove the relevant front roadwheel.

2 Remove the split-pin, then partially unscrew the castellated nut securing the track rod end to the steering arm. Using a balljoint separator tool, separate the track rod end from the steering arm **(see illustrations)**. Remove the nut. Discard the split-pin – a new one must be used on refitting.

16.2b Using a balljoint separator tool, separate the track rod end from the steering arm

3 Counterhold the track rod end using the flats provided, then loosen the track rod end locknut.

4 Counting the exact number of turns required to do so, unscrew the track rod end from the track rod.

Refitting

5 Carefully clean the track rod end and the track rod threads.

6 Renew the track rod end if the rubber dust cover is cracked, split or perished, or if the movement of the balljoint is either sloppy or too stiff. Also check for other signs of damage such as worn threads.

7 Screw the track rod end onto the track rod by the number of turns noted before removal.

8 Counterhold the track rod end using the flats provided, then securely tighten the track rod end locknut.

9 Ensure that the balljoint taper is clean, then engage the taper with the steering arm on the hub carrier.

10 Refit the castellated nut, and tighten to the specified torque.

11 If necessary, tighten the nut further (ensure that the maximum torque for the nut is not exceeded) until the nearest grooves in the nut are aligned with the split-pin hole in the track rod end, then fit a new split-pin.

12 Refit the roadwheel, and lower the vehicle to the ground.

13 Check the front wheel alignment with reference to Chapter 1.

Chapter 11
Bodywork and fittings

Contents

Degrees of difficulty

Easy, suitable for novice with little experience	Fairly easy, suitable for beginner with some experience	Fairly difficult, suitable for competent DIY mechanic	Difficult, suitable for experienced DIY mechanic	Very difficult, suitable for expert DIY or professional

Specifications

Torque wrench settings	Nm	lbf ft
Door hinge bolts	25	18
Door lock striker bolts	15	11
Front seat securing bolts	50	37
Seat belt anchor bolts	50	37

1 General information

The bodyshell is made of pressed-steel sections, and is available in four-door Saloon, and three- and five-door Hatchback versions. Most components are welded together, but some use is made of structural adhesives; the front wings are bolted on.

The bonnet, door, and some other vulnerable panels, are made of zinc-coated metal, and are further protected by being coated with an anti-chip primer, prior to being sprayed.

Extensive use is made of plastic materials, mainly in the interior, but also in exterior components. The outer sections of the front and rear bumpers are injection-moulded from a synthetic material which is very strong, and yet light. Plastic components such as wheel arch liners are fitted to the underside of the vehicle, to improve the body's resistance to corrosion.

2 Maintenance – bodywork and underframe
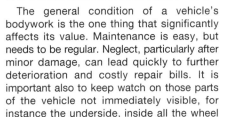

The general condition of a vehicle's bodywork is the one thing that significantly affects its value. Maintenance is easy, but needs to be regular. Neglect, particularly after minor damage, can lead quickly to further deterioration and costly repair bills. It is important also to keep watch on those parts of the vehicle not immediately visible, for instance the underside, inside all the wheel arches, and the lower part of the engine compartment.

The basic maintenance routine for the bodywork is washing – preferably with a lot of water, from a hose. This will remove all the loose solids which may have stuck to the vehicle. It is important to flush these off in such a way as to prevent grit from scratching the finish. The wheel arches and underframe need washing in the same way, to remove any accumulated mud, which will retain moisture and tend to encourage rust. Paradoxically enough, the best time to clean the underframe and wheel arches is in wet weather, when the mud is thoroughly wet and soft. In very wet weather, the underframe is usually cleaned of large accumulations automatically, and this is a good time for inspection.

Periodically, except on vehicles with a wax-based underbody protective coating, it is a good idea to have the whole of the underframe of the vehicle steam-cleaned, engine compartment included, so that a thorough inspection can be carried out to see what minor repairs and renovations are necessary. Steam-cleaning is available at many garages, and is necessary for the removal of the accumulation of oily grime, which sometimes is allowed to become thick in certain areas. If steam-cleaning facilities are not available, there are some excellent grease solvents available which can be brush-applied; the dirt can then be simply hosed off. Note that these methods should not be used on vehicles with wax-based underbody protective coating, or the coating will be removed. Such vehicles should be inspected annually, preferably just prior to Winter, when the underbody should be washed down, and any damage to the wax coating repaired. Ideally, a completely fresh coat should be applied. It would also be worth considering the use of such wax-based protection for injection into door panels, sills, box sections, etc, as an additional safeguard against rust damage, where such protection is not provided by the vehicle manufacturer.

After washing paintwork, wipe off with a chamois leather to give an unspotted clear finish. A coat of clear protective wax polish will give added protection against chemical pollutants in the air. If the paintwork sheen has dulled or oxidised, use a cleaner/polisher combination to restore the brilliance of the shine. This requires a little effort, but such dulling is usually caused because regular washing has been neglected. Care needs to be taken with metallic paintwork, as special non-abrasive cleaner/polisher is required to avoid damage to the finish. Always check that the door and ventilator opening drain holes and pipes are completely clear, so that water can be drained out. Brightwork should be treated in the same way as paintwork. Windscreens and windows can be kept clear of the smeary film which often appears, by the use of proprietary glass cleaner. Never use any form of wax or other body or chromium polish on glass.

3 Maintenance –
upholstery and carpets

Mats and carpets should be brushed or vacuum-cleaned regularly, to keep them free of grit. If they are badly stained, remove them from the vehicle for scrubbing or sponging, and make quite sure they are dry before refitting. Seats and interior trim panels can be kept clean by wiping with a damp cloth. If they do become stained (which can be more apparent on light-coloured upholstery), use a little liquid detergent and a soft nail brush to scour the grime out of the grain of the material. Do not forget to keep the headlining clean in the same way as the upholstery. When using liquid cleaners inside the vehicle, do not over-wet the surfaces being cleaned. Excessive damp could get into the seams and padded interior, causing stains, offensive odours or even rot.

HAYNES HiNT *If the inside of the vehicle gets wet accidentally, it is worthwhile taking some trouble to dry it out properly, particularly where carpets are involved. Do not leave oil or electric heaters inside the vehicle for this purpose.*

4 Minor body damage –
repair

Minor scratches in bodywork

If the scratch is very superficial, and does not penetrate to the metal of the bodywork, repair is very simple. Lightly rub the area of the scratch with a paintwork renovator, or a very fine cutting paste, to remove loose paint from the scratch, and to clear the surrounding bodywork of wax polish. Rinse the area with clean water.

Apply touch-up paint to the scratch using a fine paint brush; continue to apply fine layers of paint until the surface of the paint in the scratch is level with the surrounding paintwork. Allow the new paint at least two weeks to harden, then blend it into the surrounding paintwork by rubbing the scratch area with a paintwork renovator or a very fine cutting paste. Finally, apply wax polish.

Where the scratch has penetrated right through to the metal of the bodywork, causing the metal to rust, a different repair technique is required. Remove any loose rust from the bottom of the scratch with a penknife, then apply rust-inhibiting paint to prevent the formation of rust in the future. Using a rubber or nylon applicator, fill the scratch with bodystopper paste. If required, this paste can be mixed with cellulose thinners to provide a very thin paste which is ideal for filling narrow scratches. Before the stopper-paste in the scratch hardens, wrap a piece of smooth cotton rag around the top of a finger. Dip the finger in cellulose thinners, and quickly sweep it across the surface of the stopper-paste in the scratch; this will ensure that the surface of the stopper-paste is slightly hollowed. The scratch can now be painted over as described earlier in this Section.

Dents in bodywork

When deep denting of the vehicle's bodywork has taken place, the first task is to pull the dent out, until the affected bodywork almost attains its original shape. There is little point in trying to restore the original shape completely, as the metal in the damaged area will have stretched on impact, and cannot be reshaped fully to its original contour. It is better to bring the level of the dent up to a point which is about 3 mm below the level of the surrounding bodywork. In cases where the dent is very shallow anyway, it is not worth trying to pull it out at all. If the underside of the

---

dent is accessible, it can be hammered out gently from behind, using a mallet with a wooden or plastic head. Whilst doing this, hold a suitable block of wood firmly against the outside of the panel, to absorb the impact from the hammer blows and thus prevent a large area of the bodywork from being 'belled-out'.

Should the dent be in a section of the bodywork which has a double skin, or some other factor making it inaccessible from behind, a different technique is called for. Drill several small holes through the metal inside the area – particularly in the deeper section. Then screw long self-tapping screws into the holes, just sufficiently for them to gain a good purchase in the metal. Now the dent can be pulled out by pulling on the protruding heads of the screws with a pair of pliers.

The next stage of the repair is the removal of the paint from the damaged area, and from an inch or so of the surrounding 'sound' bodywork. This is accomplished most easily by using a wire brush or abrasive pad on a power drill, although it can be done just as effectively by hand, using sheets of abrasive paper. To complete the preparation for filling, score the surface of the bare metal with a screwdriver or the tang of a file, or alternatively, drill small holes in the affected area. This will provide a really good 'key' for the filler paste.

To complete the repair, see the Section on filling and respraying.

Rust holes/gashes in bodywork

Remove all paint from the affected area, and from an inch or so of the surrounding 'sound' bodywork, using an abrasive pad or a wire brush on a power drill. If these are not available, a few sheets of abrasive paper will do the job most effectively. With the paint removed, you will be able to judge the severity of the corrosion, and therefore decide whether to renew the whole panel (if this is possible) or to repair the affected area. New body panels are not as expensive as most people think, and it is often quicker and more satisfactory to fit a new panel than to attempt to repair large areas of corrosion.

Remove all fittings from the affected area, except those which will act as a guide to the original shape of the damaged bodywork (eg headlight shells etc). Then, using tin snips or a hacksaw blade, remove all loose metal and any other metal badly affected by corrosion. Hammer the edges of the hole inwards, in order to create a slight depression for the filler paste.

Wire-brush the affected area to remove the powdery rust from the surface of the remaining metal. Paint the affected area with rust-inhibiting paint, if the back of the rusted area is accessible, treat this also.

Before filling can take place, it will be necessary to block the hole in some way. This can be achieved by the use of aluminium or plastic mesh, or aluminium tape.

Aluminium or plastic mesh, or glass-fibre matting, is probably the best material to use for a large hole. Cut a piece to the approximate size and shape of the hole to be filled, then position it in the hole so that its edges are below the level of the surrounding bodywork. It can be retained in position by several blobs of filler paste around its periphery.

Aluminium tape should be used for small or very narrow holes. Pull a piece off the roll, trim it to the approximate size and shape required, then pull off the backing paper (if used) and stick the tape over the hole; it can be overlapped if the thickness of one piece is insufficient. Burnish down the edges of the tape with the handle of a screwdriver or similar, to ensure that the tape is securely attached to the metal underneath.

Filling and respraying

Before using this Section, see the Sections on dent, deep scratch, rust holes and gash repairs.

Many types of bodyfiller are available, but generally speaking, those proprietary kits which contain a tin of filler paste and a tube of resin hardener are best for this type of repair. A wide, flexible plastic or nylon applicator will be found invaluable for imparting a smooth and well-contoured finish to the surface of the filler.

Mix up a little filler on a clean piece of card or board – measure the hardener carefully (follow the maker's instructions on the pack), otherwise the filler will set too rapidly or too slowly. Using the applicator, apply the filler paste to the prepared area; draw the applicator across the surface of the filler to achieve the correct contour and to level the surface. As soon as a contour that approximates to the correct one is achieved, stop working the paste – if you carry on too long, the paste will become sticky and begin to 'pick-up' on the applicator. Continue to add thin layers of filler paste at 20-minute intervals, until the level of the filler is just proud of the surrounding bodywork.

Once the filler has hardened, the excess can be removed using a metal plane or file. From then on, progressively-finer grades of abrasive paper should be used, starting with a 40-grade production paper, and finishing with a 400-grade wet-and-dry paper. Always wrap the abrasive paper around a flat rubber, cork, or wooden block – otherwise the surface of the filler will not be completely flat. During the smoothing of the filler surface, the wet-and-dry paper should be periodically rinsed in water. This will ensure that a very smooth finish is imparted to the filler at the final stage.

At this stage, the 'dent' should be surrounded by a ring of bare metal, which in turn should be encircled by the finely 'feathered' edge of the good paintwork. Rinse the repair area with clean water, until all of the dust produced by the rubbing-down operation has gone.

Spray the whole area with a light coat of primer – this will show up any imperfections in the surface of the filler. Repair these imperfections with fresh filler paste or bodystopper, and once more smooth the surface with abrasive paper. Repeat this spray-and-repair procedure until you are satisfied that the surface of the filler, and the feathered edge of the paintwork, are perfect. Clean the repair area with clean water, and allow to dry fully.

If bodystopper is used, it can be mixed with cellulose thinners to form a really thin paste which is ideal for filling small holes.

The repair area is now ready for final spraying. Paint spraying must be carried out in a warm, dry, windless and dust-free atmosphere. This condition can be created artificially if you have access to a large indoor working area, but if you are forced to work in the open, you will have to pick your day very carefully. If you are working indoors, dousing the floor in the work area with water will help to settle the dust which would otherwise be in the atmosphere. If the repair area is confined to one body panel, mask off the surrounding panels; this will help to minimise the effects of a slight mis-match in paint colours. Bodywork fittings (eg chrome strips, door handles etc) will also need to be masked off. Use genuine masking tape, and several thicknesses of newspaper, for the masking operations.

Before commencing to spray, agitate the aerosol can thoroughly, then spray a test area (an old tin, or similar) until the technique is mastered. Cover the repair area with a thick coat of primer; the thickness should be built up using several thin layers of paint, rather than one thick one. Using 400-grade wet-and-dry paper, rub down the surface of the primer until it is really smooth. While doing this, the work area should be thoroughly doused with water, and the wet-and-dry paper periodically rinsed in water. Allow to dry before spraying on more paint.

Spray on the top coat, again building up the thickness by using several thin layers of paint. Start spraying at one edge of the repair area, and then, using a side-to-side motion, work until the whole repair area and about 2 inches of the surrounding original paintwork is covered. Remove all masking material 10 to 15 minutes after spraying on the final coat of paint.

Allow the new paint at least two weeks to harden, then, using a paintwork renovator, or a very fine cutting paste, blend the edges of the paint into the existing paintwork. Finally, apply wax polish.

Plastic components

With the use of more and more plastic body components by the vehicle manufacturers (eg bumpers, spoilers, and in some cases major body panels), rectification of more serious damage to such items has become a matter of either entrusting repair work to a specialist in this field, or renewing complete components. Repair of such damage by the DIY owner is not really feasible, owing to the cost of the equipment and materials required for effecting such repairs. The basic technique involves making a groove along the line of the crack in the plastic, using a rotary burr in a power drill. The damaged part is then welded back together, using a hot-air gun to heat up and fuse a plastic filler rod into the groove. Any excess plastic is then removed, and the area rubbed down to a smooth finish. It is important that a filler rod of the correct plastic is used, as body components can be made of a variety of different types (eg polycarbonate, ABS, polypropylene).

Damage of a less serious nature (abrasions, minor cracks etc) can be repaired by the DIY owner using a two-part epoxy filler repair material. Once mixed in equal proportions, this is used in similar fashion to the bodywork filler used on metal panels. The filler is usually cured in twenty to thirty minutes, ready for sanding and painting.

If the owner is renewing a complete component himself, or if he has repaired it with epoxy filler, he will be left with the problem of finding a suitable paint for finishing which is compatible with the type of plastic used. At one time, the use of a universal paint was not possible, owing to the complex range of plastics encountered in body component applications. Standard paints, generally speaking, will not bond to plastic or rubber satisfactorily. However, it is now possible to obtain a plastic body parts finishing kit which consists of a pre-primer treatment, a primer and coloured top coat. Full instructions are normally supplied with a kit, but basically, the method of use is to first apply the pre-primer to the component concerned, and allow it to dry for up to 30 minutes. Then the primer is applied, and left to dry for about an hour before finally applying the special-coloured top coat. The result is a correctly-coloured component, where the paint will flex with the plastic or rubber, a property that standard paint does not normally possess.

5 Major body damage – repair

Where serious damage has occurred, or large areas need renewal due to neglect, it means that complete new panels will need welding-in, and this is best left to professionals. If the damage is due to impact, it will also be necessary to check completely the alignment of the bodyshell, and this can only be carried out accurately by a Nissan dealer using special jigs. If the body is left misaligned, it is primarily dangerous, as the car will not handle properly, and secondly, uneven stresses will be imposed on the steering, suspension and possibly transmission, causing abnormal wear, or complete failure, particularly to such items as the tyres.

6.3 Unscrew the front bumper lower securing bolts

6.4a Remove the clip securing the wheel arch liner to the lower corner of the bumper . . .

6.4b . . . then reach up and undo the bolt securing the upper corner of the bumper to the front wing

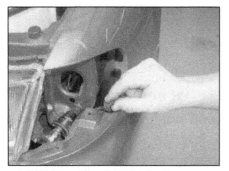

6.5 Remove the clip securing the upper edge of the bumper to the body

6.6 Undo the two outer bolts (A) and extract the centre clip (B) securing the upper edge of the bumper to the body

6.8 Withdraw the bumper forwards and remove it from the vehicle

6 Front bumper – removal and refitting

Removal

1 Remove the front grille panel as described in Section 24.
2 Remove both front direction indicator light units as described in Chapter 12, Section 7.
3 Working under the front of the vehicle, unscrew the bumper lower securing bolts **(see illustration)**.
4 Working under one of the wheel arches, undo the retaining clip centre screw then extract the clip securing the wheel arch liner to the lower corner of the bumper. Ease the

liner away then reach up and undo the bolt securing the upper corner of the bumper to the front wing **(see illustrations)**. Repeat the procedure on the remaining side of the vehicle.
5 From within the direction indicator aperture each side, undo the retaining clip centre screw then extract the clip securing the outer edge of the bumper to the body **(see illustration)**.
6 Working in the grille panel aperture, undo the two outer bolts and extract the centre clip securing the upper edge of the bumper to the body **(see illustration)**.
7 On models with front foglights, disconnect the foglight wiring connectors from under the bumper.
8 Withdraw the bumper forwards and remove it from the vehicle **(see illustration)**.

Refitting

9 Refitting is a reversal of removal.

7 Rear bumper – removal and refitting

Removal

1 Working at the rear of the wheel arches, unscrew the two bolts each side securing the upper and lower corners of the bumper to the rear wing and body brackets **(see illustrations)**.
2 Working at the lower rear corner of the bumper each side, undo the retaining clip centre screw then extract the clip **(see illustration)**.

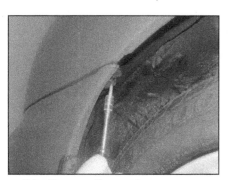

7.1a Unscrew the rear bumper upper wheel arch securing bolts . . .

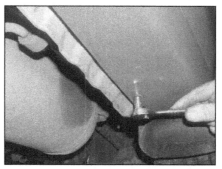

7.1b . . . and lower wheel arch securing bolts each side

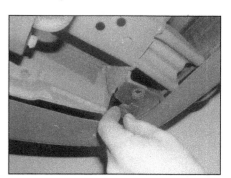

7.2 Remove the bumper lower rear corner retaining clips

3 Open the boot lid or tailgate and undo the four retaining clip centre screws then extract the clips securing the upper edge of the bumper to the body **(see illustration)**.

4 From underneath the centre of the bumper, undo the two lower retaining bolts **(see illustration)**.

5 Release the bumper from the locating lugs each side, then withdraw the bumper rearwards and remove it from the vehicle **(see illustrations)**.

Refitting

6 Refitting is a reversal of removal.

8 Bonnet –
removal, refitting and adjustment

Removal

1 Open the bonnet and have an assistant support it. Using a pencil or felt tip pen, mark the outline of each bonnet hinge relative to the bonnet, to use as a guide on refitting.

2 Disconnect the windscreen washer fluid supply hose from the connector under the bonnet, then release the hose from the clips under the bonnet **(see illustration)**.

3 Unscrew the bolts securing the hinges to the bonnet **(see illustration)** and, with the help of an assistant, carefully lift the bonnet clear. Store the bonnet out of the way in a safe place.

4 Inspect the bonnet hinges for signs of wear and free play at the pivots, and if necessary renew. Each hinge is secured to the body by two bolts.

Refitting

5 With the aid of an assistant, offer up the bonnet, and loosely fit the retaining bolts. Align the hinges with the marks made on removal, then tighten the retaining bolts securely.

6 Reconnect the windscreen washer fluid supply hose, and clip it into position under the bonnet.

7 Adjust the alignment of the bonnet as follows.

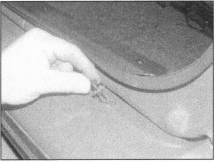

7.3 Remove the clips securing the upper edge of the bumper to the body

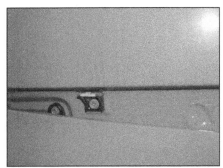

7.5a Release the bumper from the locating lugs each side . . .

Adjustment

8 Close the bonnet, and check for alignment with the adjacent panels. If necessary, slacken the hinge bolts and re-align the bonnet to suit. Once the bonnet is correctly aligned, tighten the relevant hinge bolts securely.

9 Once the bonnet is correctly aligned, check that the bonnet fastens and releases in a satisfactory manner. If adjustment is necessary, slacken the bonnet lock retaining bolts, and adjust the position of the lock to suit. Once the lock is operating correctly, securely tighten its retaining bolts.

10 If necessary, align the front edge of the bonnet with the wing panels by turning the rubbers screwed into the body front panel, to raise or lower the front edge as required.

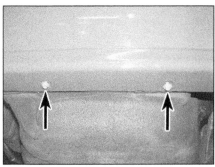

7.4 Undo the two lower centre bumper retaining bolts (arrowed)

7.5b . . . then withdraw the bumper rearwards and remove it from the vehicle

9 Bonnet release cable –
removal and refitting

Removal

1 Open and support the bonnet.

2 Remove the front grille panel as described in Section 24.

3 Unscrew the three securing bolts, and remove the lock assembly from the body panel.

4 Pull the return spring to one side, then unhook the end of the bonnet release cable from the lock lever **(see illustration)**. Withdraw the lock assembly from the vehicle.

5 Where applicable, unscrew the cable securing clip from the front body panel.

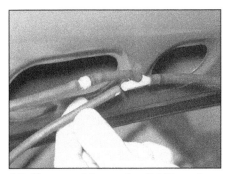

8.2 Disconnect the washer fluid hose from the bonnet connector

8.3 Unscrew the bolts (arrowed) securing the hinges to the bonnet

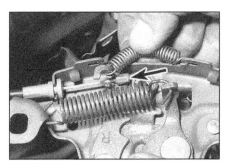

9.4 Pull the return spring to one side, and unhook the end of the bonnet release cable

6 Working in the driver's footwell, remove the securing screw from the footwell side trim panel. Release the plastic securing clip, then pull the weatherstrip from the edge of the panel. Pull the panel from the footwell to release the remaining securing clips.

7 Unscrew the two now-exposed securing bolts, and withdraw the bonnet release lever.

8 Note the routing of the cable, and release it from any clips in the engine compartment, then feed the cable through the bulkhead grommet into the passenger compartment. On some models, it may be necessary to move certain components in the engine compartment to one side, to gain access to the cable clips. It is advisable to tie a length of string to the end of the cable before removal, to aid refitting. Pull the cable through into the passenger compartment, then untie the string and leave it place until the cable is to be refitted.

Refitting

9 Refitting is a reversal of removal, but use the string to pull the cable into position, and ensure that the bulkhead grommet is securely located. Make sure that the cable is routed as noted before removal, and reposition the cable in its securing clips in the engine compartment. Check the bonnet release mechanism for correct operation on completion.

10 Bonnet lock – removal and refitting

Removal

1 Open and support the bonnet.

2 Remove the front grille panel as described in Section 24.

3 Unscrew the three securing bolts, and remove the lock assembly from the body panel **(see illustration)**.

4 Pull the return spring to one side, then unhook the end of the bonnet release cable from the lock lever, and withdraw the assembly from the vehicle.

11.5 Remove the roll-pin (arrowed) securing the door check strap to the body bracket

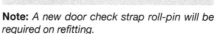

10.3 Unscrew the three bolts (arrowed), and remove the lock assembly from the body panel

Refitting

5 Refitting is a reversal of removal. If necessary, adjust the position of the lock, as described in Section 8.

11 Door – removal, refitting and adjustment

Note: *A new door check strap roll-pin will be required on refitting.*

Removal

1 Disconnect the battery negative terminal (refer to *Disconnecting the battery* in the Reference Chapter).

2 Remove the door inner trim panel, as described in Section 12.

3 Disconnect all relevant wiring from the components inside the door, and unclip the wiring harnesses from inside the door. Take careful note of the way the wire is routed, to aid refitting.

4 Carefully feed the wiring through the aperture in the front edge of the door (pull out the grommet if necessary).

5 Using a suitable punch, drive out the roll-pin securing the door check strap to the body bracket **(see illustration)**.

6 Mark the positions of the hinges on the door, to aid alignment of the door on refitting.

7 Have an assistant support the door, then unscrew the bolts securing the door hinges to

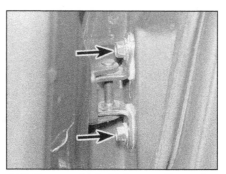

11.7 Unscrew the bolts (arrowed) securing the door hinges to the door

the door, and lift the door from the vehicle **(see illustration)**.

8 Examine the hinges for wear and damage. If necessary, the hinges can be unbolted from the body and renewed.

Refitting

9 Refitting is a reversal of removal, but align the hinges with the marks made on the body before removal, and before finally tightening the hinge securing bolts, check the door adjustment as described in the following paragraphs. Use a new roll-pin to secure the door check strap to the body bracket.

Adjustment

10 Close the door (**carefully**, in case the alignment is incorrect, which may cause scratching on the door or the body as the door is closed), and check the fit of the door with the surrounding panels.

11 If adjustment is required, loosen the hinge securing bolts (the hinge-to-door and the hinge-to-body bolt holes are elongated), and move the hinges as required to achieve satisfactory alignment. Tighten the securing bolts to the specified torque when the alignment is satisfactory.

12 Check the operation of the door lock. If necessary, slacken the securing bolts, and adjust the position of the lock striker on the body pillar to achieve satisfactory alignment. Tighten the bolts to the specified torque on completion.

12 Door inner trim panel – removal and refitting

Front door trim panel

Removal

1 Disconnect the battery negative terminal (refer to *Disconnecting the battery* in the Reference Chapter).

2 On models with manual windows, use a length of bent wire to hook out the regulator handle wire clip (pass the wire down behind the handle, and press back the plastic trim plate) until the handle can be withdrawn. Recover the plastic trim plate.

3 On early models with a detachable full-length arm rest finisher, carefully prise up the finisher to release the retaining clips. Withdraw the finisher and, where applicable, disconnect the wiring connector from the electric window switch assembly. Undo the trim panel retaining bolt now exposed, located towards the rear of the arm rest finisher location.

4 On later models prise up the front end of the electric window switch panel then disengage the hook at the rear. Withdraw the switch panel and disconnect the wiring connector **(see illustrations)**.

5 Carefully prise up the front end of the interior door handle surround and remove the surround **(see illustrations)**.

12.4a Prise up the front end of the electric window switch panel then disengage the hook at the rear

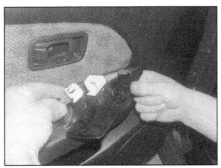

12.4b Withdraw the panel and disconnect the wiring connectors

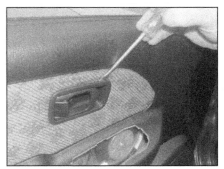

12.5a Prise up the front end of the interior door handle surround . . .

12.5b . . . then remove the surround

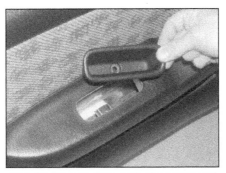

12.6 On later models, undo the screw and lift out the door pull handle

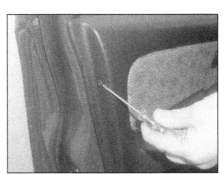

12.8 Prise out the trim panel retaining clips . . .

6 On later models, undo the screw located at the base of the door pull handle and lift out the handle **(see illustration)**.

7 Undo the screws securing the lower edge of the trim panel to the door.

8 Carefully prise out the remaining clips securing the trim panel to the door. Two clips are located along the front edge of the panel and two along the rear edge **(see illustration)**.

9 Using a suitable forked tool, release the internal securing clips around the edge of the trim panel, then lift the panel upwards to release the central locating pawl **(see illustration)**. Note that the lower window aperture weatherstrip is integral with the trim panel, and must be released from the door as the panel is withdrawn.

10 If work is to be carried out on the door internal components, it will be necessary to remove the plastic sealing sheet and stiffener panel from the inside of the door, as follows.

11 Undo the screws and remove the trim panel mounting bracket from the stiffener panel **(see illustration)**.

12 Remove the loudspeaker from the door, with reference to Chapter 12 if necessary.

13 Using a sharp knife, carefully release the sealant bead and pull the plastic sealing sheet from the door **(see illustration)**. Try to keep the sealant intact as far as possible, to ease refitting.

14 Undo the four bolts and remove the stiffener panel from the door aperture **(see illustration)**. Where applicable. Release the wiring from the clips on the panel.

Refitting

15 Refitting is a reversal of removal, bearing in mind the following points:

a) *Ensure that the sealing sheet is correctly*

12.9 . . . then release the internal clips and lift the trim panel from the door

12.13 Carefully release the sealant bead and pull the plastic sealing sheet from the door

refitted, and sealed around its edge. It should be possible to use the original mastic sealant, but if necessary, new sealant can be obtained from a Nissan dealer.

12.11 Undo the screws (arrowed) and remove the trim panel mounting bracket

12.14 Remove the stiffener panel from the door aperture

12.17 Extract the retaining clip and pull off the manual window regulator handle

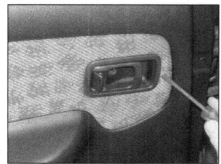

12.19 Carefully prise off the interior door handle surround

12.20 Undo the screw and lift out the door pull handle

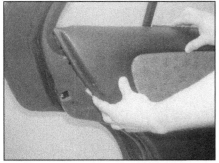

12.23 Release the internal clips and lift the trim panel from the door

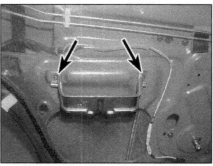

12.25 Undo the screws (arrowed) and withdraw the trim panel mounting bracket

12.26 Carefully release the sealant bead and pull the plastic sealing sheet from the door

b) *Make sure that the trim panel central securing pawl engages correctly with the bracket, and that the weatherstrip engages securely with the edge of the door and the trim panel as the panel is refitted.*

c) *On models with manual windows, refit the wire retaining clip to the regulator handle, then push the handle onto the regulator shaft.*

Rear door trim panel

Removal

16 Disconnect the battery negative terminal (refer to *Disconnecting the battery* in the Reference Chapter).
17 On models with manual windows, use a length of bent wire to hook out the regulator handle wire clip (pass the wire down behind the handle and press back the plastic trim plate) until the handle can be withdrawn **(see illustration)**. Recover the plastic trim plate.
18 On models with electric windows, prise up the front end of the electric window switch panel then disengage the hook at the rear. Withdraw the switch panel and disconnect the wiring connector.
19 Carefully prise up the front end of the interior door handle surround, disengage the hook at the rear and remove the surround **(see illustration)**.
20 Undo the screw located at the base of the door pull handle and lift out the handle **(see illustration)**.
21 Undo the screws securing the lower edge of the trim panel to the door.

22 Carefully prise out the remaining clips securing the trim panel to the door. Two clips are located along the front edge of the panel and two along the rear edge.
23 Using a suitable forked tool, release the internal securing clips around the edge of the trim panel, then lift the panel upwards to release the central locating pawl **(see illustration)**. Note that the lower window aperture weatherstrip is integral with the trim panel, and must be released from the door as the panel is withdrawn.
24 If work is to be carried out on the door internal components, it will be necessary to remove the plastic sealing sheet as follows.
25 Remove the two securing screws, then withdraw the trim panel mounting bracket from the door **(see illustration)**.
26 Using a sharp knife, carefully release the sealant bead and pull the plastic sealing sheet from the door **(see illustration)**. Try to keep

13.3 Undo the retaining screw, then slide the interior handle towards the front of the door to remove

the sealant intact as far as possible, to ease refitting.

Refitting

27 Refitting is a reversal of removal, bearing in mind the following points:
a) *Ensure that the sealing sheet is correctly refitted, and sealed around its edge. It should be possible to use the original mastic sealant, but if necessary, new sealant can be obtained from a Nissan dealer.*
b) *Make sure that the weatherstrip engages securely with the edge of the door and the trim panel as the panel is refitted.*
c) *On models with manual windows, refit the wire retaining clip to the regulator handle, then push the handle onto the regulator shaft.*

---

13 Door handle and lock components – removal and refitting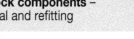

Interior door handle

Removal

1 Remove the door inner trim panel and plastic sealing sheet as described in Section 12.
2 Release the plastic clip securing the lock operating rods to the door panel.
3 Remove the screw securing the front of the interior handle to the door, then slide the handle towards the front of the door to release the rear securing lugs **(see illustration)**.

4 Withdraw the handle assembly from the door, then release the securing clips, and disconnect the lock operating rods, noting their routing.

Refitting

5 Refitting is a reversal of removal, bearing in mind the following points:
a) *Ensure that the lock operating rods are routed as noted before removal.*
b) *Check the operation of the handle/lock mechanism before refitting the door inner trim panel.*
c) *Refit the door inner trim panel with reference to Section 12.*

Front door exterior handle

Removal

6 With the window fully raised, remove the door inner trim panel, plastic sealing sheet and stiffener panel as described in Section 12.
7 Remove the bolt securing the window rear guide channel to the rear edge of the door. Pull the channel down and remove it from the door to improve access to the exterior handle rear securing nut.
8 Unscrew the two bolts securing the metal shield to the inside of the door, and remove the shield for improved access to the handle assembly **(see illustrations)**.
9 Disconnect the lock operating rod from the exterior handle and lift the rod from the door.
10 Working through the door aperture, unscrew the two handle securing nuts **(see illustration)**.
11 Recover the metal shield plate from the handle rear stud, then lift handle assembly from the outside of the door **(see illustration)**.

Refitting

12 Refitting is a reversal of removal, bearing in mind the following points:
a) *Ensure that the lock operating rod is correctly reconnected, and that the metal shield and shield plate are correctly refitted.*
b) *Check the operation of the handle/lock mechanism before refitting the door inner trim panel.*
c) *Refit the door inner trim panel with reference to Section 12.*

Rear door exterior handle

Removal

13 With the window fully raised, remove the door inner trim panel and the plastic sealing sheet as described in Section 12.
14 Remove the interior handle and the door lock assembly, as described elsewhere in this Section.
15 Working through the door aperture, remove the two securing nuts, then withdraw the exterior handle assembly from the door.

Refitting

16 Refitting is a reversal of removal, but refit the interior handle and the door lock assembly

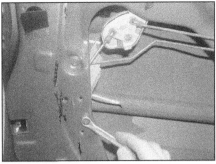

13.8a Unscrew the two bolts . . .

13.10 Unscrew the two securing nuts . . .

as described in the relevant paragraphs of this Section, and refit the door inner trim panel with reference to Section 12.

Front door lock cylinder

Removal

17 With the window fully raised, remove the door inner trim panel, plastic sealing sheet and stiffener panel, as described in Section 12.
18 Disconnect the operating rod from the lock cylinder lever.
19 Using a suitable screwdriver or a pair of pliers, prise off the lock cylinder securing plate, then withdraw the lock cylinder from outside the door **(see illustration)**. Note that the lock cylinder securing plate has raised tangs which lock against the inner door skin – this means that some force is required to prise the plate free.

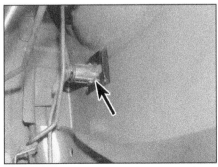

13.19 Front door lock cylinder securing plate (arrowed)

13.8b . . . and remove the metal shield

13.11 . . . and remove the front door exterior handle

Refitting

20 Refitting is a reversal of removal, bearing in mind the following points:
a) *Ensure that the lock operating rod is correctly reconnected.*
b) *Refit the door inner trim panel as described in Section 12.*

Front door lock

Removal

21 With the window fully raised, remove door inner trim panel, plastic sealing sheet and stiffener panel, as described in Section 12.
22 Remove the bolt securing the window rear guide channel to the rear edge of the door. Pull the channel down and remove it from the door **(see illustrations)**.
23 Unscrew the two bolts securing the metal shield to the inside of the door, and remove the shield.

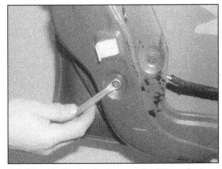

13.22a Undo the bolt securing the window rear guide channel to the door . . .

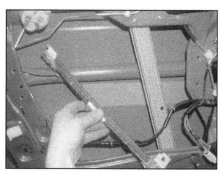

13.22b ... then pull the channel down and remove it from the door

13.25a Unscrew the lock assembly securing screws and central locking motor retaining bolt (arrowed) ...

13.25b ... disconnect the wiring and remove the front door lock assembly

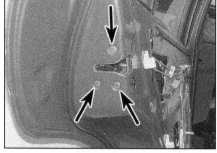

13.31 Rear door lock securing screws (arrowed) ...

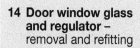

13.32 ... and central locking motor retaining bolt (arrowed)

24 Reach in through the door aperture, and disconnect the lock operating rods from the lock, noting their locations and routing.
25 Working at the rear edge of the door, unscrew the three lock securing screws, then unscrew the central locking motor retaining bolt on the inside of the door panel. Disconnect the lock motor wiring plug and withdraw the lock assembly through the door aperture **(see illustrations)**.

Refitting

26 Refitting is a reversal of removal, bearing in mind the following points:
a) *Ensure that the lock operating rods are correctly reconnected and routed, as noted before removal.*
b) *Check the operation of the handle/lock mechanism before refitting the door inner trim panel.*
c) *Refit the door inner trim panel as described in Section 12.*

Rear door lock

Removal

27 With the window fully raised, remove the door inner trim panel and the plastic sealing sheet, as described in Section 12.
28 Release the plastic clip securing the lock operating rods to the door panel.
29 Remove the screw securing the front of the interior handle to the door, then slide the handle towards the front of the door to release the rear securing lugs.
30 Withdraw the handle assembly from the door, then release the securing clips, and

disconnect the lock operating rods, noting their routing.
31 Working at the rear edge of the door, unscrew the three lock securing screws **(see illustration)**.
32 Working inside the door, remove the screw securing the central locking motor to the door panel **(see illustration)**. Disconnect the wiring plug from the central locking motor, then manipulate the lock assembly, complete with operating rods, out through the door aperture.

Refitting

33 Refitting is a reversal of removal, bearing in mind the following points:
a) *Ensure that the lock operating rods are correctly reconnected.*
b) *Check the operation of the handle/lock mechanism before refitting the door inner trim panel.*

14.3 Undo the front door window glass retaining bolts (arrowed)

c) *Refit the door inner trim panel as described in Section 12.*

14 Door window glass and regulator – removal and refitting

Front window glass

Removal

1 Remove the door inner trim panel and plastic sealing sheet as described in Section 12.
2 Temporarily reconnect the electric window switch, and the battery negative terminal, or refit the window regulator handle, as applicable.
3 Lower the window until the two bolts securing the lower edge of the window glass to the regulator mechanism are accessible through the door aperture. Support the glass, then undo the two bolts **(see illustration)**.
4 Disengage the window glass from the regulator, raise the glass at the rear and withdraw it from the outside of the door **(see illustrations)**.

Refitting

5 Refitting is a reversal of removal, bearing in mind the following points:
a) *Take care not to dislodge the weatherstrips when fitting the glass.*
b) *Check the operation of the window mechanism before refitting the door inner trim panel.*
c) *Refit the door inner trim panel with reference to Section 12.*

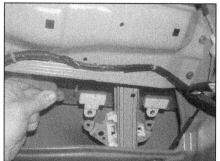

14.4a Disengage the window glass from the regulator ...

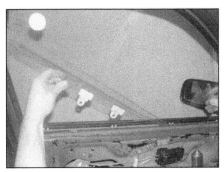

14.4b . . . raise the glass at the rear and withdraw it from outside the door

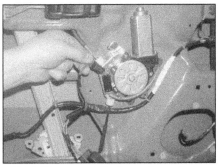

14.10 Disconnect the front door window regulator motor wiring connector

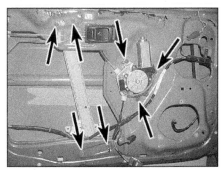

14.11a Unscrew the regulator mechanism securing bolts (arrowed) . . .

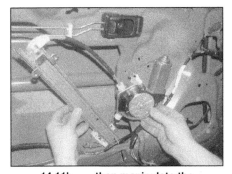

14.11b . . . then manipulate the motor/regulator assembly out through the door aperture

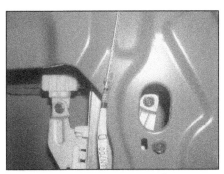

14.16 Undo the two bolts securing the rear door sliding window glass to the regulator

14.17 Lift the glass up and out through the inside of the door

Front window regulator

Removal

6 Remove the door inner trim panel, plastic sealing sheet, and stiffener panel as described in Section 12.

7 Temporarily reconnect the electric window switch, and the battery negative terminal, or refit the window regulator handle, as applicable.

8 Lower the window until the two bolts securing the lower edge of the window glass to the regulator mechanism are accessible through the door aperture. Support the glass, undo the two bolts and disengage the glass from the regulator.

9 Fully raise the window glass, and secure the glass in position using suitable tape, or by wedging the glass in position using rags between the glass and the edge of the door – ensure that the glass cannot drop into the door. Alternatively, lift the glass panel out through the window aperture.

10 Where applicable, disconnect the regulator motor wiring connector **(see illustration)**.

11 Unscrew the three bolts securing the motor assembly to the door **(see illustration)**.

12 Unscrew the two upper and the two lower regulator mechanism securing bolts, then manipulate the complete motor/regulator assembly out through the aperture in the door.

13 Refitting is a reversal of removal, bearing in mind the following points:

a) Check the operation of the window mechanism before refitting the door inner trim panel.

b) Refit the door inner trim panel with reference to Section 12.

Rear sliding window glass

Removal

14 Remove the door inner trim panel and plastic sealing sheet as described in Section 12.

15 Temporarily reconnect the electric window switch, and the battery negative terminal, or refit the window regulator handle, as applicable.

16 Lower the window until the two bolts securing the lower edge of the window glass to the regulator mechanism are accessible through the door aperture. Support the glass, then undo the two bolts **(see illustration)**.

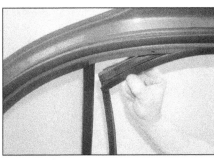

14.20 Pull the weatherstrip from the corner of the window aperture and rear guide channel

17 Disengage the window glass from the regulator and rear guide channel, then raise the glass and withdraw it from the inside of the door **(see illustration)**.

Refitting

18 Refitting is a reversal of removal, bearing in mind the following points:

a) Take care not to dislodge the weatherstrips when fitting the glass.

b) Check the operation of the window mechanism before refitting the door inner trim panel.

c) Refit the door inner trim panel with reference to Section 12.

Rear fixed window glass

Removal

19 Remove the sliding window glass, as described previously in this Section.

20 Carefully pull the weatherstrip from the top rear corner of the window aperture and pull it up and out of the rear guide channel **(see illustration)**.

21 Ease the door sealing weather strip from the top of the door frame in the area around above the rear guide channel. Undo the screw now exposed securing the rear guide channel to the top of the door frame **(see illustration)**.

22 Undo the two bolts securing the lower part of the rear guide channel to the door **(see illustration)**.

23 Lower the rear guide channel to disengage the upper mounting, then withdraw the guide channel from the door **(see illustrations)**.

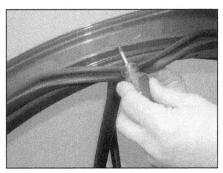

14.21 Undo the screw securing the rear guide channel to the top of the door frame

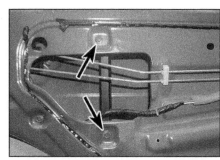

14.22 Undo the two bolts (arrowed) securing the lower part of the rear guide channel to the door

14.23a Disengage the guide channel upper mounting . . .

14.23b . . . then withdraw the guide channel from the door

14.24 Withdraw the fixed window glass from the door aperture

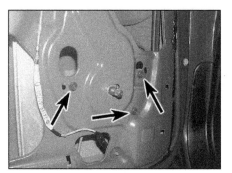

14.32 Rear door window regulator motor securing bolts (arrowed)

24 Carefully pull the fixed window glass, complete with its surrounding weatherstrip, from the door aperture **(see illustration)**.
25 If desired, pull the weatherstrip from the edge of the glass.

Refitting

26 Refitting is a reversal of removal, but ensure that the weatherstrip seals securely on both the window aperture and the window glass, and refit the sliding window glass as described previously in this Section.

Rear window regulator

Removal

27 Remove the door inner trim panel and the plastic sealing sheet as described in Section 12.
28 Temporarily reconnect the electric window switch, and the battery negative terminal, or refit the window regulator handle, as applicable.

29 Lower the window until the two bolts securing the lower edge of the window glass to the regulator mechanism are accessible through the door aperture. Support the glass, undo the two bolts and disengage the glass from the regulator.
30 Fully raise the window glass, and secure the glass in position using suitable tape, or by wedging the glass in position using rags between the glass and the edge of the door – ensure that the glass cannot drop into the door. Alternatively, lift the glass panel out through the window aperture.
31 Where applicable, disconnect the regulator motor wiring connector.
32 Unscrew the three bolts securing the motor assembly to the door **(see illustration)**.
33 Unscrew the two upper and the two lower regulator mechanism securing bolts, then manipulate the complete motor/regulator

assembly out through the aperture in the door **(see illustrations)**.

Refitting

34 Refitting is a reversal of removal, bearing in mind the following points:
a) *Where applicable, refit the window glass as described previously in this Section.*
b) *Check the operation of the window mechanism before refitting the door inner trim panel.*
c) *Refit the door inner trim panel with reference to Section 12.*

15 Boot lid –
removal, refitting
and adjustment

Removal

1 Disconnect the battery negative terminal (refer to *Disconnecting the battery* in the Reference Chapter).
2 Working inside the luggage compartment, disconnect the boot lid wiring harness connector, located on the right-hand side of the luggage compartment. Release the cable ties securing the harness to the boot lid hinge.
3 Using a pencil or felt tip pen, mark the outline of each boot hinge relative to the boot lid, to use as a guide on refitting.
4 Have an assistant support the boot lid, then unscrew the bolts securing the hinges to the boot lid **(see illustration)**. Lift the boot lid from the vehicle – take care not to scratch the bodywork as the boot lid is removed.

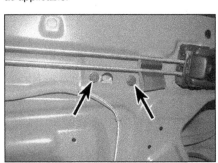

14.33a Rear door window regulator mechanism upper securing bolts (arrowed) . . .

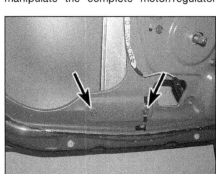

14.33b . . . and lower securing bolts (arrowed)

Refitting

5 With the aid of an assistant, offer up the boot lid, and loosely fit the retaining bolts. Align the hinges with the marks made on removal, then tighten the retaining bolts securely.

6 Reconnect the wiring harness and secure it to the hinges using new cable ties.

7 Adjust the alignment of the boot lid as follows.

Adjustment

8 Close the boot lid (**carefully**, in case the alignment is incorrect, which may cause scratching on the lid or the body as the boot lid is closed), and check for alignment with the adjacent panels. If necessary, slacken the hinge bolts and re-align the boot lid to suit. Once the boot lid is correctly aligned, tighten the hinge bolts securely.

9 Once the boot lid is correctly aligned, check that the boot lid fastens and releases in a satisfactory manner. If adjustment is necessary, slacken the boot lid striker retaining bolts, and adjust the position of the striker to suit. Once the lock is operating correctly, securely tighten the striker retaining bolts.

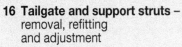

16 Tailgate and support struts – removal, refitting and adjustment

Tailgate

Removal

1 Disconnect the battery negative terminal (refer to *Disconnecting the battery* in the Reference Chapter).

2 Remove the tailgate interior trim panels as described in Section 27.

3 Working inside the tailgate, disconnect the wiring plugs from the tailgate wiper motor, the heated rear window element, the rear number plate lights, the rear light clusters, and the high-level stop-light. Also unbolt the earth lead(s). Check for any other wiring connectors which must be disconnected to facilitate tailgate removal, then release the harness securing cable ties.

4 Tie a length of string to the wiring harness, then prise the wiring harness grommet from the front corner of the tailgate. Feed the wiring harness through the aperture in the tailgate. Untie the string from the wiring harness, and leave the string in place in the tailgate, to aid refitting.

5 Remove the tailgate washer nozzle, as described in Chapter 12, Section 15, then tie a length of string to the fluid hose, and repeat the procedure carried out on the wiring harness.

6 Have an assistant support the tailgate in the open position.

7 Unscrew two bolts each side securing the support strut upper mounting plates to the tailgate (**see illustration**).

8 Unscrew the bolts securing the hinges to

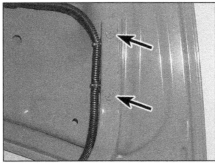

15.4 Boot lid hinge-to-boot lid retaining bolts (arrowed)

the tailgate (**see illustration**), then lift the tailgate from the vehicle.

Refitting

9 Refitting is a reversal of removal, bearing in mind the following points.

10 Tie the string to the wiring harness, and use the string to pull the wiring harness through the aperture and into the tailgate.

11 Do not fully tighten the hinge bolts until the tailgate adjustment has been checked, as described in the following paragraphs.

Adjustment

12 Close the tailgate (**carefully**, in case the alignment is incorrect, which may cause scratching on the tailgate or the body as the tailgate is closed), and check for alignment with the adjacent panels. If adjustment is required, it will be necessary to prise off the rear plastic end trim from the headlining to gain access to the tailgate hinge-to-body

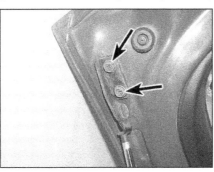

16.7 Tailgate support strut upper mounting plate bolts (arrowed)

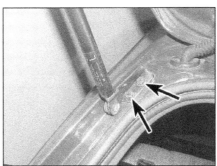

16.16 Tailgate support strut lower mounting plate bolts (arrowed)

retaining nuts. The end trim is secured internally by three metal clips.

13 Slacken the hinge retaining nuts and re-align the tailgate to suit. Once the tailgate is correctly aligned, tighten the hinge retaining nuts fully and refit the trim.

14 Once the tailgate is correctly aligned, check that the tailgate fastens and releases in a satisfactory manner. If adjustment is necessary, slacken the tailgate lock retaining bolts, and adjust the position of the lock to suit. Once the lock is operating correctly, securely tighten its retaining bolts.

Support struts

Removal

15 To remove a strut, first ensure that the tailgate is adequately supported.

16 Unscrew two bolts securing the support strut upper mounting plates to the tailgate, and the two bolts securing the lower mounting plates to the body (**see illustration**). Lift away the support strut.

Refitting

17 Refitting is a reversal of removal.

17 Boot lid lock components – removal and refitting

Boot lid lock

Removal

1 Open the boot lid and unclip the plastic cover from the lock (**see illustration**).

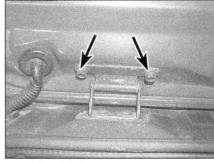

16.8 Tailgate hinge-to-tailgate retaining bolts (arrowed)

17.1 Unclip the cover from the boot lid lock . . .

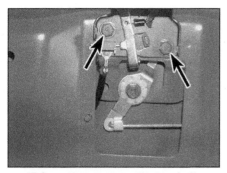

17.2 ... then unscrew the two bolts (arrowed) and remove the lock

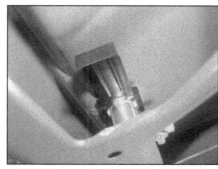

17.5a Remove the boot lid lock cylinder securing plate ...

17.5b ... and remove the lock cylinder

17.7a Extract the retaining clips ...

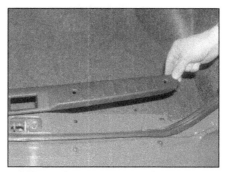

17.7b ... then release the trim panel and remove it from the luggage compartment

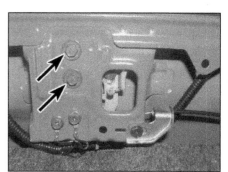

17.8 Boot lid lock striker securing bolts (arrowed)

2 Unscrew the two securing bolts, and withdraw the lock from the boot lid **(see illustration)**. Disconnect the wiring connectors and remove the lock.

Refitting

3 Refitting is a reversal of removal.

Boot lid lock cylinder

Removal

4 Open the boot lid, release the securing clip, and disconnect the lock operating rod from the lock cylinder lever.
5 Using a suitable pair of pliers, pull out the lock cylinder securing plate, then withdraw the lock cylinder from outside the boot lid **(see illustrations)**.

Refitting

6 Refitting is a reversal of removal, but ensure that the securing clip is securely refitted.

Boot lid lock striker

Removal

7 Open the boot lid, undo the retaining clip centre screws then extract the clips securing the plastic trim panel at the rear of the luggage compartment. Release the trim panel and remove it from the luggage compartment **(see illustrations)**.
8 Unscrew the two securing bolts **(see illustration)**, withdraw the lock striker assembly, and disconnect the boot lid release cable from the lever on the striker assembly.

Refitting

9 Refitting is a reversal of removal, but check the operation of the boot lid release mechanism, as follows, before refitting the luggage compartment trim panel.
10 Check that the boot lid fastens and releases in a satisfactory manner. If adjust-

ment is necessary, slacken the striker retaining bolts, and adjust the position of the striker to suit. Once the lock is operating correctly, securely tighten the striker retaining bolts.
11 Check that the boot lid opens satisfactorily using the interior release handle. If necessary, slacken the retaining bolt and reposition the release cable support bracket located alongside the striker in the luggage compartment. When the release mechanism is operating correctly, securely tighten the release cable support bracket bolt.

18 Tailgate lock components – removal and refitting

Tailgate lock

Removal

1 Disconnect the battery negative terminal (refer to *Disconnecting the battery* in the Reference Chapter).
2 Remove the tailgate inner trim panel as described in Section 27.
3 Detach the luggage compartment light switch wiring connector from the tailgate, and separate the two halves of the connector (the light switch is integral with the lock assembly) **(see illustration)**.
4 Disconnect the lock operating rod from the lock lever.
5 Unscrew the two lock securing bolts and manipulate the lock out through the aperture in the tailgate **(see illustrations)**.

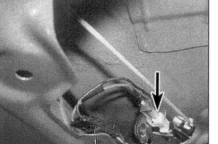

18.3 Disconnect and detach the wiring connector (arrowed) from the tailgate

18.5a Unscrew the tailgate lock securing bolts (arrowed) ...

Refitting

6 Refitting is a reversal of removal, but before refitting the tailgate trim panel, check that the tailgate fastens and releases in a satisfactory manner. If adjustment is necessary, slacken the tailgate lock retaining bolts, and adjust the position of the lock to suit. Once the lock is operating correctly, securely tighten its retaining bolts.

Tailgate lock cylinder

Removal

7 Remove the tailgate inner trim panel as described in Section 27.
8 Disconnect the lock operating rod from the lock cylinder **(see illustration)**.
9 Using a suitable pair of pliers, pull out the lock cylinder securing plate, then withdraw the lock cylinder from outside the tailgate.

Refitting

10 Refitting is a reversal of removal.

Tailgate lock striker

Removal

11 Remove the luggage compartment parcel shelf support panels and the tailgate aperture trim panel as described in Section 27.
12 Using a suitable forked tool, prise out the plastic clips securing the rear carpet trim to the body and move the carpet trim clear of the striker assembly.
13 Unscrew the two securing bolts, then withdraw the lock striker assembly, and disconnect the tailgate release cable from the lever on the striker assembly.

Refitting

14 Refitting is a reversal of removal, but check the operation of the tailgate release mechanism, as follows, before refitting the trim panel.
15 Check that the tailgate fastens and releases in a satisfactory manner. If adjustment is necessary, slacken the striker retaining bolts, and adjust the position of the striker to suit. Once the lock is operating correctly, securely tighten the striker retaining bolts.
16 Check that the tailgate opens satisfactorily using the interior release handle. If necessary, slacken the retaining bolt and reposition the release cable support bracket

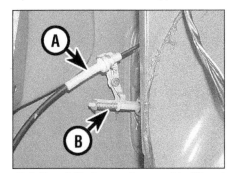

19.3 Boot lid/tailgate release cable (A) and fuel filler flap release rod (B)

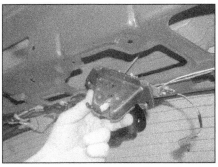

18.5b ... and manipulate the lock out through the tailgate aperture

located alongside the striker in the luggage compartment. When the release mechanism is operating correctly, securely tighten the release cable support bracket bolt and refit the trim panels.

---

19 Boot lid/tailgate and fuel filler flap release cable – removal and refitting

Note: *If the release cable breaks, or the mechanism is faulty, and the fuel filler flap cannot be released, the flap can be released manually. Proceed as described in paragraphs 1 and 2, then reach in through the body aperture, and pull the release rod to compress the spring and release the filler flap.*

Removal

1 Disconnect the end of the cable from the boot lid/tailgate lock striker, as described during removal of the lock striker in Section 17 or 18.
2 Release the rear portion of the left-hand side luggage compartment carpet trim panel, with reference to Section 27.
3 Disconnect the cable from the fuel filler flap release assembly, then feed the cable through from the boot lid/tailgate lock striker to the fuel filler flap **(see illustration)**.
4 Fold forward, or remove, the rear seat cushion for access to the cable.
5 Working on the driver's side of the vehicle, for access to the tailgate/fuel filler flap release lever, pull the sill trim panel from the sill, then pull back the carpet panel to expose the lever securing bolt.
6 Unscrew the securing bolt, then manipulate the lever assembly out through the aperture in the carpet, and disconnect the cable from the lever.
7 The cable can now be withdrawn from the vehicle, but take careful note of the cable routing, to aid refitting.

Refitting

8 Refitting is a reversal of removal, but ensure that the cable is routed as noted before removal, and check the operation of the release mechanism before refitting the carpet and trim panels.

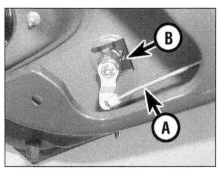

18.8 Tailgate lock cylinder operating rod (A) and securing plate (B)

---

20 Central locking system components – removal and refitting

Electronic control unit

Removal

1 The electronic control unit is located under the facia, adjacent to the steering column.
2 Disconnect the battery negative terminal (refer to *Disconnecting the battery* in Reference).
3 Remove the driver's side lower facia panel as described in Section 29.
4 Remove the two securing screws, then withdraw the unit, complete with the mounting bracket, and disconnect the wiring plugs.

Refitting

5 Refitting is a reversal of removal.

Door lock switches

Removal

6 Disconnect the battery negative terminal (refer to *Disconnecting the battery* in Reference).
7 Undo the retaining screw and withdraw the relevant switch from the door pillar **(see illustration)**.
8 Disconnect the wiring connector and remove the switch.

> **HAYNES HiNT** *Tape the wiring to the door pillar, to prevent it falling back into the pillar. Alternatively, tie a piece of string to the wiring to retrieve it.*

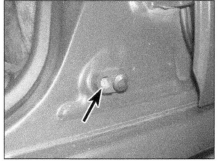

20.7 Door lock switch retaining screw (arrowed)

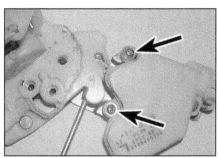

20.11 Undo the two screws (arrowed) to separate the central locking motor from the door lock

Refitting

9 Refitting is a reversal of removal.

Door lock motor

Removal

10 Remove the door lock as described in Section 13.
11 The motor is secured to the lock assembly by two screws **(see illustration)**.

Refitting

12 Refit the door lock as described in Section 13.

Remote control battery renewal

13 Undo the retaining screw and carefully prise the two halves of the transmitter apart and remove the two batteries.
14 Fit the new batteries, observing the correct polarity, and clip the transmitter back together. Refit the retaining screw.

21.3 Carefully prise off the door mirror interior trim panel

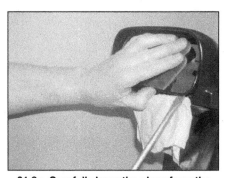

21.8a Carefully lever the glass from the mirror . . .

21 Exterior mirror and associated components – removal and refitting

Mirror assembly

Removal

1 If working on an electric mirror, disconnect the battery negative terminal (refer to *Disconnecting the battery* in Reference).
2 Remove the door inner trim panel, as described in Section 12.
3 Carefully prise off the mirror interior trim panel to release the three internal retaining clips **(see illustration)**.
4 Peel back the plastic sealing sheet from the front upper corner of the door.
5 Reach through the aperture in the door, and disconnect the mirror wiring connector. Unclip the mirror wiring from the door, noting its routing.
6 Remove the three securing screws, and withdraw the mirror from the outside of the door **(see illustration)**.

Refitting

7 Refitting is a reversal of removal, ensuring that the mirror wiring is routed as noted before removal.

Mirror glass

Removal

8 Insert a suitable thin plastic or wooden tool between the mirror glass and the mirror body, and lever the glass to release the securing

21.6 Remove the three screws (arrowed), and withdraw the mirror from the door

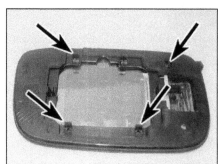

21.8b . . . to release the internal clips (arrowed)

clips **(see illustrations)**. Once the clips are released, remove the mirror glass from the mirror body.

Refitting

9 Align the retaining clips on the glass with the lugs on the mirror body, then push the glass fully into position.

22 Windscreen, tailgate glass and fixed windows – general information

These areas of glass are secured by the tight fit of the weatherstrip in the body aperture, and are bonded in position with a special adhesive. Renewal of such fixed glass is a difficult, messy and time-consuming task, which is considered beyond the scope of the home mechanic. It is difficult, unless one has plenty of practice, to obtain a secure, waterproof fit. Furthermore, the task carries a high risk of breakage; this applies especially to the laminated glass windscreen. In view of this, owners are strongly advised to have this sort of work carried out by one of the many specialist windscreen fitters.

For those possessing the necessary skills and equipment to carry out this task, some preliminary removal of the vehicle interior trim and associated components is necessary, as follows, referring to the procedures contained in the Sections and Chapters indicated.

Windscreen

1 Remove both wiper arms (Chapter 12).
2 Remove the windscreen scuttle grille panel (Section 24 of this Chapter).
3 Remove the front pillar trim on both sides (Section 27 of this Chapter).
4 Remove the sun visors.

Rear fixed side window glass

5 Remove the relevant luggage compartment trim panels (Section 27 of this Chapter).

Tailgate window glass

6 Remove the tailgate trim panels (Section 27 of this Chapter).

23 Sunroof – general information

1 Due to the complexity of the sunroof mechanism, considerable expertise is needed to repair, renew or adjust the sunroof components successfully. Removal of the roof first requires the headlining to be removed, which is a complex and tedious operation, and not a task to be undertaken lightly. Therefore, any problems with the sunroof should be referred to a Nissan dealer.
2 On models with an electric sunroof, if the sunroof motor fails to operate, first check the relevant fuse. If the fault cannot be traced and rectified, the sunroof can be opened and

closed manually, using the special crank handle supplied in the vehicle tool kit, to turn the motor spindle.

3 To gain access to the motor spindle, ensure that the ignition key is in the 'off' position, then carefully prise the overhead console from its location. Disconnect the wiring connector and remove the console. Engage the crank handle with the spindle, and turn the handle clockwise to close the sunroof.

4 Once the roof is closed, remove the crank handle, reconnect the wiring connector and clip the overhead console back into place.

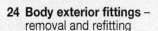

24 Body exterior fittings – removal and refitting

Front grille panel

1 Open the bonnet and support the bonnet.

2 Using a screwdriver, depress the retaining tangs of the three grille panel retaining clips, one at a time, while at the same time pulling the grille forward **(see illustration)**.

3 Once the three retaining clips are released, lift the grille panel up and remove it from its location **(see illustration)**.

4 To refit, locate the panel in position and push the retaining clips into their locations until they lock into place.

Windscreen scuttle grille panel

5 Open and support the bonnet.

6 Remove the windscreen wiper arms as described in Chapter 12.

7 Pull the weatherstrip from the front edge of the scuttle grille panel to release the retaining clips **(see illustration)**.

8 Remove the securing screws and plastic clips, and lift off the two halves of the scuttle grille panel **(see illustration)**.

9 Refitting is a reversal of removal.

Wheel arch liners

10 The wheel arch liners are secured by a combination of self-tapping screws and plastic clips, and removal is self-evident.

Body trim strips and badges

11 The various body trim strips and badges are held in position with a special adhesive

25.4 Remove the trim cover for access to the seat rail rear securing bolts

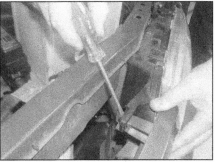

24.2 Depress the retaining tangs of the grille panel retaining clips . . .

24.7 Pull the weatherstrip from the front edge of the windscreen scuttle grille panel

tape. Removal requires the trim/badge to be heated, to soften the adhesive, and then cut away from the surface. Due to the high risk of damage to the vehicle paintwork during this operation, it is recommended that this task should be entrusted to a Nissan dealer.

25 Seats – removal and refitting

Front seat

> **Warning: Certain later models are equipped with side airbags built into the outer sides of the front seats. Refer to Chapter 12 for the precautions which should be observed when dealing with an airbag system. Do not tamper with the airbag unit in any way, and do not attempt to test any airbag system components. Note that the airbag is triggered if the mechanism is supplied with an electrical current (including via an ohmmeter), or if the assembly is subjected to a temperature of greater than 100°C.**

Removal

1 On models with side airbags in the front seats, de-activate the airbag system as described in Chapter 12 before attempting to remove the seat.

2 Move the seat fully forwards.

3 Unclip the trim cover from the rear corner of the seat outer rail.

24.3 . . . then lift the grille panel up and remove it from its location

24.8 Remove the screws and clips, then lift off the two halves of the scuttle grille panel

4 Remove the securing screw, and remove the trim cover from the rear corner of the seat inner rail **(see illustration)**.

5 Unscrew the seat rail rear securing bolts.

6 Slide the seat fully rearwards, then unscrew the seat rail front securing bolts **(see illustration)**.

7 Where applicable, disconnect the wiring connectors, then lift the seat, complete with the rails, from the vehicle.

Refitting

8 Refitting is a reversal of removal, but tighten the securing bolts to the specified torque.

Rear seat back

Removal

9 Fold the relevant rear seat back section forwards, then lift the trim at the base of the

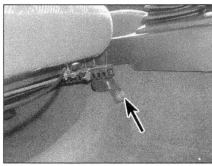

25.6 Seat rail front securing bolt (arrowed)

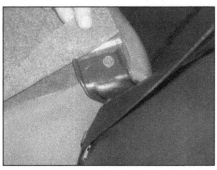

25.9 Rear seat-to-hinge retaining bolt

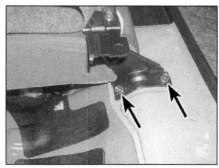

25.10 Rear seat hinge-to-floor retaining bolts (arrowed)

seat back for access to the seat-to-hinge retaining bolt (see illustration).

10 Undo the seat-to-hinge retaining bolt, or the four nuts or bolts, as applicable (two on each side) securing the hinges to the floor, as desired, then remove the seat back from the vehicle (see illustration).

Refitting

11 Refitting is a reversal of removal.

Rear seat cushion

Removal

12 On Hatchback models, unclip the trim cover from the base of the cushion hinges, unscrew the bolts securing the hinges to the floor, and remove the seat cushion, complete with the hinges, from the vehicle.

13 On Saloon models, lift the seat cushion upwards to disengage the locating pegs and remove the cushion from the vehicle.

26.4a Immobilise the front seat belt mechanical pretensioner unit by moving the safety lever . . .

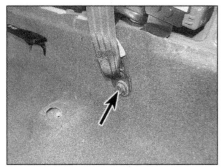

26.5 Front seat belt lower anchor bolt (arrowed)

Refitting

14 Refitting is a reversal of removal.

26 Seat belt components – removal and refitting

Note: *Record the positions of the washers and spacers on the seat belt anchors, and ensure they are refitted in their original positions.*

Front seat belt removal

> ⚠ **Warning: On certain models, the front seat belt inertia reels are equipped with a mechanical or pyrotechnic pretensioner mechanism. Refer to the airbag system precautions contained in Chapter 12 which apply equally to the seat belt pretensioners. Do**

26.4b . . . to the vertical position

26.6 Front seat belt inertia reel upper mounting bolt (arrowed)

not tamper with the inertia reel pre-tensioner unit in any way, and do not attempt to test the unit.

Saloon and 5-door Hatchback

1 De-activate the airbag system (which will also de-activate the pyrotechnic pretensioner mechanism, where fitted) as described in Chapter 12 before attempting to remove the seatbelt.

2 Remove the relevant front seat as described in Section 25.

3 Remove the centre pillar trim panels as described in Section 27.

4 On models with mechanical seat belt pretensioners, immobilise the pretensioner unit by moving the safety lever to the vertical position. Ensure that the safety lever engages with the detent and locks in position before proceeding (see illustrations).

5 Unscrew the seat belt lower anchor bolt, noting that the bolt also secures the inertial reel (see illustration).

6 Undo the inertia reel upper mounting bolt and remove the seat belt assembly from the vehicle (see illustration).

3-door Hatchback

7 De-activate the airbag system (which will also de-activate the pyrotechnic pretensioner mechanism, where fitted) as described in Chapter 12 before attempting to remove the seatbelt.

8 Remove the relevant front and rear seat as described in Section 25.

9 Remove the lower side trim panel as described in Section 27.

10 On models with mechanical seat belt pretensioners, immobilise the pretensioner unit by moving the safety lever to the vertical position. Ensure that the safety lever engages with the detent and locks in position before proceeding (see illustrations 26.4a and 26.4b).

11 Undo the two bolts and remove the seat belt lower anchor rail.

12 Undo the two bolts securing the upper belt guide to the door pillar.

13 Undo the two inertia reel retaining bolts and remove the seat belt assembly from the vehicle.

Front seat belt refitting

14 Refitting is a reversal of removal, ensuring that all mounting bolts are tightened to the specified torque. On models with mechanical seat belt pretensioners, prior to refitting the trim panel(s), return the safety lever to its original position to unlock the pretensioner unit.

Front seat belt stalk

Removal

15 Carefully prise off the plastic trim panel from the inner edge of the seat to expose the seat belt stalk anchor bolt.

16 Unscrew the bolt and withdraw the stalk assembly.

Refitting

17 Refitting is a reversal of removal, but tighten the stalk anchor bolt to the specified torque.

Rear seat belt and buckles

Removal

18 Remove the relevant rear seat as described in Section 25.

19 Working as described in Section 27, on Hatchback models, remove the relevant parcel shelf support panel, then locally pull away the luggage compartment carpet trim for access to the inertia reel. On Saloon models, remove the rear pillar trim and rear parcel shelf, then locally pull away the front of the luggage compartment carpet trim panel.

20 Unscrew the inertia reel securing bolt, the belt upper and lower anchor bolts, and withdraw the seat belt assembly from the vehicle **(see illustrations)**.

21 Undo the seat belt buckle anchor bolt(s) and remove the relevant seat belt buckles **(see illustration)**.

Refitting

22 Refitting is a reversal of removal, but tighten the seat belt anchor bolts to the specified torque.

---

27 Interior trim panels –
removal and refitting

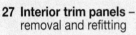

General

1 The interior trim panels are secured by a combination of metal and plastic clips and screws. When releasing certain types of securing clips, a suitable forked tool will prove invaluable to avoid damage to the panel and clips. A degree of force will be necessary to pull some of the panels from their locations, especially where numerous internal retaining clips are used. Be prepared for some of the plastic clips to break when their relevant panel is being removed.

Door inner trim

2 Refer to Section 12.

Footwell trim

3 Extract the plastic clip securing the footwell trim panel to the front door pillar. Pull the weatherstrip from the edge of the sill trim panel, lift up the sill trim panel then remove the footwell trim panel by pulling it outward to release the metal retaining clip **(see illustration)**.

4 Refitting is a reversal of removal.

Sill trim

5 Carefully pull the panel upwards from the sill to release the securing clips **(see illustration)**.

6 To refit the panel, simply push it back into position, ensuring that the retaining clips engage.

26.20a Rear sear belt inertia reel mounting bolt (arrowed) . . .

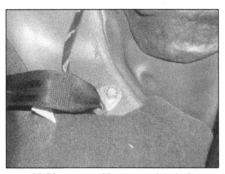

26.20c . . . and lower anchor bolt

Front pillar trim

7 Carefully prise the front door weatherstrip from the edge of the panel, then pull the upper part of the panel from the pillar. Lift the panel

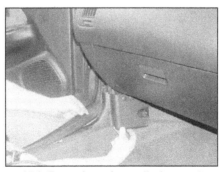

27.3 Removing a footwell trim panel

27.7a Pull the upper part of the front pillar trim panel from the pillar . . .

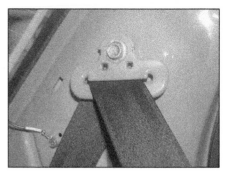

26.20b . . . upper anchor bolt (Saloon model shown) . . .

26.21 Rear seat belt buckle anchor bolts

up to disengage the lower lugs from the facia and remove the trim panel **(see illustrations)**.

8 Refitting is a reversal of removal, ensuring the weatherstrip is correctly seated.

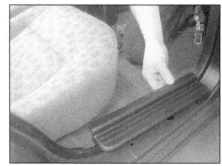

27.5 Removing a sill trim panel

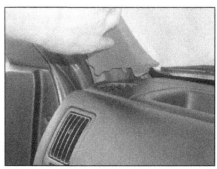

27.7b . . . then lift it up to disengage the lower lugs from the facia

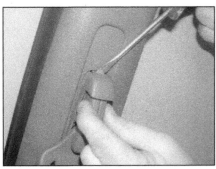

27.10a Depress the seat belt height adjuster release buttons, and depress the upper pawl of the trim cap with a screwdriver

27.10b Release the lower pawl and lift off the trim cap

27.13 Pull the centre pillar lower trim panel from the pillar to release the retaining clips

Centre pillar trim

Saloon and 5-door Hatchback

9 Remove the relevant front seat as described in Section 25.

10 Depress the seat belt height adjuster release buttons, while at the same time depressing the upper retaining pawl of the height adjuster trim cap with a small screwdriver. Release the lower pawl and lift off the trim cap **(see illustrations)**. Undo the seat belt upper anchor bolt and release the seat belt from the height adjuster. Record the positions of the washers and spacers on the upper anchor bolt to ensure correct refitting.

11 Remove the sill trim panels from the front and rear door apertures by carefully pulling the panels upwards from the sill to release the securing clips.

12 Carefully prise the front and rear door weatherstrip from the edges of the centre pillar upper trim panel.

13 Pull the centre pillar lower trim panel from the pillar to release the retaining clips then remove the panel **(see illustration)**.

14 Pull the top of the centre pillar upper trim panel from the pillar to release the upper retaining clip, then lift it up to disengage the lower retaining hook.

15 Refitting is a reversal of removal, ensuring that all retaining clips are fully engaged. Refit the seat belt upper anchor bolt with the washers and spacers correctly positioned as noted during removal. Tighten the seat belt anchor bolt to the specified torque.

Rear pillar trim

Saloon

16 If the trim panel is to be completely removed it will be necessary to remove the rear seat cushion, as described in Section 25, for access to the rear seat belt lower anchor bolt. If the panel is being removed for access to other components only, it can be moved to one side without disengaging the seat belt.

17 Fold down the rear seat back, then carefully prise the rear door weatherstrip from the edge of the panel. Pull the upper and rear part of the panel from the pillar, then lift the panel up to disengage the lower lug from the rear pillar **(see illustrations)**.

18 If the panel is to be removed completely, undo the rear seat belt lower anchor bolt. Record the positions of the washers and spacers on the anchor bolt to ensure correct refitting.

19 Feed the seat belt through the trim panel and remove the panel.

20 Refitting is a reversal of removal, ensuring that all retaining clips are fully engaged and the weatherstrip is correctly seated. Refit the seat belt lower anchor bolt with the washers and spacers correctly positioned as noted during removal. Tighten the seat belt anchor bolt to the specified torque.

Rear parcel shelf

Saloon

21 Remove the rear pillar trim on both sides as described previously in this Section. Note that it is not necessary to completely remove the panels, just position them clear of the parcel shelf.

22 Lift the front edge of the parcel shelf upwards to disengage the front retaining clips, then reach under the shelf and push it up to disengage the rear clips **(see illustrations)**. Remove the parcel shelf from its location.

23 Refitting is a reversal of removal but ensure that all clips are securely engaged.

Rear seat back finisher

Saloon

Note: *The seat back finisher panel is in three sections, the two side sections can be removed singularly, but both must be removed to allow removal of the centre section.*

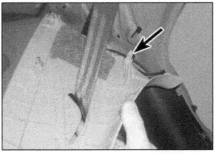
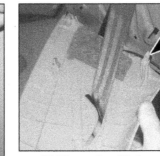

27.17a Pull the rear pillar trim panel from the pillar . . .

27.17b . . . then lift the panel up to disengage the lower lug (arrowed) from the rear pillar – Saloon models

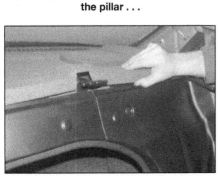

27.22a Lift the front edge of the parcel shelf upwards to disengage the front clips . . .

27.22b . . . then reach under the shelf and push it up to disengage the rear clips – Saloon models

24 Remove the rear parcel shelf as described previously in this Section.

25 Remove the rear seat back(s) as described in Section 25.

26 Pull the edge trim from its location around the central aperture in the seat back finisher panel **(see illustration)**.

27 Carefully prise both rear door weatherstrips from the edges of the seat back finisher panel.

28 Remove the seat belt side retaining clip from the edge of the door aperture, noting its position for refitting **(see illustration)**.

29 Extract the retaining clips securing the relevant side section to the body **(see illustration)**.

30 Pull the side section forward to release the edge retaining clips from the door aperture, then remove the side section **(see illustration)**.

31 To remove the centre section, remove both side sections then extract the retaining clips and remove the centre section from the body.

32 Refitting is a reversal of removal but ensure that all clips are securely engaged.

Luggage area – Saloon

Boot lid aperture trim

33 The rear plastic trim panel at the base of the boot lid aperture can be removed by undoing the retaining clip centre screws then extracting the clips. Release the trim panel and remove it from the luggage compartment.

34 Refitting is a reversal of removal but ensure that all clips are securely engaged.

Carpet trim

35 The luggage compartment carpet trim panels are secured at the rear by an upper and lower plastic clip which can by withdrawn by turning half a turn. Release the remaining clips using a suitable forked tool and lift out the panel **(see illustrations)**. Where applicable, remove the luggage strap brackets, to enable the panels to be removed.

36 Refitting is a reversal of removal but ensure that all clips are securely engaged.

Luggage area – 3-door Hatchback

Parcel shelf support

37 Lift out the parcel shelf, then lower the rear seat back. Undo the two retaining bolts now exposed at the front edge of the parcel shelf support panel.

38 Disconnect the loudspeaker and luggage compartment light wiring connectors, then fold down the carpet and undo the central retaining bolt located under the loudspeaker.

39 Undo the retaining bolt securing the rear of the support panel in the tailgate aperture.

40 Lift the panel up to disengage the upper pillar trim, and withdraw it from its location. Release the seat belt guide plate and feed the seat belt through the slit in the edge of the panel. Remove the panel from the vehicle.

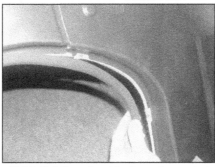

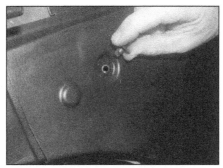

27.26 Pull the edge trim from the central aperture in the seat back finisher panel – Saloon models

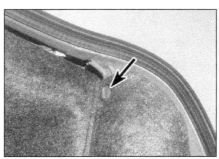

27.29 Extract the seat back finisher panel retaining clips . . .

41 Refitting is a reversal of removal but ensure that all clips are securely engaged.

Lower side trim

42 Remove the rear seat back(s) as described in Section 25.

43 Remove the relevant parcel shelf support panel as described previously.

44 Carefully prise the door weatherstrip from the edge of the lower side trim panel.

45 At the upper rear corner of the panel, undo the retaining clip centre screw and extract the clip.

46 Pull the panel forward to release the retaining clips, then remove the panel.

47 Refitting is a reversal of removal but ensure that all clips are securely engaged.

Upper pillar trim

48 Remove the relevant parcel shelf support panel as described previously.

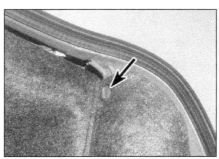

27.35a Release the luggage compartment trim panel rear clips (arrowed) by turning them half a turn . . .

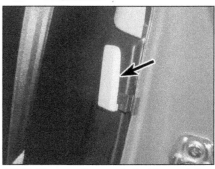

27.28 Remove the seat belt side retaining clip (arrowed) from the door aperture – Saloon models

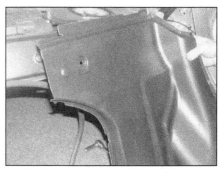

27.30 . . . and remove the seat back finisher side section – Saloon models

49 Remove the trim covers from the seat belt upper mountings and undo the anchor bolts. Record the positions of the washers and spacers on the anchor bolts to ensure correct refitting.

50 At the rear of the panel prise out the trim cap and undo the retaining screw now exposed.

51 Pull the panel from its location to disengage the retaining clips and remove the panel from the vehicle.

52 Refitting is a reversal of removal, ensuring that all retaining clips are fully engaged. Refit the seat belt anchor bolts with the washers and spacers correctly positioned as noted during removal. Tighten the seat belt anchor bolts to the specified torque.

Tailgate aperture trim

53 The rear plastic trim panel at the base of the tailgate aperture can be withdrawn after

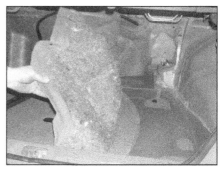

27.35b . . . release the remaining clips and lift out the panel – Saloon models

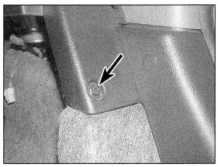

27.58 Undo the retaining bolt (arrowed) at the front edge of the parcel shelf support panel – 5-door Hatchback models

27.59 Undo the two central parcel shelf support retaining bolts (arrowed) – 5-door Hatchback models

27.60 Undo the bolt (arrowed) securing the parcel shelf support panel in the tailgate aperture – 5-door Hatchback models

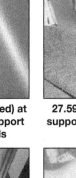

27.61a Withdraw the parcel shelf support panel from the upper pillar trim . . .

27.61b . . . and feed the seat belt through the slit in the panel – 5-door Hatchback models

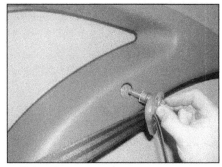

27.64 Remove the rear seat belt upper anchor bolt . . .

removing both parcel shelf support panels as described previously. Pull the panel from the tailgate aperture to release the metal retaining clips and remove the panel from the vehicle.
54 Refitting is a reversal of removal but ensure that all clips are securely engaged.

Carpet trim

55 Remove the relevant rear seat side trim panel as described previously.
56 Remove the plastic clip at the rear of the panel by turning the clip half a turn. Release the remaining clips using a suitable forked tool and lift out the panel. Where applicable, remove the luggage strap brackets, to enable the panels to be removed.
57 Refitting is a reversal of removal but ensure that all clips are securely engaged.

Luggage area – 5-door Hatchback

Parcel shelf support

58 Lift out the parcel shelf, then lower the rear seat back. Undo the retaining bolt now exposed at the front edge of the parcel shelf support panel (see illustration).
59 Disconnect the loudspeaker and luggage compartment light wiring connectors, then fold down the carpet and undo the two central retaining bolts located under the loudspeaker (see illustration).
60 Undo the retaining bolt securing the rear of the support panel in the tailgate aperture (see illustration).
61 Lift the panel up to disengage the upper pillar trim, and withdraw it from its location.

Feed the seat belt through the slit in the edge of the panel and remove the panel from the vehicle (see illustrations).
62 Refitting is a reversal of removal but ensure that all clips are securely engaged.

Upper pillar trim

63 Remove the relevant parcel shelf support panel as described previously.
64 Remove the trim cover from the rear seat belt upper mounting and undo the anchor bolt (see illustration). Record the positions of the washers and spacers on the anchor bolt to ensure correct refitting.
65 At the rear of the panel, prise out the trim cap and undo the retaining screw now exposed (see illustration).
66 Pull the panel from its location to disengage the retaining clips and remove the panel from the vehicle (see illustration).

27.65 . . . prise out the trim cap and undo the retaining screw . . .

67 Refitting is a reversal of removal, ensuring that all retaining clips are fully engaged. Refit the seat belt upper anchor bolt with the washers and spacers correctly positioned as noted during removal. Tighten the seat belt anchor bolt to the specified torque.

Rear seat side trim

68 Remove the rear seat back(s) as described in Section 25.
69 Remove the relevant parcel shelf support panel as described previously.
70 Carefully prise the rear door weatherstrip from the edge of the rear seat side trim panel.
71 Extract the upper retaining clip securing the panel to the body (see illustration).
72 Pull the panel forward to release the edge retaining clips from the door aperture, then remove the panel (see illustration).
73 Refitting is a reversal of removal but ensure that all clips are securely engaged.

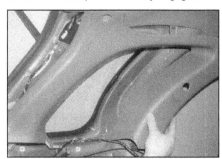

27.66 . . . then pull the upper pillar trim panel from its location – 5-door Hatchback models

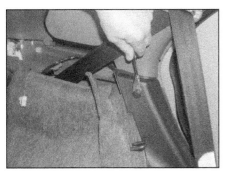

27.71 Extract the upper retaining clip securing the rear seat side trim panel . . .

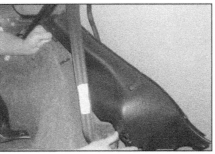

27.72 . . . then release the edge retaining clips from the door aperture and remove the panel – 5-door Hatchback models

27.79 Removing the tailgate inner side trim panel

Tailgate aperture trim

74 The rear plastic trim panel at the base of the tailgate aperture can be withdrawn after removing both parcel shelf support panels as described previously. Pull the panel from the tailgate aperture to release the metal retaining clips and remove the panel from the vehicle.
75 Refitting is a reversal of removal but ensure that all clips are securely engaged.

Carpet trim

76 Remove the relevant rear seat side trim panel as described previously.
77 Remove the plastic clip at the rear of the panel by turning the clip half a turn. Release the remaining clips using a suitable forked tool and lift out the panel. Where applicable, remove the luggage strap brackets, to enable the panels to be removed.
78 Refitting is a reversal of removal but ensure that all clips are securely engaged.

Tailgate inner trim

79 To remove the side trim panels, pull the lower part (nearest to the main trim panel) from its location to disengage the retaining clips, then slide it down to disengage the upper locating clip **(see illustration)**.
80 To remove the main panel, undo the retaining screw and remove the pull handle **(see illustration)**.
81 Pull the panel away from the tailgate to release the retaining clips around the edge, then remove the panel from the tailgate **(see illustration)**.
82 Refitting is a reversal of removal, but ensure that all clips are securely engaged.

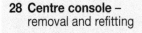

28 Centre console –
removal and refitting

Removal

1 Disconnect the battery negative terminal (refer to *Disconnecting the battery* in Reference).
2 Pull up the handbrake lever as far as it will go.
3 Slide the front seats forward, then undo the screws on each side of the console, at the rear **(see illustration)**.
4 Similarly, undo the screws on each side of

27.80 Undo the retaining screw and remove the pull handle . . .

the console at the front, below the facia **(see illustration)**.
5 Carefully prise up the cover plate, located in front of the handbrake lever, and undo the two screws now exposed **(see illustration)**.

28.3 Undo the screws (arrowed) each side securing the centre console at the rear . . .

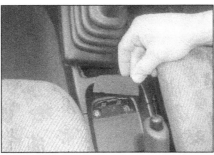

28.5 Carefully prise up the cover plate in front of the handbrake lever and undo the two screws now exposed

27.81 Then pull off the tailgate main trim panel

6 Carefully prise the gear lever boot, or automatic transmission selector lever trim from its location, twist it down and push it through the console aperture **(see illustrations)**.

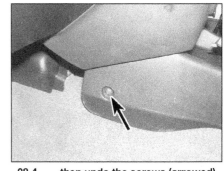

28.4 . . . then undo the screws (arrowed) securing the console at the front

28.6 Prise up the gear lever boot, twist it down, and push it through the console aperture

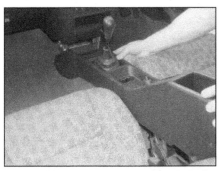

28.7 Lift the console up at the rear and manipulate it over the handbrake lever

29.1a Pull out the plastic hinge pin each side . . .

29.1b . . . and remove the glovebox from the facia

7 Lift the console up at the rear, manipulate it over the handbrake lever and remove the console from the vehicle **(see illustration)**.

Refitting

8 Refitting is a reversal of removal.

29 Facia panels –
removed and refitting

> ⚠ **Warning: Certain models are equipped with an airbag system. The driver's airbag is mounted in the steering wheel centre pad and, where fitted, the passenger's airbag is mounted in the passenger's side of the facia. Make sure that the safety recommendations given in Chapter 12 are followed, to prevent personal injury.**

Glovebox

Removal

1 Working under the glovebox, pull out the plastic hinge pin each side and remove the glovebox from the facia **(see illustrations)**.

Refitting

2 Refitting is a reversal of removal.

Driver's side lower panel

Removal

3 Undo the three screws securing the panel to the facia, one at each lower corner, and one in the upper centre **(see illustration)**.
4 Carefully pull the panel from the facia to release the upper metal clips, and remove the panel **(see illustration)**.

Refitting

5 Refitting is a reversal of removal.

Centre switch/ventilation nozzle

Removal

6 Disconnect the battery negative terminal (refer to *Disconnecting the battery* in the Reference Chapter).
7 Undo the two screws located above the heater/ventilation control unit **(see illustration)**.
8 Using a spatula or similar tool, carefully release the panel upper metal retaining clips **(see illustration)**.
9 Pull the panel away from the facia to release the remaining internal plastic clips **(see illustration)**.
10 Disconnect the switch wiring connectors at the rear of the panel and remove the panel from the vehicle **(see illustration)**.

Refitting

11 Refitting is a reversal of removal.

29.3 Undo the driver's side lower facia panel securing screws . . .

29.4 . . . then pull the panel from the facia to release the upper metal clips, and remove the panel

29.7 Undo the two facia switch/ventilation nozzle housing retaining screws (arrowed) above the heater/ventilation control unit

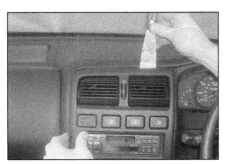

29.8 Carefully release the switch/ventilation nozzle housing upper metal retaining clips

29.9 Pull the panel away from the facia to release the remaining internal clips . . .

29.10 . . . then disconnect the switch wiring connectors at the rear of the panel

Steering column shrouds

Removal

12 Unclip the trim plate from the ignition switch/steering column lock **(see illustration)**.

13 Working under the steering column lower shroud, undo the shroud securing screws, then manipulate the upper shroud from the steering column (on models with an adjustable steering column, this is aided by fully lowering the column) **(see illustration)**.

14 Release the lower shroud from the steering column and withdraw the shroud **(see illustration)**.

Refitting

15 Refitting is a reversal of removal, but ensure that the shroud halves engage correctly with each other.

Instrument panel surround

Removal

16 Release the facia switch/ventilation nozzle housing as described previously, noting that it is not necessary to completely remove the housing, but just release it from the facia.

17 Undo the two screws located at the top of the surround **(see illustration)**.

18 Ease the surround away from the facia and disengage the two lower retaining clips. Manipulate the surround around the steering wheel (on models with an adjustable steering column, this is aided by fully lowering the column) and remove the surround from the vehicle **(see illustration)**.

Refitting

19 Refitting is a reversal of removal.

Complete assembly

Note: *This is an involved procedure, and it is suggested that this complete Section is read through thoroughly before beginning the operation. It is advisable to make careful note of all wiring connections, and the routing of all wiring, to aid refitting.*

Removal

20 Disconnect the battery negative terminal (refer to *Disconnecting the battery* in the Reference Chapter).

21 Remove the footwell trim panels as described in Section 27.

29.12 Unclip the trim plate from the ignition switch/steering column lock . . .

29.14 . . . then withdraw the lower shroud

22 Remove the following facia panels, as described previously in this Section. Note the routing of all wiring, and keep all securing screws and fixings with the relevant panels to avoid confusion on refitting.
  a) *Glovebox.*
  b) *Driver's side lower facia panel.*
  c) *Facia centre switch/ventilation nozzle housing.*
  d) *Steering column shrouds.*
  e) *Instrument panel surround.*

23 Undo the two screws and remove the oddments box located below the heater/ventilation control unit.

24 Remove the heater/ventilation control unit, as described in Chapter 3.

25 Remove the radio/cassette player and the instrument panel as described in Chapter 12.

26 Carefully prise out the switch panel on the driver's side, disconnect the switch wiring connectors and remove the panel.

29.13 . . . undo the shroud securing screws, then manipulate the upper shroud from the steering column . . .

29.17 Undo the two screws located at the top of the instrument panel surround . . .

27 Where fitted, remove the passenger's airbag as described in Chapter 12.

28 Remove the centre console as described in Section 28.

29 Remove the front pillar trim panels as described in Section 27.

30 Referring to Chapter 10, undo the steering column mounting nuts/bolts and lower the column to allow the steering wheel to rest on the driver's seat.

31 Working on the upper edge of the facia next to the windscreen, prise up the blanking plate, and the anti-theft system LED plate, from their locations. Disconnect the wiring connector and remove the anti-theft system LED plate **(see illustration)**.

32 Undo the facia mounting nuts/bolts from the following locations:
  a) *One bolt each side securing the base of the facia to the mounting brackets on the transmission tunnel **(see illustration)**.*

29.18 . . . ease the surround away from the facia, disengage the two lower retaining clips, and withdraw the surround

29.31 Remove the blanking plate and the anti-theft LED plate from each side of the facia at the top

29.32a Undo the bolts (arrowed) securing the base of the facia to the mounting brackets . . .

b) *One nut each side securing the lower outer edge of the facia to the body* **(see illustration)**.

c) *One bolt each side securing the top edge of the facia to the body (behind the previously removed blanking plates or anti-theft warning light plate).*

33 Make a final check to ensure that all relevant wiring has been disconnected, then pull the upper part of the facia panel towards the rear of the car to disengage it from the scuttle – some manipulation may be required **(see illustration)**.

34 Once the facia panel has been released, withdraw it through the passenger's door aperture.

Refitting

35 Refitting is essentially a reversal of the removal procedure, bearing in mind the following points:

a) *Ensure that all wiring is correctly reconnected, and routed.*

b) *Refit all surrounding facia panels with reference to the relevant paragraphs of this Section.*

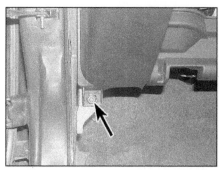

29.32b ... and the nuts (arrowed) securing the lower edge of the facia to the body

29.33 Pull the upper part of the facia panel towards the rear of the car to disengage it from the scuttle

Chapter 12
Body electrical systems

Contents

Degrees of difficulty

Easy, suitable for novice with little experience	Fairly easy, suitable for beginner with some experience	Fairly difficult, suitable for competent DIY mechanic	Difficult, suitable for experienced DIY mechanic	Very difficult, suitable for expert DIY or professional 

Specifications

Bulb ratings

	Watts
Courtesy light	10
Front direction indicator light	21
Front direction indicator repeater light	5
Front driving light	55
Front foglight	55
Front sidelight	5
Headlights	60/55
High-level stop-light:	
Saloon models	5
Hatchback models	21
Luggage compartment light:	
Saloon models	3.4
Hatchback models	5
Map reading light	10
Rear direction indicator light	21
Rear foglight	21
Rear number plate light	5
Reversing light	21
Stop/tail light	21/5

Torque wrench setting

	Nm	lbf ft
Airbag unit retaining bolts	20	15

1 General information and precautions

General information

The electrical system is of 12 volt negative earth type. Power for the lights and all electrical accessories is supplied by a lead-acid type battery, which is charged by the alternator.

This Chapter covers repair and service procedures for the various electrical components not associated with the engine. Information on the battery, alternator and starter motor can be found in Chapter 5A, and information on the ignition system is contained in Chapter 5B.

It should be noted that, prior to working on any component in the electrical system, the battery negative terminal should first be disconnected, to prevent the possibility of electrical short-circuits and/or fires (refer to *Disconnecting the battery* in the Reference Chapter).

Precautions

⚠️ *Warning: Before carrying out any work on the electrical system, read through the precautions given in 'Safety first!' at the beginning of this manual, and in Chapter 5A.*

⚠️ *Warning: Certain later models are equipped with an airbag system and pyrotechnic seat belt pretensioners. When working on the electrical system, refer to the precautions given in Section 21, to avoid the possibility of personal injury.*

2 Electrical fault finding – general information

Note: *Refer to the precautions given in 'Safety first!' and in Section 1 of this Chapter before starting work. The following tests relate to testing of the main electrical circuits, and should not be used to test delicate electronic circuits (such as engine management systems or anti-lock braking systems), particularly where an electronic control unit is used.*

General

1 A typical electrical circuit consists of an electrical component, any switches, relays, motors, fuses, fusible links or circuit breakers related to that component, and the wiring and connectors which link the component to both the battery and the chassis. To help to pinpoint a problem in an electrical circuit, wiring diagrams are included at the end of this chapter.

2 Before attempting to diagnose an electrical fault, first study the appropriate wiring diagram, to obtain a more complete understanding of the components included in the particular circuit concerned. The possible sources of a fault can be narrowed down by noting whether other components related to the circuit are operating properly. If several components or circuits fail at one time, the problem is likely to be related to a shared fuse or earth connection.

3 Electrical problems usually stem from simple causes, such as loose or corroded connections, a faulty earth connection, a blown fuse, a melted fusible link, or a faulty relay (refer to Section 3 for details of testing relays). Visually inspect the condition of all fuses, wires and connections in a problem circuit before testing the components. Use the wiring diagrams to determine which terminal connections will need to be checked, in order to pinpoint the trouble spot.

4 The basic tools required for electrical fault finding include a circuit tester or voltmeter (a 12 volt bulb with a set of test leads can also be used for certain tests); a self-powered test light (sometimes known as a continuity tester); an ohmmeter (to measure resistance); a battery and set of test leads; and a jumper wire, preferably with a circuit breaker or fuse incorporated, which can be used to bypass suspect wires or electrical components. Before attempting to locate a problem with test instruments, use the wiring diagram to determine where to make the connections.

5 To find the source of an intermittent wiring fault (usually due to a poor or dirty connection, or damaged wiring insulation), a 'wiggle' test can be performed on the wiring. This involves wiggling the wiring by hand, to see if the fault occurs as the wiring is moved. It should be possible to narrow down the source of the fault to a particular section of wiring. This method of testing can be used in conjunction with any of the tests described in the following sub-Sections.

6 Apart from problems due to poor connections, two basic types of fault can occur in an electrical circuit – open-circuit, or short-circuit.

7 Open-circuit faults are caused by a break somewhere in the circuit, which prevents current from flowing. An open-circuit fault will prevent a component from working, but will not cause the relevant circuit fuse to blow.

8 Short-circuit faults are normally caused by a breakdown in wiring insulation, which allows a feed wire to touch either another wire, or an earthed component such as the bodyshell. This allows the current flowing in the circuit to 'escape' along an alternative route, usually to earth. As the circuit does not now follow its original complete path, it is known as a 'short' circuit. A short-circuit fault will normally cause the relevant circuit fuse to blow.

Finding an open-circuit

9 To check for an open-circuit, connect one lead of a circuit tester or voltmeter to either the negative battery terminal or a known good earth.

10 Connect the other lead to a connector in the circuit being tested, preferably nearest to the battery or fuse.

11 Switch on the circuit, bearing in mind that some circuits are live only when the ignition switch is moved to a particular position.

12 If voltage is present (indicated either by the tester bulb lighting or a voltmeter reading, as applicable), this means that the section of the circuit between the relevant connector and the battery is problem-free.

13 Continue to check the remainder of the circuit in the same fashion.

14 When a point is reached at which no voltage is present, the problem must lie between that point and the previous test point with voltage. Most problems can be traced to a broken, corroded or loose connection.

Finding a short-circuit

15 To check for a short-circuit, first disconnect the load(s) from the circuit (loads are the components which draw current from a circuit, such as bulbs, motors, heating elements, etc).

16 Remove the relevant fuse from the circuit, and connect a circuit tester or voltmeter to the fuse connections.

17 Switch on the circuit, bearing in mind that some circuits are live only when the ignition switch is moved to a particular position.

18 If voltage is present (indicated either by the tester bulb lighting or a voltmeter reading, as applicable), this means that there is a short-circuit.

19 If no voltage is present, but the fuse still blows with the load(s) connected, this indicates an internal fault in the load(s).

Finding an earth fault

20 The battery negative terminal is connected to 'earth' – the metal of the engine/transmission and the car body – and most systems are wired so that they only receive a positive feed, the current returning via the metal of the car body. This means that the component mounting and the body form part of that circuit. Loose or corroded mountings can therefore cause a range of electrical faults, ranging from total failure of a circuit, to a puzzling partial fault. In particular, lights may shine dimly (especially when another circuit sharing the same earth point is in operation), motors (eg, wiper motors or the radiator cooling fan motor) may run slowly, and the operation of one circuit may have an apparently-unrelated effect on another. Note that on many vehicles, earth straps are used between certain components, such as the engine/transmission and the body, usually where there is no metal-to-metal contact between components, due to flexible rubber mountings, etc.

21 To check whether a component is properly earthed, disconnect the battery, and connect one lead of an ohmmeter to a known good earth point. Connect the other lead to the wire or earth connection being tested. The resistance reading should be zero; if not, check the connection as follows.

22 If an earth connection is thought to be faulty, dismantle the connection, and clean back to bare metal both the bodyshell and the wire terminal or the component earth connection mating surface. Be careful to remove all traces of dirt and corrosion, then use a knife to trim away any paint, so that a clean metal-to-metal joint is made. On reassembly, tighten the joint fasteners securely; if a wire terminal is being refitted, use serrated washers between the terminal and the bodyshell, to ensure a clean and secure connection. When the connection is remade, prevent the onset of corrosion in the future by applying a coat of petroleum jelly or silicone-based grease, or by spraying on (at regular intervals) a proprietary ignition sealer.

3 Fuses and relays – general information

Fuses

1 Fuses are designed to break a circuit when a predetermined current is reached, in order to protect the components and wiring which could be damaged by excessive current flow. Any excessive current flow will be due to a fault in the circuit, usually a short-circuit (see Section 2).

2 The main fuses are located in the fusebox on the driver's side of the facia.

3 For access to the fuses, pull open the cover flap **(see illustration)**.

4 Additional fuses and circuit-breakers are located in an auxiliary fusebox in the engine compartment, attached to a bracket, next to the battery **(see illustration)**.

5 A blown fuse can be recognised from its melted or broken wire.

6 To remove a fuse, first ensure that the relevant circuit is switched off.

7 Using the plastic tool clipped to the main fusebox lid, pull the fuse from its location **(see illustration)**.

8 Spare fuses are provided in the main fusebox.

9 Before renewing a blown fuse, trace and rectify the cause, and always use a fuse of the correct rating (fuse ratings are specified on the inside of the fusebox cover flap). Never substitute a fuse of a higher rating, or make temporary repairs using wire or metal foil; more serious damage, or even fire, could result.

10 Note that the fuses are colour-coded as follows.

Colour	Rating
Orange	*5A*
Red	*10A*
Blue	*15A*
Yellow	*20A*
Clear or White	*25A*
Green	*30A*

3.3 For access to the main fuses, pull open the fusebox cover flap

11 The radio/cassette player fuse is located in the rear of the unit, and can be accessed after removing the radio/cassette player.

Relays

12 A relay is an electrically-operated switch, which is used for the following reasons:
  a) *A relay can switch a heavy current remotely from the circuit in which the current is flowing, therefore allowing the use of lighter-gauge wiring and switch contacts.*
  b) *A relay can receive more than one control input, unlike a mechanical switch.*
  c) *A relay can have a timer function – for example, the intermittent wiper relay.*

13 Various relays are located behind the facia, next to the fusebox, and in the relay box on the right-hand side of the engine compartment **(see illustration)**.

14 The direction indicator/hazard warning flasher unit is located on a bracket behind the facia, adjacent to the fusebox, and can be accessed once the driver's side lower facia panel has been removed (see Chapter 11, Section 29).

15 The 'headlights/rear foglight on' warning buzzer is located under the centre of the facia, adjacent to the steering column.

16 Additional relays may be located in various other locations, depending on model – for example, the fuel injection/engine management relay is located adjacent to the engine management system ECU, just in front of the centre console, mounted on the transmission tunnel floor.

3.7 Using the plastic tool clipped to the fusebox lid, pull the fuse from its location

3.4 Additional fuses and circuit-breakers are located in an auxiliary fusebox (arrowed) next to the battery

17 If a circuit or system controlled by a relay develops a fault, and the relay is suspect, operate the system. If the relay is functioning, it should be possible to hear it 'click' as it is energised, If this is the case, the fault lies with the components or wiring of the system. If the relay is not being energised, then either the relay is not receiving a main supply or a switching voltage, or the relay itself is faulty. Testing is by the substitution of a known good unit, but be careful – while some relays are identical in appearance and in operation, others look similar but perform different functions.

18 To remove a relay, first ensure that the relevant circuit is switched off. The relay can then simply be pulled out from the socket, and pushed back into position.

4 Switches – removal and refitting

Note: *Disconnect the battery negative terminal (refer to 'Disconnecting the battery' in the Reference Chapter), before removing any switch, and reconnect the terminal after refitting the switch.*

Ignition switch/steering lock

1 Refer to Chapter 10.

Steering column switches

2 Remove the steering column shrouds, as described in Chapter 11, Section 29.

3.13 The main relays are located in the engine compartment relay box (arrowed)

4.3a Undo the retaining screw . . .

4.3b . . . then slide the relevant combination switch from the bracket on the steering column

4.6 Disconnect the wiring connector from the relevant centre facia-mounted switch . . .

3 Undo the switch retaining screw, then slide the relevant switch from the bracket on the steering column, and disconnect the wiring connectors **(see illustrations)**.
4 Refitting is a reversal of removal.

Facia-mounted switches

5 Remove the facia centre switch/ventilation nozzle housing as described in Chapter 11, Section 29.
6 Working at the rear of the panel, disconnect the wiring connector from the relevant switch **(see illustration)**.
7 Push the switch forwards from the panel to release the securing clips **(see illustration)**.
8 Refitting is a reversal of removal.

Heater/ventilation switches

9 Remove the heater/ventilation control unit as described in Chapter 3.

10 If the blower motor switch is being removed, carefully pull off the switch control knob **(see illustration)**. If using pliers, wrap a suitable piece of cloth or card around the knob to protect it.
11 Release the securing clips, using a screwdriver if necessary, then pull the switch from the control panel **(see illustration)**.
12 Refit the switch using a reversal of the removal procedure, then refit the heater/ventilation control unit as described in Chapter 3.

Driver's side/ventilation panel switches

13 These switches include the electric door mirror switch, the instrument panel illumination switch, the front foglight switch and the headlight beam adjustment switch, according to model.

14 If removing the instrument panel illumination switch, the front foglight switch or the headlight beam adjustment switch, prise the switch assembly from the facia, and disconnect the wiring connector **(see illustration)**.
15 To remove the electric door mirror switch, first remove the switch assembly above the mirror switch, as described in the previous paragraph. Reach through the switch aperture and push the electric window switch out of the facia from behind **(see illustration)**. Disconnect the wiring and remove the switch.
16 Refit the relevant switch using a reversal of the removal procedure.

Centre console switches

17 Remove the centre console, as described in Chapter 11.
18 Working at the rear of the centre console, release the securing clips, then push the switch out through the top of the console.
19 Refit the relevant switch using a reversal of the removal procedure, then refit the centre console.

Electric window switches

Front door switch

20 On early models with a detachable full-length arm rest finisher, carefully prise up the finisher to release the retaining clips. Withdraw the finisher and disconnect the wiring connector from the electric window switch assembly.
21 On later models, prise up the front end of the electric window switch panel then

4.7 . . . then push the switch forwards from the panel to release the securing clips

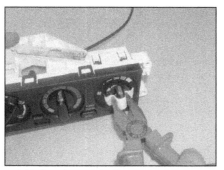

4.10 Carefully pull off the blower motor switch control knob . . .

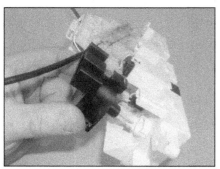

4.11 . . . then release the securing clips and pull the switch from the control panel

4.14 Removing the upper switches from the driver's side switch/ventilation panel

4.15 Push the lower switch from the panel by reaching through the upper aperture

4.21a Prise up the front end of the electric window switch panel then disengage the hook at the rear

4.21b Disconnect the wiring connector and remove the switch panel

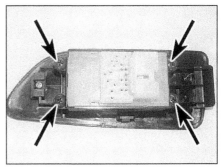

4.22 Undo the four screws (arrowed) to remove the driver's door electric window switch from the panel

disengage the hook at the rear. Withdraw the switch panel and disconnect the wiring connector **(see illustrations)**.

22 If working on the driver's door switch, from the underside of the arm rest finisher or switch panel, as applicable, undo the four screws and remove the electric window switch assembly **(see illustration)**.

23 If working on the passenger's door switch, release the retaining clips and withdraw the switch from the arm rest finisher or switch panel, as applicable.

24 Refit the switch assembly then refit the arm rest finisher or switch panel to the door trim panel.

Electric sunroof switch

25 Carefully prise the right-hand side of the overhead console from its location, disengage the retaining lug on the left-hand side and withdraw the console from the roof **(see illustration)**.

26 From the rear of the console, undo the retaining screw and lift off the relevant switch **(see illustration)**. Disconnect the wiring connector and remove the switch from the console.

27 Refitting is a reversal of removal.

Courtesy light/door ajar warning switches

28 Open the door to expose the switch in the door pillar.

29 Remove the securing screw, then withdraw the switch from the door pillar. Disconnect the wiring connector as it becomes accessible.

 HAYNES HiNT *Tape the wiring to the door pillar, or tie a length of string to the wiring, to retrieve it if it falls back into the door pillar.*

30 Refitting is a reversal of removal, but ensure that the rubber gaiter is correctly seated on the switch.

Luggage area light switch

31 The switch is integral with the boot lid/tailgate lock. Removal and refitting details for the boot lid/tailgate lock are provided in Chapter 11.

4.25 Carefully prise the overhead console from its location . . .

Map reading light switches

32 The switches are integral with the light assembly and cannot be renewed independently.

---

5 Bulbs (exterior lights) – renewal

General

1 Whenever a bulb is renewed, note the following points:

a) *Disconnect the battery negative terminal (refer to 'Disconnecting the battery' in the Reference Chapter).*

b) *Remember that, if the light has just been in use, the bulb may be extremely hot.*

c) *Always check the bulb contacts and holder, ensuring that there is clean metal-*

5.3 Turn the plastic cover anti-clockwise and remove it from the rear of the headlight

4.26 . . . then undo the retaining screws (arrowed) securing the electric sunroof switches

to-metal contact between the bulb and its live contact(s) and earth. Clean off any corrosion or dirt before fitting a new bulb.

d) *Wherever bayonet-type bulbs are fitted, ensure that the live contact(s) bear firmly against the bulb contact.*

e) *Always ensure that the new bulb is of the correct rating (see Specifications), and that it is completely clean before fitting it; this applies particularly to headlight/foglight bulbs (see following paragraphs).*

Headlight

2 Open and support the bonnet.

3 Turn the plastic cover anti-clockwise and remove it from the rear of the headlight **(see illustration)**.

4 Squeeze the securing lugs, and disconnect the wiring connector from the rear of the bulb **(see illustration)**.

5.4 Squeeze the securing lugs, and disconnect the wiring connector

5.5 Pull the rubber boot from the rear of the headlight

5.7 Release the spring clip and withdraw the headlight bulb

5.12 Withdraw the sidelight bulbholder and remove the push-fit bulb

5 Pull the rubber boot from the rear of the headlight **(see illustration)**.
6 Squeeze the retaining spring-clip lugs, and release the clip from the rear of the bulb.
7 Withdraw the bulb **(see illustration)**.
8 When handling the new bulb, use a tissue or clean cloth, to avoid touching the glass with the fingers; moisture and grease from the skin can cause blackening and rapid failure of this type of bulb. If the glass is accidentally touched, wipe it clean using methylated spirit.
9 Install the new bulb, ensuring that its locating tabs are correctly located in the light unit cut-outs. Secure the bulb in position with the retaining clip, then refit the rubber boot. Refit the plastic cover, turning it clockwise to lock, and reconnect the wiring connector.

Sidelight

10 Open and support the bonnet.
11 Working at the rear of the headlight unit, twist the sidelight bulbholder anti-clockwise, and withdraw it from the light unit.
12 The bulb is a push-fit in the bulbholder **(see illustration)**.
13 Fit the new bulb using a reversal of the removal procedure.

Front indicator

14 Open and support the bonnet.
15 Undo the direction indicator light unit securing screw (located at the top corner of the light unit) **(see illustration)**.
16 Withdraw the light unit forwards from the wing panel, then twist the bulbholder anti-

clockwise and withdraw it from the light unit **(see illustration)**.
17 The bulb is a bayonet fit in the bulbholder.
18 Fit the new bulb, then refit the light assembly using a reversal of the removal procedure.

Front indicator side repeater

19 Push the light unit towards the front of the vehicle and release it from the wing panel.
20 Withdraw the light unit, then twist the bulbholder anti-clockwise to release it from the light unit.
21 The bulb is a push-fit in the bulbholder.
22 Fit the new bulb using a reversal of the removal procedure.

Front foglight

23 Remove the two lens surround securing screws, then withdraw the surround, lens, and reflector.
24 Carefully prise off the trim cover from the inner end of the foglight lens.
25 Undo the two screws now exposed and withdraw the light unit from the bumper.
26 Turn the plastic cover anti-clockwise and remove it from the rear of the foglight.
27 Disconnect the wiring from the bulb, then release the spring clip, and withdraw the bulb from the rear of the light unit.
28 Fit the new bulb using a reversal of the removal procedure.

Stop/tail light and rear indicator

29 Open the boot lid or tailgate.
30 Working at the side of the luggage compartment, turn the plastic retaining clip anti-clockwise and pull back the carpet trim **(see illustration)**.
31 Turn the relevant bulbholder anti-clockwise and withdraw it from the light unit.
32 The bulbs are a bayonet fit in the bulbholder **(see illustration)**.
33 Fit the new bulb using a reversal of the removal procedure. Note that the stop/tail light bulb has offset pins, to ensure correct installation.

Reversing light and rear foglight

34 Open the boot lid or tailgate.
35 Undo the three fasteners (Saloon models) or depress the retaining tag (Hatchback models) and lift off the boot lid/tailgate

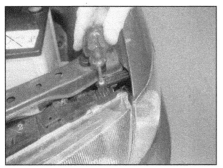

5.15 Undo the direction indicator light unit securing screw

5.16 Withdraw the direction indicator light unit then twist the bulbholder anti-clockwise to remove

5.30 Turn the retaining clip anti-clockwise and pull back the carpet for access to the light unit

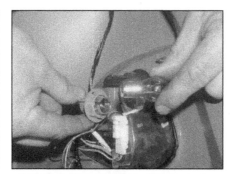

5.32 Turn the bulbholder anti-clockwise then remove the bayonet fit bulb

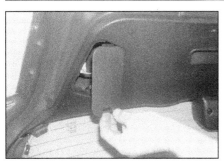

5.35 On Hatchback models, remove the tailgate aperture panel for access to the light unit

5.37 Turn the bulbholder anti-clockwise then remove the bayonet fit bulb

5.39 Undo the two screws (arrowed) and withdraw the number plate light unit

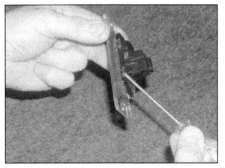

5.40 Release the retaining lugs and unclip the lens from the light unit

5.41 Remove the push-fit bulb from the light unit base

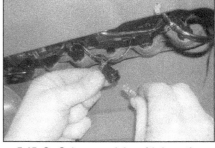

5.45 On Saloon models, withdraw the relevant high-level stoplight bulbholder and remove the push-fit bulb

aperture access panel behind the rear of the light unit **(see illustration)**.

36 Working through the aperture in the boot lid/tailgate, turn the relevant bulbholder anti-clockwise and withdraw it from the light unit.

37 The bulbs are a bayonet fit in the bulbholder **(see illustration)**.

38 Fit the new bulb using a reversal of the removal procedure.

Number plate light

39 Undo the two securing screws, and withdraw the light unit from the boot lid/tailgate **(see illustration)**. Disconnect the wiring connector and remove the light unit.

40 Release the retaining lugs and unclip the lens from the light unit for access to the bulb **(see illustration)**.

41 The bulb is a push-fit in the light unit **(see illustration)**.

42 Fit the new bulb using a reversal of removal procedure.

High-level stop-light

Saloon models

43 Open the boot lid.

44 Turn the relevant bulbholder anti-clockwise and withdraw it from the light unit.

45 The bulb is a push-fit in the bulbholder **(see illustration)**.

46 Fit the new bulb using a reversal of the removal procedure.

Hatchback models

47 Remove the main tailgate inner trim panel as described in Chapter 11, Section 27.

48 Undo the two screws securing the plastic cover to the light unit. Disengage the locating clips and lift off the cover **(see illustration)**.

49 Turn the bulbholder anti-clockwise and withdraw it from the light unit.

50 The bulb is a bayonet fit in the bulbholder **(see illustration)**.

51 Fit the new bulb using a reversal of the removal procedure.

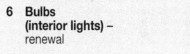

6   Bulbs (interior lights) – renewal

General

1 Refer to Section 5, paragraph 1.

Map reading light

2 Carefully prise the lens from the light unit (if necessary, carefully use a flat-bladed screwdriver) **(see illustration)**.

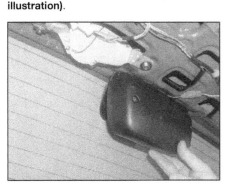

5.48 On Hatchback models, undo the two screws, disengage the locating clips and lift off the high-level stop-light cover

5.50 Withdraw the bulbholder and remove the bayonet fit bulb

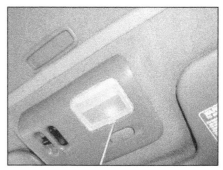

6.2 Carefully prise the lens from the map reading light unit . . .

6.3 . . . then remove the bayonet fit bulb

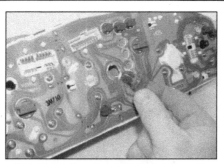

6.11 Twist the relevant bulbholder anti-clockwise to remove it from the instrument panel

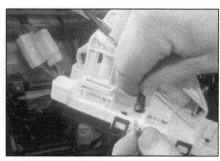

6.17 Twist the heater control panel bulbholder anti-clockwise then remove the push-fit bulb

3 The bulb is a bayonet fit in the light unit **(see illustration)**.
4 Fit the new bulb using a reversal of the removal procedure.

Courtesy light

5 Proceed as described previously for the map reading light, but note that the bulb is a festoon type.

Luggage area light

6 Open the tailgate or boot lid, as applicable.
7 On Saloon models, turn the lens anti-clockwise and remove it from the light unit. On Hatchback models, unclip the inner edge of the lens and pivot the lens down.
8 The bulb is a push-fit in the light assembly.
9 Fit the new bulb using a reversal of the removal procedure.

Instrument panel lights

10 Remove the instrument panel, as described in Section 9.
11 Twist the relevant bulbholder anti-clockwise to remove it from the rear of the instrument panel **(see illustration)**.
12 The bulb is a push-fit in the bulbholder.
13 Fit the new bulb, refit the bulbholder and twist it clockwise to lock it into position.
14 Refit the instrument panel as described in Section 9.

Heater control illumination

15 Withdraw the heater/ventilation control unit from the facia as described in Chapter 3. Note that it is not necessary to disconnect the control cables.

16 Twist the relevant bulbholder anti-clockwise to remove it from the rear of the control unit.
17 The bulb is a push-fit in the bulbholder **(see illustration)**.
18 Fit the new bulb, refit the bulbholder and twist it clockwise to lock it into position.
19 Refit the heater/ventilation control unit as described in Chapter 3.

Switch illumination bulbs

20 All of the switches are fitted with illuminating bulbs, and some are also fitted with a bulb to show when the circuit concerned is operating.
21 In most cases, if a bulb blows, the complete switch must be renewed, but on certain models, some of the switch bulbs can be renewed. Check with a Nissan dealer for information on the availability of spare bulbs.
22 On switches where the bulbs can be renewed, remove the relevant switch as described in Section 4, then pull the bulbholder from the top of the switch. The bulbs are integral with the bulbholders.

---

7 Exterior light units – removal and refitting

Note: *Disconnect the battery negative terminal (refer to 'Disconnecting the battery' in the Reference Chapter), before removing any light unit, and reconnect the terminal after refitting the light.*

Headlight

Removal

1 Remove the front grille panel as described in Chapter 11, Section 24.
2 Remove the relevant direction indicator light as described later in this Section.
3 Unscrew the two headlight side securing bolts **(see illustration)**.
4 Similarly, unscrew the two headlight rear securing nuts **(see illustration)**.
5 Withdraw the headlight unit from the vehicle and disconnect the wiring connectors **(see illustration)**.

Refitting

6 Refitting is a reversal of removal. On completion, it is wise to have the headlight beam alignment checked with reference to Section 8.

Front indicator

7 The procedure is described as part of the bulb renewal procedure in Section 5.

Front indicator side repeater

8 The procedure is described as part of the bulb renewal procedure in Section 5.

Front foglight

9 The procedure is described as part of the bulb renewal procedure in Section 5.

Stop/tail and rear indicator

Note: *A new light unit sealing strip will be required on refitting.*

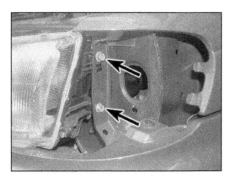

7.3 Unscrew the two headlight side securing bolts (arrowed) . . .

7.4 . . . and rear securing nuts (arrowed) . . .

7.5 . . . then withdraw the headlight unit and disconnect the wiring connectors

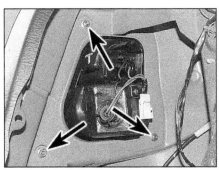

7.13a Unscrew the three securing nuts (arrowed) . . .

7.13b . . . then withdraw the stop/tail and rear direction indicator light unit

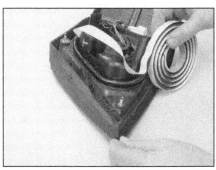

7.15 Apply a new sealing strip to the light unit prior to refitting

Removal

10 Open the boot lid or tailgate.

11 Working at the side of the luggage compartment, turn the plastic retaining clip anti-clockwise and pull back the carpet trim.

12 Disconnect the wiring connector at the rear of the light unit.

13 Working inside the luggage compartment, unscrew the three securing nuts, then withdraw the light unit from the rear of the vehicle **(see illustrations)**. Note that the light unit is held in place with a sealing compound.

Refitting

14 Before refitting the light unit, clean all traces of sealant from the light unit and the rear wing panel.

15 Apply the new sealing strip to the rear of the light unit (the sealing strip is supplied in rolls, and may have to be trimmed to the required length) **(see illustration)**.

16 Refit the light unit using a reversal of the removal procedure.

Reversing light and rear foglight

Note: *A new light unit sealing strip will be required on refitting.*

Removal

17 Open the boot lid or tailgate.

18 On Saloon models, where fitted, undo the three fasteners and lift off the boot lid aperture access panel behind the rear of the light unit. On Hatchback models, remove the main tailgate inner trim panel as described in Chapter 11, Section 27.

19 Disconnect the wiring connector at the rear of the light unit.

20 Unscrew the four securing nuts, then pull the light unit from the outside of the boot lid/tailgate **(see illustration)**. Note that the light unit is held in place with a sealing compound.

Refitting

21 Before refitting the light unit, clean all traces of sealant from the light unit and the boot lid/tailgate.

22 Apply the new sealing strip to the rear of the light unit (the sealing strip is supplied in rolls, and may have to be trimmed to the required length).

23 Refit the light unit using a reversal of the removal procedure.

Rear number plate light

24 The procedure is described as part of the bulb renewal procedure in Section 5.

High-level stop-light

Saloon models

25 Open the boot lid.

26 Undo the two retaining nuts and withdraw the light unit from the boot lid.

27 Disconnect the wiring connector and remove the light unit.

28 Refit the light unit using a reversal of the removal procedure.

Hatchback models

29 Remove the main tailgate inner trim panel as described in Chapter 11, Section 27.

30 Undo the two screws securing the plastic cover to the light unit. Disengage the locating clips and lift off the cover.

31 Turn the bulbholder anti-clockwise and withdraw it from the light unit.

32 Undo the two bolts and remove the light unit from the tailgate **(see illustration)**.

33 Refit the light unit using a reversal of the removal procedure.

8 Headlight beam alignment – general information

Accurate adjustment of the headlight beam

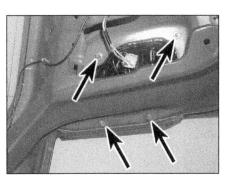

7.20 Reversing light/rear foglight unit securing nuts (arrowed)

is only possible using optical beam-setting equipment, and this work should therefore be carried out by a Nissan dealer or suitably-equipped workshop.

For reference, the outer adjusting screw (nearest the vehicle wing) is used to adjust the vertical alignment, and the inner screw (located nearest to the radiator) is used to adjust the horizontal alignment. Note that on models with electric headlight beam adjustment, the adjustment switch must be set to position 0 when carrying out beam alignment.

Certain models are equipped with a headlight beam adjustment system, controlled by a switch located on the facia, which allows the aim of the headlights to be adjusted to compensate for the varying loads carried in the vehicle. The switch should be positioned according to the load being carried in the vehicle – eg, position 0 for driver with no passengers or luggage, up to position 3 for maximum load, or towing.

9 Instrument panel – removal and refitting

Removal

1 Disconnect the battery negative terminal (refer to *Disconnecting the battery* in the Reference Chapter).

2 Remove the instrument panel surround as described in Chapter 11, Section 29.

7.32 High-level stop-light retaining bolts (arrowed) – Hatchback models

9.3a Remove the two upper instrument panel securing screws . . .

9.3b . . . and the two lower screws

9.4a Pull the panel free . . .

9.4b . . . and disconnect the wiring connectors

3 Remove the two upper and the two lower instrument panel securing screws **(see illustrations)**.
4 Pull the instrument panel forwards, and disconnect the wiring connectors from the rear of the panel **(see illustrations)**. Withdraw the instrument panel from the facia.

Refitting

5 Refitting is a reversal of removal.

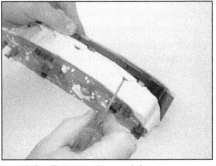

10.2a Release the retaining tags . . .

10.2b . . . and withdraw the lens assembly and facia plate from the front of the instrument panel

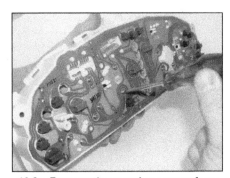

10.3a Remove the securing screws from the rear of the instrument panel . . .

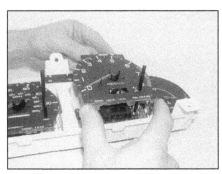

10.3b . . . then withdraw the relevant instrument from the front of the panel

10 Instrument panel components – removal and refitting

General

1 Remove the instrument panel as described in Section 9, then proceed as described under the relevant sub-heading.

Instruments

2 Using a small screwdriver, unclip the lens assembly and facia plate from the front of the instrument panel **(see illustrations)**.
3 Remove the securing screws from the rear of the instrument panel, then withdraw the relevant instrument from the front of the panel **(see illustrations)**.
4 Refitting is a reversal of removal, but refit the instrument panel with reference to Section 9.

Illumination and warning bulbs

5 Twist the relevant bulbholder anti-clockwise to remove it from the rear of the instrument panel.
6 The bulbs are a push-fit in the bulbholders.

Printed circuit

7 The printed circuit assembly can be removed from the rear of the instrument panel after removing all the instruments, as described previously in this Section, and the instrument panel illumination and warning light bulbs.

11 Horn – removal and refitting

Removal

1 Disconnect the battery negative terminal (refer to *Disconnecting the battery* in the Reference Chapter).
2 Remove the front grille panel as described in Chapter 11, Section 24.
3 Disconnect the wiring from the relevant horn.
4 Unscrew the securing bolt, and withdraw the horn complete with its mounting bracket.

Refitting

5 Refitting is a reversal of removal.

12 Wiper arm – removal and refitting

Removal

1 Operate the wiper motor, then switch it off so that the wiper arm returns to the at-rest/parked position.
2 If a windscreen or tailgate wiper is being removed, stick a piece of tape alongside the edge of the wiper blade, to use as an alignment aid on refitting.

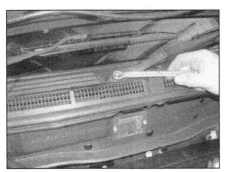

12.3a Undo the wiper arm spindle nut . . .

12.3b . . . then lift the blade off the glass, and pull the wiper arm off its spindle

3 Where applicable, lift off the wiper arm spindle nut cover, then slacken and remove the spindle nut. Lift the blade off the glass, and pull the wiper arm off its spindle **(see illustrations)**. If necessary, the arm can be levered off the spindle using a suitable flat-bladed screwdriver. If both windscreen wiper arms are removed, note their locations, as different arms are fitted to the driver's and passenger's sides.

Refitting

4 Ensure that the wiper arm and spindle splines are clean and dry.
5 When refitting a windscreen or tailgate wiper arm, refit the arm to the spindle, aligning the wiper blade with the tape fitted before removal. If both windscreen wiper arms have been removed, ensure that the arms are refitted to their correct positions as noted before removal.

6 When refitting a headlight wiper arm, hold the arm below the stops at the bottom of the headlight unit, until the spindle nut has been refitted and tightened.
7 Refit the spindle nut, tighten it securely, and where applicable, clip the nut cover back into position. If refitting a headlight wiper arm, once the spindle nut has been tightened, position the wiper blade against the upper surfaces of the stops on the headlight.

---

13 Windscreen wiper motor and linkage – removal and refitting

Motor

Removal

1 Disconnect the battery negative terminal

13.3 Disconnect the windscreen wiper motor wiring connector

13.4a Unscrew the four bolts (arrowed) and withdraw the motor . . .

13.4b . . . then prise the crank arm from the motor drive balljoint

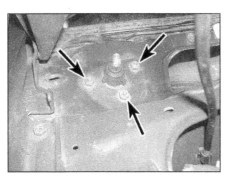

13.8 Windscreen wiper linkage spindle securing nuts (arrowed)

(refer to *Disconnecting the battery* in the Reference Chapter).
2 Remove the windscreen scuttle grille panel as described in Chapter 11, Section 24.
3 Disconnect the motor wiring connector **(see illustration)**.
4 Unscrew the four motor securing bolts, and withdraw the motor. When sufficient clearance exists, prise the crank arm from the motor drive balljoint **(see illustrations)**.

Refitting

5 Refitting is a reversal of removal, but ensure that the motor drive is in the 'parked' position before reconnecting the crank arm. Reach in through the aperture in the scuttle to connect the crank arm to the motor drive balljoint.

Linkage

Removal

6 Remove the wiper arms as described in Section 12.
7 Remove the wiper motor as described previously in this Section.
8 Undo the three nuts securing each spindle unit to the scuttle and withdraw the linkage out through the scuttle aperture **(see illustration)**.

Refitting

9 Refitting is a reversal of removal, but ensure that motor drive is in the 'parked' position before reconnecting the crank arm.

---

14 Tailgate wiper motor – removal and refitting

Removal

1 Disconnect the battery negative terminal (refer to *Disconnecting the battery* in the Reference Chapter).
2 Remove the tailgate inner trim panels as described in Chapter 11, Section 27.
3 Remove the wiper arm with reference to Section 12.
4 Disconnect the tailgate wiper motor wiring connector.
5 Unscrew the three bolts securing the motor mounting bracket to the tailgate and withdraw the motor assembly **(see illustrations)**.

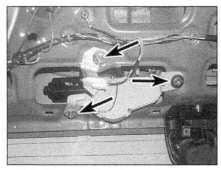

14.5a Unscrew the tailgate wiper motor securing bolts (arrowed) . . .

14.5b ... and withdraw the motor assembly

Refitting

6 Refitting is a reversal of removal.

15 Windscreen/tailgate washer system components – removal and refitting

Washer fluid reservoir

Removal

1 Working in the engine compartment, twist the reservoir filler neck clockwise, then pull it from the top of the reservoir.
2 Disconnect the battery negative terminal (refer to *Disconnecting the battery* in the Reference Chapter).
3 Firmly apply the handbrake, then jack up the front of the vehicle and support it securely on axle stands (see *Jacking and vehicle*

16.4a Lift off the radio/cassette player surround ...

17.2a Undo the securing screws (arrowed) ...

support). Remove the front right-hand roadwheel.
4 From underneath the front of the car, undo the retaining screws and remove the wheel arch liner from underneath the wing to gain access to the washer reservoir. If necessary, also undo the retaining screws and remove the engine undershield to improve access.
5 Disconnect the wiring connector(s) from the washer pump(s), and from the fluid level sensor, where applicable.
6 Disconnect the fluid hose(s) from the washer pump(s) – if the reservoir still contains fluid, be prepared for fluid spillage.
7 Where applicable, release the wiring harness from its clips, and move the harness to one side to allow sufficient clearance to remove the reservoir.
8 Working under the wheel arch, remove the three reservoir securing screws, then lower the reservoir from under the wheel arch.

Refitting

9 Refitting is a reversal of removal.

Washer pump(s)

Removal

10 Proceed as described in paragraphs 2 to 4.
11 Disconnect the wiring connector and the fluid hose from the relevant washer pump.
12 Pull the washer pump from the reservoir and, where applicable, recover the grommet. If the reservoir still contains fluid, be prepared for fluid spillage.

Refitting

13 Refitting is a reversal of removal.

16.4b ... then withdraw the radio/cassette player from the facia

17.2b ... then withdraw the loudspeaker from the door and disconnect the wiring connector

Windscreen washer nozzle

Removal

14 Open and support the bonnet.
15 Working under the bonnet, release the securing tabs using a suitable screwdriver, then push the nozzle from the bonnet. Disconnect the fluid hose, and withdraw the nozzle.

Refitting

16 Refitting is a reversal of removal.

Tailgate washer nozzle

Removal

17 Open the tailgate.
18 Prise off the rubber cover, then push out the washer nozzle though the outside of the tailgate, and disconnect the fluid hose.

Refitting

19 Refitting is a reversal of removal.

16 Radio/cassette player – removal and refitting

Removal

1 Disconnect the battery negative terminal (refer to *Disconnecting the battery* in the Reference Chapter).
2 Remove the facia centre switch/ventilation nozzle housing as described in Chapter 11, Section 29.
3 Remove the four now-exposed radio/ cassette player securing screws.
4 Lift off the surround, then pull the unit forwards from the facia **(see illustrations)**. Disconnect the wiring connectors and the aerial lead from the rear of the unit.
5 Note the bayonet fuse, which is a push-fit in the rear of the unit.

Refitting

6 Refitting is a reversal of removal, ensuring that the wiring is freely routed behind the unit.

17 Loudspeakers – removal and refitting

Door-mounted loudspeakers

1 Remove the door inner trim panel as described in Chapter 11.
2 Undo the three securing screws, then withdraw the loudspeaker from the door and disconnect the wiring connector **(see illustrations)**.
3 Refitting is a reversal of removal, but refit the inner door trim panel with reference to Chapter 11.

Pillar-mounted loudspeakers

4 Disconnect the battery negative terminal (refer to *Disconnecting the battery* in the Reference Chapter).

17.10 Rear loudspeaker securing screws (arrowed) – Saloon models

5 Remove the relevant front pillar trim panel as described in Chapter 11, Section 27.
6 Undo the two screws securing the loudspeaker to the pillar, then lift out the loudspeaker, and separate the two halves of the wiring connector (where necessary, pull the insulating foam from the connector).
7 Refitting is a reversal of removal.

Rear loudspeakers

Saloon models

8 Disconnect the battery negative terminal (refer to *Disconnecting the battery* in the Reference Chapter).
9 Remove the rear parcel shelf, as described in Chapter 11, Section 27.
10 Undo the four securing screws, then withdraw the loudspeaker from the body panel, and disconnect the wiring connector **(see illustration)**.
11 Refitting is a reversal of removal.

Hatchback models

12 Disconnect the battery negative terminal (refer to *Disconnecting the battery* in the Reference Chapter).
13 Remove the relevant luggage compartment parcel shelf support panel, as described in Chapter 11 Section 27.
14 Working under the parcel shelf support panel, unscrew the four loudspeaker securing screws, then withdraw the loudspeaker from under the panel. Disconnect the wiring connector and remove the loudspeaker **(see illustration)**.
15 Refitting is a reversal of removal.

17.14 Rear loudspeaker securing screws (arrowed) – Hatchback models (shown with parcel shelf support removed)

18 Radio aerial – removal and refitting

Front roof-mounted aerial

Removal

1 Working on the roof of the vehicle, remove the two securing screws, then carefully pull the aerial from the roof, and disconnect the aerial lead.

Refitting

2 Refitting is a reversal of removal, but ensure that the aerial housing is securely engaged with the roof panel.

Rear roof-mounted aerial

3 Removal of the aerial base is a complicated operation entailing partial removal of the headlining. If any problems with the aerial or aerial wiring are experienced, contact a Nissan dealer for advice.
4 The aerial stalk can be unscrewed from the base for security purposes or when using a car wash facility.

19 Anti-theft system and engine immobiliser – general information

All models in the range are equipped as standard with a central locking system incorporating an electronic engine immobiliser function.
The electronic engine immobiliser is operated by a transponder fitted to the ignition key, in conjunction with an analogue module fitted around the ignition switch.
When the ignition key is inserted in the switch and turned to the ignition 'on' position, the control module sends a preprogrammed recognition code signal to the analogue module on the ignition switch. If the recognition code signal matches that of the transponder on the ignition key, an unlocking request signal is sent to the engine management ECU allowing the engine to be started. If the ignition key signal is not recognised, the engine management system remains immobilised.
When the ignition is switched off, a locking signal is sent to the ECU and the engine is immobilised until the unlocking request signal is again received.

20 Heated front seat components – general information

Certain models may be equipped with heated front seats. The seats are heated by electrical elements built into the seat cushions.
For access to the heating elements, the

seats must be dismantled, and this work should be entrusted to a Nissan dealer.
Removal and refitting details for the heated seat (centre console) switches are given in Section 4.

21 Airbag system – general information, precautions and system de-activation

General information

A driver's and passenger's airbag are fitted as standard or optional equipment on most models. The driver's airbag is located in the steering wheel centre pad and the passenger's airbag is located above the glovebox in the facia. Side airbags are also available on certain later models and are located in the front seats.
The system is armed only when the ignition is switched on, however, a reserve power source maintains a power supply to the system in the event of a break in the main electrical supply. The steering wheel and facia airbags are activated by a 'g' sensor (deceleration sensor), and controlled by an electronic control unit located in the steering wheel (single airbag system) or under the rear of the centre console (twin airbag system). The side airbags are activated by severe side impact and operate in conjunction with the main system.
The airbags are inflated by a gas generator, which forces the bag out from its location in the steering wheel, facia or seat back frame.

Precautions

⚠ **Warning: The following precautions must be observed when working on vehicles equipped with an airbag system, to prevent the possibility of personal injury.**

General precautions

a) *Do not disconnect the battery with the engine running.*
b) *Before carrying out any work in the vicinity of the airbag, removal of any of the airbag components, or any welding work on the vehicle, de-activate the system as described in the following sub-Section.*
c) *Do not attempt to test any of the airbag system circuits using test meters or any other test equipment.*
d) *If the airbag warning light comes on, or any fault in the system is suspected, consult a Nissan dealer without delay. **Do not** attempt to carry out fault diagnosis, or any dismantling of the components.*

Precautions to be taken when handling an airbag

a) *Transport the airbag by itself, bag upward.*
b) *Do not put your arms around the airbag.*
c) *Carry the airbag close to the body, bag outward.*

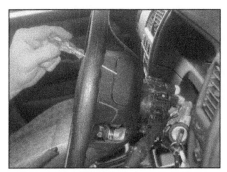

22.4a Prise off the steering wheel lower cover . . .

22.4b . . . and the two side covers. The airbag securing bolts (arrowed) are located behind the side covers

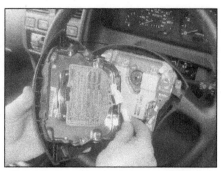

22.6 Lift the airbag unit from the steering wheel and disconnect the wiring connector

d) Do not drop the airbag or expose it to impacts.
e) Do not attempt to dismantle the airbag unit.
f) Do not connect any form of electrical equipment to any part of the airbag circuit.

Precautions to be taken when storing an airbag unit

a) Store the unit in a cupboard with the airbag upward.
b) Do not expose the airbag to temperatures above 80°C.
c) Do not expose the airbag to flames.
d) Do not attempt to dispose of the airbag – consult a Nissan dealer.
e) Never refit an airbag which is known to be faulty or damaged.

De-activation of airbag system

The system must be de-activated before carrying out any work on the airbag components or surrounding area:
a) Switch on the ignition and check the operation of the airbag warning light on the instrument panel. The light should illuminate when the ignition is switched on, then extinguish.
b) Switch off the ignition.
c) Remove the ignition key.
d) Switch off all electrical equipment.
e) Disconnect the battery negative terminal (refer to 'Disconnecting the battery' in the Reference Chapter).
f) Insulate the battery negative terminal and the end of the battery negative lead to prevent any possibility of contact.
g) Wait for at least ten minutes before carrying out any further work.

Activation of airbag system

To activate the system on completion of any work, proceed as follows:
a) Ensure that there are no occupants in the vehicle, and that there are no loose objects around the vicinity of the steering wheel. Close the vehicle doors and windows.
b) Ensure that the ignition is switched off then reconnect the battery negative terminal.

c) Open the driver's door and switch on the ignition, without reaching in front of the steering wheel. Check that the airbag warning light illuminates briefly then extinguishes.
d) Switch off the ignition.
e) If the airbag warning light does not operate as described in paragraph c), consult a Nissan dealer before driving the vehicle.

22 Airbag system components – removal and refitting

⚠️ **Warning: Refer to the precautions given in Section 21 before attempting to carry out work on any of the airbag components. Any suspected faults with the airbag system should be referred to a Nissan dealer – under no circumstances attempt to carry out any work other than removal and refitting of the front airbag unit(s) and/or the rotary connector, as described in the following paragraphs.**

Airbag electronic control units

1 On models with a driver's side airbag only, the ECU is integral with the steering wheel and must not be separated from the wheel.
2 On models with driver and passenger airbags, the ECU is located under the centre console at the rear, and is accessible after removal of the centre console as described in

22.13a Unscrew the four securing screws . . .

Chapter 11. The side airbags are also controlled by this ECU.

Driver's airbag unit

Note: A Torx T50 type tamperproof socket bit will be required to remove the airbag retaining bolts, and new bolts must be used for refitting.

Removal

3 De-activate the airbag system as described in Section 21.
4 Prise the lower cover and the two side covers from the rear of the steering wheel/airbag unit **(see illustrations)**.
5 Using a Torx tamperproof bit, undo the two airbag unit securing bolts (one each side).
6 Carefully lift the airbag unit from the steering wheel and disconnect the wiring connector **(see illustration)**.
7 If the airbag unit is to be stored for any length of time, refer to the storage precautions given in Section 21.

Refitting

8 Refitting is a reversal of removal, bearing in mind the following points:
a) Use new bolts to secure the airbag unit and tighten the bolts to the specified torque.
b) Do not strike the airbag unit, or expose it to impacts during refitting.
c) On completion of refitting, activate the airbag system as described in Section 21.

Airbag rotary connector

Removal

9 Remove the driver's airbag unit, as described previously in this Section.
10 Remove the steering wheel as described in Chapter 10.
11 Remove the steering column shrouds as described in Chapter 11, Section 29.
12 Trace the rotary connector wiring back to the connectors below the steering column and disconnect the wiring.
13 Unscrew the four securing screws, and withdraw the rotary connector from the steering column, feeding the wiring harness through the stalk switch bracket **(see illustrations)**. Note the routing of the wiring harness.

Refitting

14 Refitting is a reversal of removal, bearing in mind the following points:

a) *Centralise the rotary connector by first turning the rotating centre part fully clockwise. From this position, turn it anti-clockwise 2.5 turns (3.5 turns on later models), and align the arrow on the rotating centre part with the corresponding arrow on the connector body* **(see illustration)**. *Instructions for centralising the unit are also provided on a label on the face of the unit and should be referred to for latest information.*

b) *Ensure that the roadwheels are in the straight-ahead position before refitting the rotary connector and steering wheel.*

c) *Before refitting the steering column shrouds, ensure that the rotary connector wiring harness is correctly routed as noted before removal.*

d) *Refit the steering wheel as described in Chapter 10, and refit the airbag unit as described previously in this Section.*

Passenger's airbag unit

Note: *A Torx T50 type tamperproof socket bit will be required to remove the airbag retaining bolts, and new bolts must be used for refitting.*

Removal

15 The passenger's airbag is fitted to the

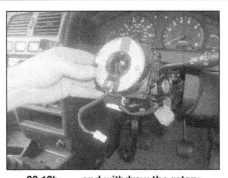

22.13b . . . and withdraw the rotary connector from the steering column

upper part of the facia, above the glovebox.

16 De-activate the airbag system as described in Section 21.

17 Remove the glovebox as described in Chapter 11, Section 29.

18 Disconnect the airbag wiring connector.

19 Undo the nuts and tamperproof bolts securing the airbag assembly to the facia support rail. Carefully withdraw the airbag from the facia.

20 If the airbag unit is to be stored for any length of time, refer to the storage precautions given in Section 21.

Refitting

21 Refitting is a reversal of removal, bearing

22.14 With the rotary connector centralised, align the arrow on the rotating part with the arrow on the connector body

in mind the following points:

a) *Use new bolts to secure the airbag unit and tighten the bolts to the specified torque.*

b) *Do not strike the airbag unit, or expose it to impacts during refitting.*

c) *On completion of refitting, activate the airbag system as described in Section 21.*

Side airbag units

22 The side airbags are located internally within the front seat back and no attempt should be made to remove them. Any suspected problems with the side airbag system should be referred to a Nissan dealer.

NISSAN ALMERA 1995 on wiring diagrams

Diagram 1

Key to symbols

Bulb	⊗
Switch	
Multiple contact switch (ganged)	
Fuse/fusible link and current rating	F5 30A
Resistor	
Variable resistor	
Connecting wires	
Plug and socket contact	
Item no.	2
Pump/motor	Ⓜ
Earth point and location	E12
Gauge/meter	
Diode	
Wire splice or soldered joint	
Solenoid actuator	
Light emitting diode (LED)	
Wire colour and wire size (brown with black tracer, 4mm²)	Br/Sw 4
Screened cable	

Dashed outline denotes part of a larger item, containing in this case an electronic or solid state device.
2 – unspecified connector pin 2.
182 1 – Connector C182, pin 1.

Earth points

E1	Battery earth	**E6**	Transmisson tunnel	**E11**	LH rear pillar
E2	Dashboard	**E7**	RH passenger compartment floor	**E12**	Engine ground
E3	Engine compartment front	**E8**	Above ABS module	**E13**	RH rear engine
E4	Passenger compartment floor	**E9**	Below ABS module	**E14**	Dashboard centre
E5	LH passenger compartment floor	**E10**	Luggage compartment centre		

Key to circuits

Diagram 1	Information for wiring diagrams
Diagram 2	Front and rear wash/wipe, airbag and cigarette lighter
Diagram 3	Central locking, sunroof and ABS
Diagram 4	Interior lights, headlights, sidelights, licence plate light and tail lights, brake light and horn
Diagram 5	Headlight leveling, direction indicator lights, hazard warning, reverse lights, foglights and heated seats
Diagram 6	Electric mirrors, electric windows, electric rear quarter windows, heated rear window and mirrors
Diagram 7	Heater blower, air conditioning and starting and charging
Diagram 8	Automatic transmission, radio and instrument module

Engine fuse box locations

Fuse table

Fuses	Rating	Circuit protected	Fuses	Rating	Circuit protected
Fa	-	Not used	F16	10A	Oxygen sensor
Fb	-	Not used	F17	15A	Fuel pump
Fc	30A	ABS	F18	-	Not used
Fd	30A	ABS	F19	20A	Wash/wipe
Fe	30A	A/C	F21	3A	Airbags
Ff	25A	Electric windows, sun roof, illumination, central locking	F22	10A	Airbags
			F23	10A	Rear wash/wipe
Fg	75A	Main power	F24	7.5A	Horn, radio, headlight levelling
Fh	-	Not used	F25	10A	Starter, A/C, engine management
Fi	30A	Ignition switch	F26	-	Not used
F1	15A	Heater, A/C	F29	-	Not used
F2	15A	Heater, A/C	F30	7.5A	A/C,
F3	20A	Heated rear window	F31	7.5A	Alternator
F4	15A	Front fog lights	F36	10A	Tail lights, fog lights, illumination
F5	-	Not used	F37	-	Not used
F6	7.5A	Heater, A/C	F39	15A	Headlights, Rear fog light
F7	20A	Electric windows, sunroof	F40	15A	Headlights
F8	10A	Instrument module, automatic transmission, reverse lights	F41	7.5A	Main power
			F42	10A	Horn
F9	10A	Rear wiper (early models)	F43	7.5A	Rear fog lights
F10	15A	Radio, illumination	F44	-	Not used
F11	7.5A	Direction indicators	F45	-	Not used
F12	7.5A	ABS			
F13	15A	Horn			
F14	10A	ABS, brake lights			
F15	10A	Direction indicators			
F20	7.5A	Interior lights			

MTS
H32652

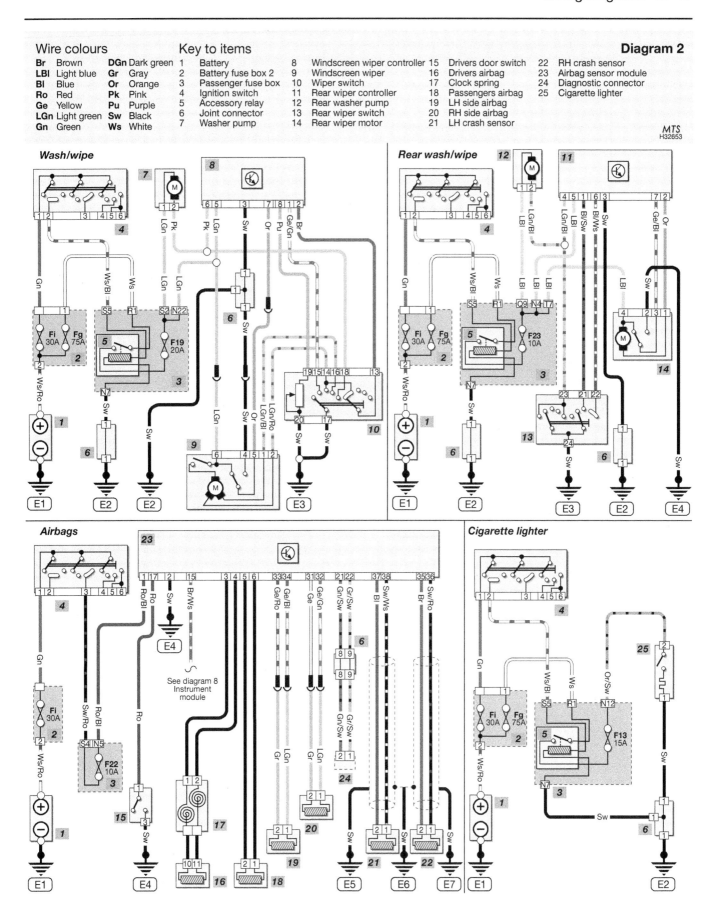

Diagram 2

Wire colours

Br	Brown	DGn	Dark green
LBl	Light blue	Gr	Gray
Bl	Blue	Or	Orange
Ro	Red	Pk	Pink
Ge	Yellow	Pu	Purple
LGn	Light green	Sw	Black
Gn	Green	Ws	White

Key to items

1	Battery	8	Windscreen wiper controller
2	Battery fuse box 2	9	Windscreen wiper
3	Passenger fuse box	10	Wiper switch
4	Ignition switch	11	Rear wiper controller
5	Accessory relay	12	Rear washer pump
6	Joint connector	13	Rear wiper switch
7	Washer pump	14	Rear wiper motor
15	Drivers door switch	22	RH crash sensor
16	Drivers airbag	23	Airbag sensor module
17	Clock spring	24	Diagnostic connector
18	Passengers airbag	25	Cigarette lighter
19	LH side airbag		
20	RH side airbag		
21	LH crash sensor		

MTS
H32653

Wash/wipe

Rear wash/wipe

Airbags

Cigarette lighter

Diagram 3

Wire colours

Br	Brown	**DGn**	Dark green
LBl	Light blue	**Gr**	Gray
Bl	Blue	**Or**	Orange
Ro	Red	**Pk**	Pink
Ge	Yellow	**Pu**	Purple
LGn	Light green	**Sw**	Black
Gn	Green	**Ws**	White

Key to items

1	Battery	28	Solenoid valve relay	36	RH rear wheel sensor
2	Battery fuse box 2	29	ABS relay box	37	Engine fuse box 3
3	Passenger fuse box	30	ABS module	38	Drivers door switch
4	Ignition switch	31	Brake light switch	39	Passenger door switch
6	Joint connector	32	ABS actuator	40	LH rear door switch
24	Diagnostic connector	33	LH front wheel sensor	41	RH rear door switch
26	Ignition relay	34	RH front wheel sensor	42	Circuit breaker
27	Motor relay	35	LH rear wheel sensor	43	Remote control unit

44	Drivers door lock
45	Passenger door lock
46	LH rear door lock
47	RH rear door lock
48	Central locking module
49	Sun roof motor module
50	Electric window relay
51	Sun roof switch

MTS H32654

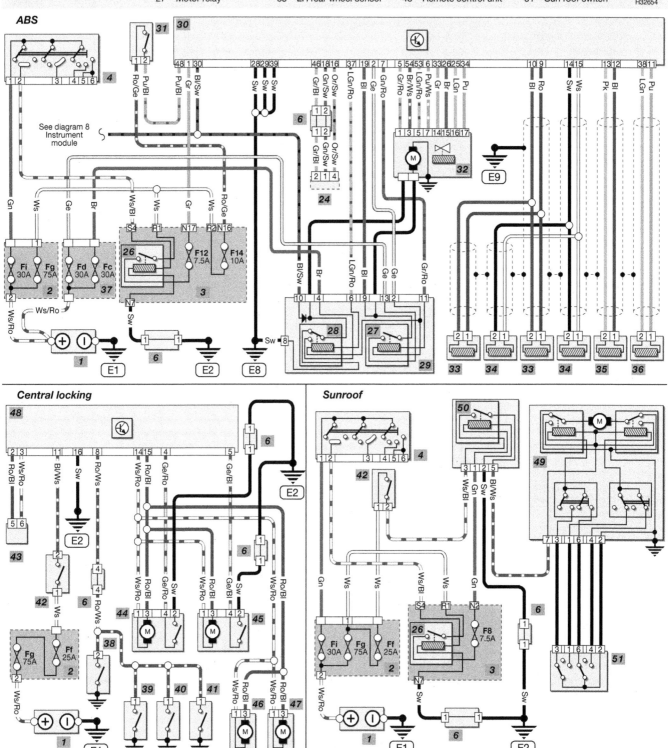

Wire colours

Br	Brown	**Gr**	Gray
LBl	Light blue	**Or**	Orange
Bl	Blue	**Pk**	Pink
Ro	Red	**Pu**	Purple
Ge	Yellow	**Sw**	Black
LGn	Light green	**Ws**	White
Gn	Green		
DGn	Dark green		

Key to items

1	Battery	52	Light switch	62	Glovebox light
2	Battery fuse box 2	53	Spot light	63	Hazard light switch
3	Passenger fuse box	54	Luggage/boot light	64	Headlight washer switch
6	Joint connector	55	Luggage/boot	65	Fan switch
17	Clock spring		light switch	66	CD player
26	Ignition relay	56	Front fog light switch	67	Horn relay
31	Brake light switch	57	Rear fog light switch	68	Dual horn
38	Drivers door switch	58	Auto gearbox lever	69	Horn switch
39	Passenger door switch	59	Ashtray illumination	70	LH tail light
40	LH rear door switch	60	Heated rear window switch		a) tail light
41	RH rear door switch	61	Glovebox light switch		b) brake light
51	Sun roof switch				

71	RH tail light (as 70)
72	Number plate light
73	LH side light
74	RH side light
75	LH headlight
	a) main beam
	b) low beam
76	RH headlight (as 75)
77	High level brake light
134	Instrument module

* Saloon
** Hatchback
*** With sunroof
**** Without sunroof

Diagram 4

MTS
H32655

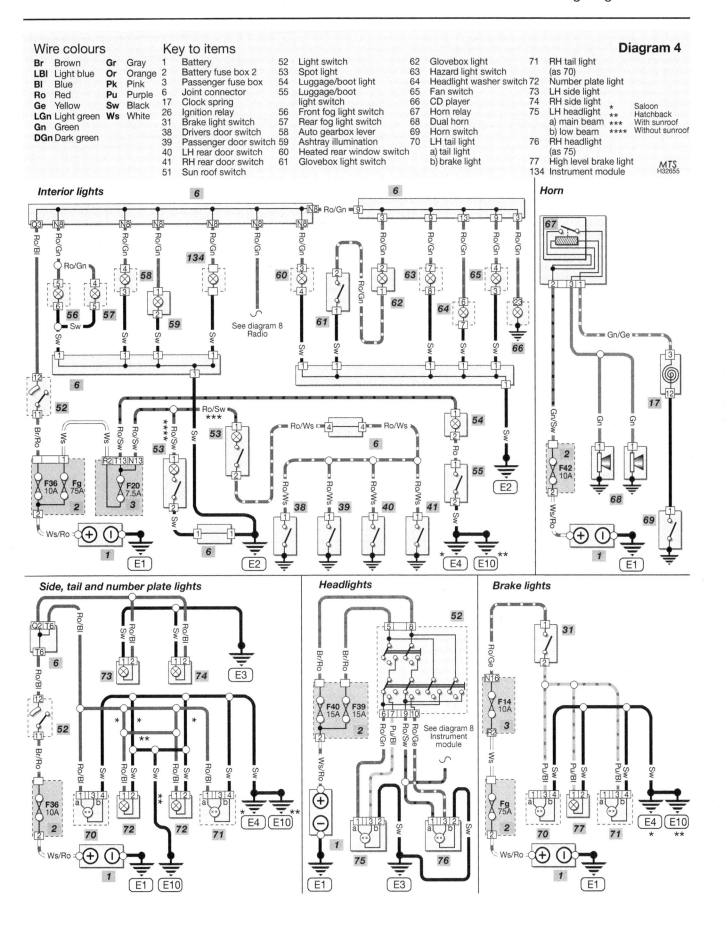

Wire colours

Br	Brown	**DGn**	Dark green
LBl	Light blue	**Gr**	Gray
Bl	Blue	**Or**	Orange
Ro	Red	**Pk**	Pink
Ge	Yellow	**Pu**	Purple
LGn	Light green	**Sw**	Black
Gn	Green	**Ws**	White

Key to items

1 Battery
2 Battery fuse box 2
3 Passenger fuse box
4 Ignition switch
6 Joint connector
26 Ignition relay
56 Front fog light switch
57 Rear fog light switch

70 LH tail light
  c) direction indicator
71 RH tail light
  (as 70)
78 Combination flasher unit
79 Indicator switch
80 LH side direction indicator
81 RH side direction indicator

82 LH front direction indicator
83 RH front direction indicator
84 Hazard warning switch
85 Headlight leveling switch
86 LH headlight motor
87 RH headlight motor
88 Reverse light switch
89 Inhibitor switch

90 LH reversing light
91 RH reversing light
92 Front fog light relay
93 Front fog lights
94 Rear fog light relay
95 Rear fog lights
* Saloon
** Hatchback
*** Manual
**** Automatic

Diagram 5

MTS
H32656

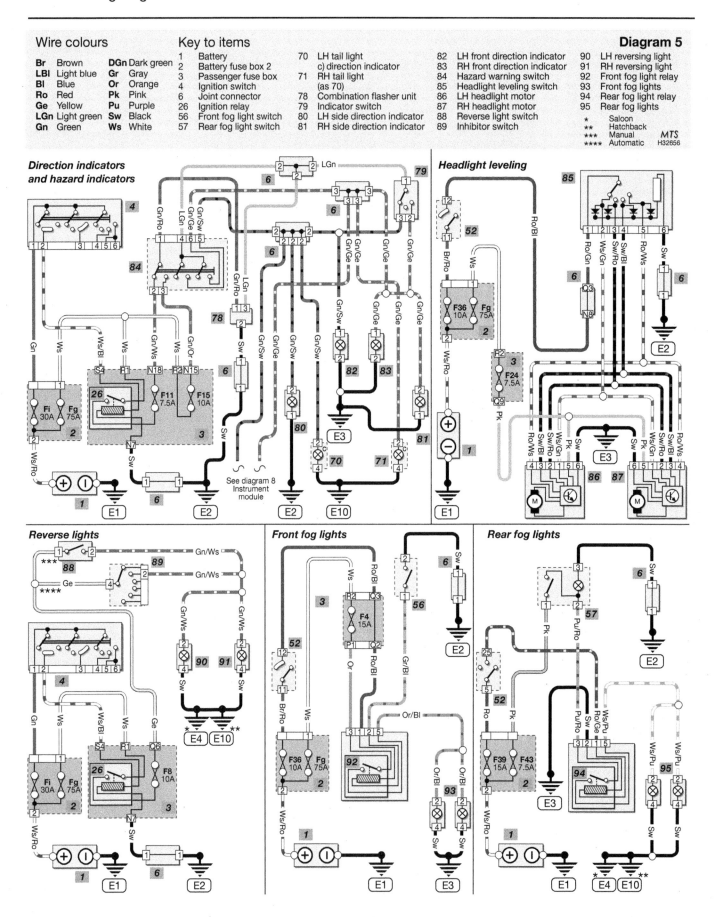

Direction indicators and hazard indicators

Headlight leveling

Reverse lights

Front fog lights

Rear fog lights

Wire colours

Br	Brown	DGn	Dark green
LBl	Light blue	Gr	Gray
Bl	Blue	Or	Orange
Ro	Red	Pk	Pink
Ge	Yellow	Pu	Purple
LGn	Light green	Sw	Black
Gn	Green	Ws	White

Key to items

1	Battery	42	Circuit breaker
2	Battery fuse box 2	50	Electric window relay
3	Passenger fuse box	60	Heated rear window switch
4	Ignition switch	96	Mirror switch
5	Accessory relay	97	LH mirror motors
6	Joint connector	98	RH mirror motors
26	Ignition relay	99	Condenser
100	Rear window heater	107	RH rear window switch
101	Driver's window switch	108	RH rear window switch
102	Drivers window motor		
103	Passenger's window switch		
104	Passenger's window motor		
105	LH rear window switch		
106	LH rear window motor		

Diagram 6

* Hatchback only   MTS
H32657

Electric mirrors

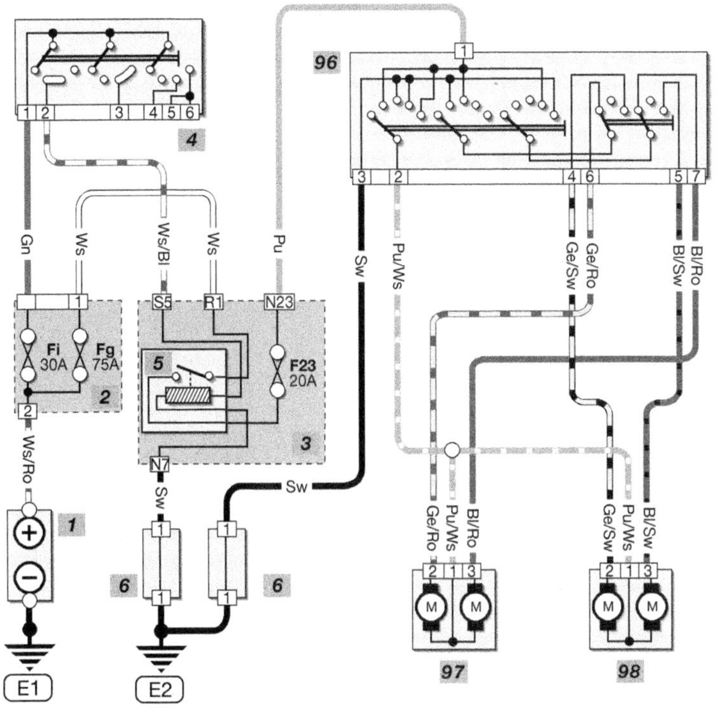

Heated rear window

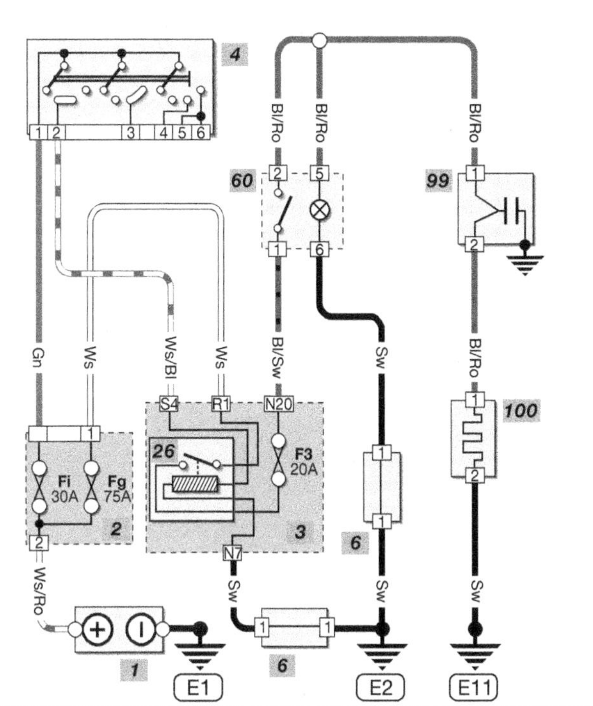

Electric windows

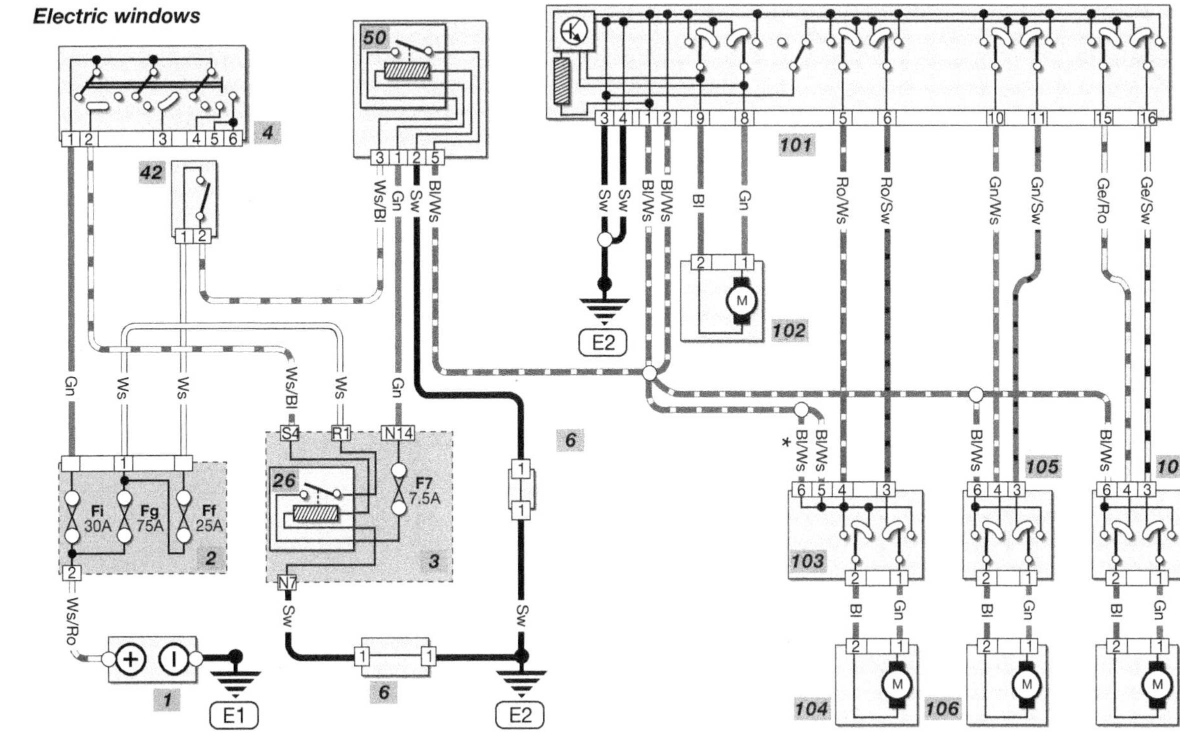

Wire colours

Br	Brown	DGn	Dark green
LBl	Light blue	Gr	Gray
Bl	Blue	Or	Orange
Ro	Red	Pk	Pink
Ge	Yellow	Pu	Purple
LGn	Light green	Sw	Black
Gn	Green	Ws	White

Key to items

1	Battery	109	Blower relay	115	A/C relay	121	Dual pressure switch
2	Battery fuse box 2	110	Intake door motor	116	A/C switch	122	Alternator
3	Passenger fuse box	111	Fan resistor	117	Thermo control amplifier	123	Starter motor
4	Ignition switch	112	Recirculation switch	118	Solenoid valve		
6	Joint connector	113	Fan resistor	119	Thermal protector switch		
65	Fan switch	114	Battery fuse box 1	120	A/C compressor		

Diagram 7

MTS
H32658

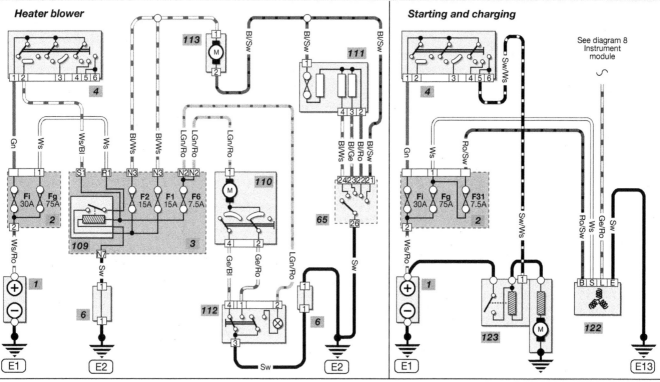

Heater blower

Starting and charging

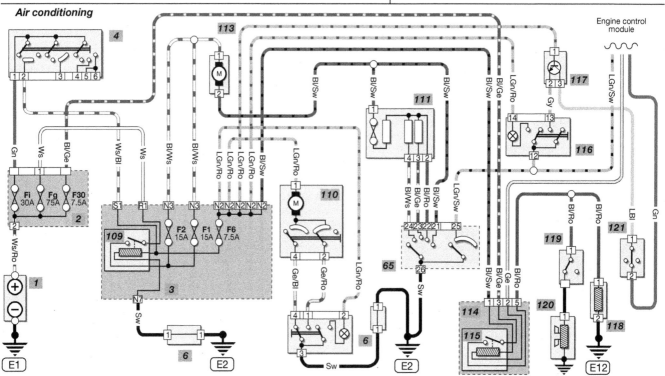

Air conditioning

Wire colours

Br	Brown	**DGn**	Dark green
LBl	Light blue	**Gr**	Gray
Bl	Blue	**Or**	Orange
Ro	Red	**Pk**	Pink
Ge	Yellow	**Pu**	Purple
LGn	Light green	**Sw**	Black
Gn	Green	**Ws**	White

MTS
H32659

Key to items

1 Battery
2 Battery fuse box 2
3 Passenger fuse box
4 Ignition switch
5 Accessory relay
6 Joint connector
26 Ignition relay
38 Drivers door switch
39 Passenger door switch
40 LH rear door switch
41 RH rear door switch
65 Fan switch
89 Inhibitor switch

124 Radio
125 Aerial
126 LH rear speaker
127 RH rear speaker
128 LH pillar speaker
129 RH pillar speaker
130 LH door speaker
131 RH door speaker
132 Solenoid valve
133 Overdrive control switch
134 Instrument module cont.
  a) speedometer
  b) tachometer

134 Instrument module cont.
  c) fuel gauge
  d) water temperature gauge
  e) engine warning
  f) ABS warning
  g) airbag warning
  h) low fuel warning
  i) oil warning
  j) brake warning
  k) charge warning
  l) washer fluid level
  m) door open warning

134 Instrument module cont.
  n) main beam warning
  o) LH direction indicator
  p) RH direction indicator
135 Vehicle speed sensor
136 Fuel tank sensor
137 Temperature sensor
138 Washer fluid level sensor
139 Brake fluid level sensor
140 Handbrake switch
141 Oil pressure switch

Diagram 8

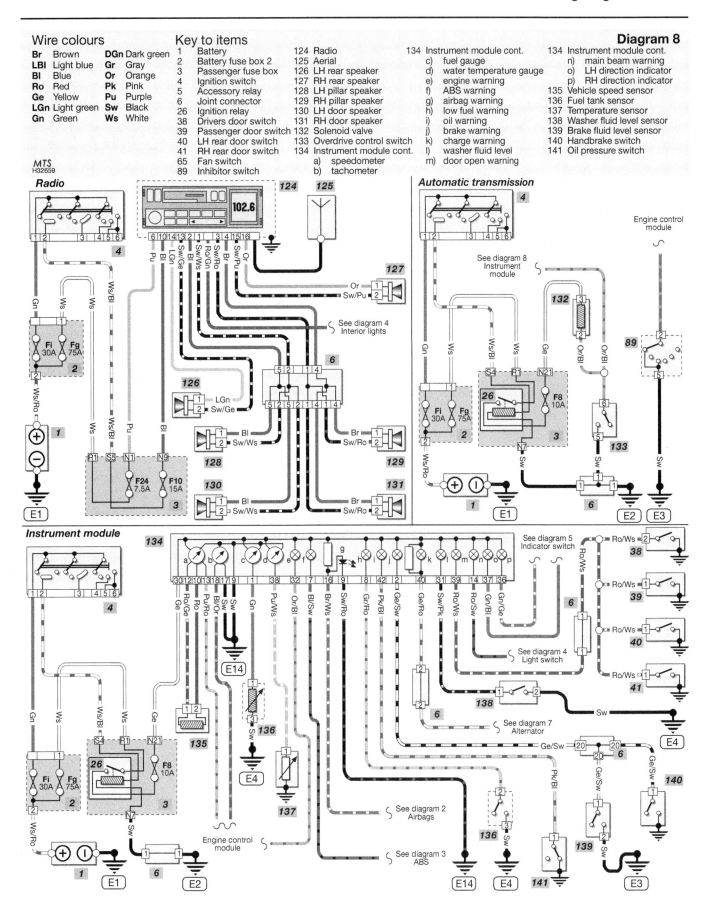

Notes

Dimensions and weights

Note: *All figures are approximate, and may vary according to model. Refer to manufacturer's data for exact figures.*

Dimensions

Overall length:
    Hatchback models . . . . 4140 mm
    Saloon models . . . . 4340 mm
Overall width . . . . 1690 mm
Overall height (unladen) . . . . 1395 mm
Wheelbase . . . . 2535 mm
Front track . . . . 1470 mm
Rear track . . . . 1435 mm

Weights

Kerb weight:
    1.4 litre models . . . . 1035 to 1090 kg
    1.6 litre manual transmission models . . . . 1070 to 1155 kg
    1.6 litre automatic transmission models . . . . 1110 to 1185 kg
Maximum towing weight (with a braked trailer):
    1.4 litre models . . . . 900 kg
    1.6 litre manual transmission models . . . . 1120 kg
    1.6 litre automatic transmission models . . . . 1160 kg

Length (distance)
Inches (in)	x 25.4	= Millimetres (mm)	x 0.0394	= Inches (in)
Feet (ft)	x 0.305	= Metres (m)	x 3.281	= Feet (ft)
Miles	x 1.609	= Kilometres (km)	x 0.621	= Miles

Volume (capacity)
Cubic inches (cu in; in³)	x 16.387	= Cubic centimetres (cc; cm³)	x 0.061	= Cubic inches (cu in; in³)
Imperial pints (Imp pt)	x 0.568	= Litres (l)	x 1.76	= Imperial pints (Imp pt)
Imperial quarts (Imp qt)	x 1.137	= Litres (l)	x 0.88	= Imperial quarts (Imp qt)
Imperial quarts (Imp qt)	x 1.201	= US quarts (US qt)	x 0.833	= Imperial quarts (Imp qt)
US quarts (US qt)	x 0.946	= Litres (l)	x 1.057	= US quarts (US qt)
Imperial gallons (Imp gal)	x 4.546	= Litres (l)	x 0.22	= Imperial gallons (Imp gal)
Imperial gallons (Imp gal)	x 1.201	= US gallons (US gal)	x 0.833	= Imperial gallons (Imp gal)
US gallons (US gal)	x 3.785	= Litres (l)	x 0.264	= US gallons (US gal)

Mass (weight)
Ounces (oz)	x 28.35	= Grams (g)	x 0.035	= Ounces (oz)
Pounds (lb)	x 0.454	= Kilograms (kg)	x 2.205	= Pounds (lb)

Force
Ounces-force (ozf; oz)	x 0.278	= Newtons (N)	x 3.6	= Ounces-force (ozf; oz)
Pounds-force (lbf; lb)	x 4.448	= Newtons (N)	x 0.225	= Pounds-force (lbf; lb)
Newtons (N)	x 0.1	= Kilograms-force (kgf; kg)	x 9.81	= Newtons (N)

Pressure
Pounds-force per square inch (psi; lbf/in²; lb/in²)	x 0.070	= Kilograms-force per square centimetre (kgf/cm²; kg/cm²)	x 14.223	= Pounds-force per square inch (psi; lbf/in²; lb/in²)
Pounds-force per square inch (psi; lbf/in²; lb/in²)	x 0.068	= Atmospheres (atm)	x 14.696	= Pounds-force per square inch (psi; lbf/in²; lb/in²)
Pounds-force per square inch (psi; lbf/in²; lb/in²)	x 0.069	= Bars	x 14.5	= Pounds-force per square inch (psi; lbf/in²; lb/in²)
Pounds-force per square inch (psi; lbf/in²; lb/in²)	x 6.895	= Kilopascals (kPa)	x 0.145	= Pounds-force per square inch (psi; lbf/in²; lb/in²)
Kilopascals (kPa)	x 0.01	= Kilograms-force per square centimetre (kgf/cm²; kg/cm²)	x 98.1	= Kilopascals (kPa)
Millibar (mbar)	x 100	= Pascals (Pa)	x 0.01	= Millibar (mbar)
Millibar (mbar)	x 0.0145	= Pounds-force per square inch (psi; lbf/in²; lb/in²)	x 68.947	= Millibar (mbar)
Millibar (mbar)	x 0.75	= Millimetres of mercury (mmHg)	x 1.333	= Millibar (mbar)
Millibar (mbar)	x 0.401	= Inches of water (inH₂O)	x 2.491	= Millibar (mbar)
Millimetres of mercury (mmHg)	x 0.535	= Inches of water (inH₂O)	x 1.868	= Millimetres of mercury (mmHg)
Inches of water (inH₂O)	x 0.036	= Pounds-force per square inch (psi; lbf/in²; lb/in²)	x 27.68	= Inches of water (inH₂O)

Torque (moment of force)
Pounds-force inches (lbf in; lb in)	x 1.152	= Kilograms-force centimetre (kgf cm; kg cm)	x 0.868	= Pounds-force inches (lbf in; lb in)
Pounds-force inches (lbf in; lb in)	x 0.113	= Newton metres (Nm)	x 8.85	= Pounds-force inches (lbf in; lb in)
Pounds-force inches (lbf in; lb in)	x 0.083	= Pounds-force feet (lbf ft; lb ft)	x 12	= Pounds-force inches (lbf in; lb in)
Pounds-force feet (lbf ft; lb ft)	x 0.138	= Kilograms-force metres (kgf m; kg m)	x 7.233	= Pounds-force feet (lbf ft; lb ft)
Pounds-force feet (lbf ft; lb ft)	x 1.356	= Newton metres (Nm)	x 0.738	= Pounds-force feet (lbf ft; lb ft)
Newton metres (Nm)	x 0.102	= Kilograms-force metres (kgf m; kg m)	x 9.804	= Newton metres (Nm)

Power
Horsepower (hp)	x 745.7	= Watts (W)	x 0.0013	= Horsepower (hp)

Velocity (speed)
Miles per hour (miles/hr; mph)	x 1.609	= Kilometres per hour (km/hr; kph)	x 0.621	= Miles per hour (miles/hr; mph)

Fuel consumption*
Miles per gallon, Imperial (mpg)	x 0.354	= Kilometres per litre (km/l)	x 2.825	= Miles per gallon, Imperial (mpg)
Miles per gallon, US (mpg)	x 0.425	= Kilometres per litre (km/l)	x 2.352	= Miles per gallon, US (mpg)

Temperature
Degrees Fahrenheit = (°C x 1.8) + 32   Degrees Celsius (Degrees Centigrade; °C) = (°F - 32) x 0.56

It is common practice to convert from miles per gallon (mpg) to litres/100 kilometres (l/100km), where mpg x l/100 km = 282

Spare parts are available from many sources, including maker's appointed garages, accessory shops, and motor factors. To be sure of obtaining the correct parts, it will sometimes be necessary to quote the vehicle identification number. If possible, it can also be useful to take the old parts along for positive identification. Items such as starter motors and alternators may be available under a service exchange scheme - any parts returned should be clean.

Our advice regarding spare parts is as follows.

Officially appointed garages

This is the best source of parts which are peculiar to your car, and which are not otherwise generally available (eg, badges, interior trim, certain body panels, etc). It is also the only place at which you should buy parts if the vehicle is still under warranty.

Accessory shops

These are very good places to buy materials and components needed for the maintenance of your car (oil, air and fuel filters, light bulbs, drivebelts, greases, brake pads, touch-up paint, etc). Components of this nature sold by a reputable shop are usually of the same standard as those used by the car manufacturer.

Besides components, these shops also sell tools and general accessories, usually have convenient opening hours, charge lower prices, and can often be found close to home. Some accessory shops have parts counters where components needed for almost any repair job can be purchased or ordered.

Motor factors

Good factors will stock the more important components which wear out comparatively quickly, and can sometimes supply individual components needed for the overhaul of a larger assembly (eg, brake seals and hydraulic parts, bearing shells, pistons, valves). They may also handle work such as cylinder block reboring, crankshaft regrinding, etc.

Tyre and exhaust specialists

These outlets may be independent, or members of a local or national chain. They frequently offer competitive prices when compared with a main dealer or local garage, but it will pay to obtain several quotes before making a decision. When researching prices, also be sure to ask what "extras" may be added - for instance, fitting a new valve, balancing the wheel, and checking the tracking (front wheels) are all commonly charged on top of the price of a new tyre.

Other sources

Beware of parts or materials obtained from market stalls, car boot sales or similar outlets. Such items are not invariably sub-standard, but there is little chance of compensation if they do prove unsatisfactory. in the case of safety-critical components such as brake pads, there is the risk not only of financial loss, but also of an accident causing injury or death.

Second-hand components or assemblies obtained from a car breaker can be a good buy in some circumstances, but this sort of purchase is best made by the experienced DIY mechanic.

Vehicle identification

Modifications are a continuing and unpublicised process in vehicle manufacture, quite apart from major model changes. Spare parts manuals and lists are compiled upon a numerical basis, the individual vehicle identification numbers being essential to correct identification of the component concerned. When ordering spare parts, always give as much information as possible. Quote the car model, year of manufacture, body and engine numbers as appropriate.

The *Vehicle Identification Number (VIN)* plate is riveted to the engine compartment bulkhead, and can be viewed once the bonnet is open. The plate carries the VIN, vehicle weight information, and paint and trim colour codes. The vehicle identification number is also stamped into the bulkhead by the side of the plate **(see illustration)**.

The *engine number* is stamped on a machined surface on the front side of the cylinder block, at the flywheel end. The first part of the engine number gives the engine code – eg, GA14 DE.

The *transmission number* is stamped on the clutch release lever on manual transmission models, and on the governor cap on the top of the transmission casing on automatic transmission models.

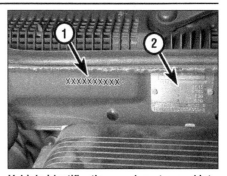

Vehicle identification number stamped into bulkhead (1) and VIN plate (2)

Whenever servicing, repair or overhaul work is carried out on the car or its components, observe the following procedures and instructions. This will assist in carrying out the operation efficiently and to a professional standard of workmanship.

Joint mating faces and gaskets

When separating components at their mating faces, never insert screwdrivers or similar implements into the joint between the faces in order to prise them apart. This can cause severe damage which results in oil leaks, coolant leaks, etc upon reassembly. Separation is usually achieved by tapping along the joint with a soft-faced hammer in order to break the seal. However, note that this method may not be suitable where dowels are used for component location.

Where a gasket is used between the mating faces of two components, a new one must be fitted on reassembly; fit it dry unless otherwise stated in the repair procedure. Make sure that the mating faces are clean and dry, with all traces of old gasket removed. When cleaning a joint face, use a tool which is unlikely to score or damage the face, and remove any burrs or nicks with an oilstone or fine file.

Make sure that tapped holes are cleaned with a pipe cleaner, and keep them free of jointing compound, if this is being used, unless specifically instructed otherwise.

Ensure that all orifices, channels or pipes are clear, and blow through them, preferably using compressed air.

Oil seals

Oil seals can be removed by levering them out with a wide flat-bladed screwdriver or similar implement. Alternatively, a number of self-tapping screws may be screwed into the seal, and these used as a purchase for pliers or some similar device in order to pull the seal free.

Whenever an oil seal is removed from its working location, either individually or as part of an assembly, it should be renewed.

The very fine sealing lip of the seal is easily damaged, and will not seal if the surface it contacts is not completely clean and free from scratches, nicks or grooves. If the original sealing surface of the component cannot be restored, and the manufacturer has not made provision for slight relocation of the seal relative to the sealing surface, the component should be renewed.

Protect the lips of the seal from any surface which may damage them in the course of fitting. Use tape or a conical sleeve where possible. Lubricate the seal lips with oil before fitting and, on dual-lipped seals, fill the space between the lips with grease.

Unless otherwise stated, oil seals must be fitted with their sealing lips toward the lubricant to be sealed.

Use a tubular drift or block of wood of the appropriate size to install the seal and, if the seal housing is shouldered, drive the seal down to the shoulder. If the seal housing is unshouldered, the seal should be fitted with its face flush with the housing top face (unless otherwise instructed).

Screw threads and fastenings

Seized nuts, bolts and screws are quite a common occurrence where corrosion has set in, and the use of penetrating oil or releasing fluid will often overcome this problem if the offending item is soaked for a while before attempting to release it. The use of an impact driver may also provide a means of releasing such stubborn fastening devices, when used in conjunction with the appropriate screwdriver bit or socket. If none of these methods works, it may be necessary to resort to the careful application of heat, or the use of a hacksaw or nut splitter device.

Studs are usually removed by locking two nuts together on the threaded part, and then using a spanner on the lower nut to unscrew the stud. Studs or bolts which have broken off below the surface of the component in which they are mounted can sometimes be removed using a stud extractor. Always ensure that a blind tapped hole is completely free from oil, grease, water or other fluid before installing the bolt or stud. Failure to do this could cause the housing to crack due to the hydraulic action of the bolt or stud as it is screwed in.

When tightening a castellated nut to accept a split pin, tighten the nut to the specified torque, where applicable, and then tighten further to the next split pin hole. Never slacken the nut to align the split pin hole, unless stated in the repair procedure.

When checking or retightening a nut or bolt to a specified torque setting, slacken the nut or bolt by a quarter of a turn, and then retighten to the specified setting. However, this should not be attempted where angular tightening has been used.

For some screw fastenings, notably cylinder head bolts or nuts, torque wrench settings are no longer specified for the latter stages of tightening, "angle-tightening" being called up instead. Typically, a fairly low torque wrench setting will be applied to the bolts/nuts in the correct sequence, followed by one or more stages of tightening through specified angles.

Locknuts, locktabs and washers

Any fastening which will rotate against a component or housing during tightening should always have a washer between it and the relevant component or housing.

Spring or split washers should always be renewed when they are used to lock a critical component such as a big-end bearing retaining bolt or nut. Locktabs which are folded over to retain a nut or bolt should always be renewed.

Self-locking nuts can be re-used in non-critical areas, providing resistance can be felt when the locking portion passes over the bolt or stud thread. However, it should be noted that self-locking stiffnuts tend to lose their effectiveness after long periods of use, and should then be renewed as a matter of course.

Split pins must always be replaced with new ones of the correct size for the hole.

When thread-locking compound is found on the threads of a fastener which is to be re-used, it should be cleaned off with a wire brush and solvent, and fresh compound applied on reassembly.

Special tools

Some repair procedures in this manual entail the use of special tools such as a press, two or three-legged pullers, spring compressors, etc. Wherever possible, suitable readily-available alternatives to the manufacturer's special tools are described, and are shown in use. In some instances, where no alternative is possible, it has been necessary to resort to the use of a manufacturer's tool, and this has been done for reasons of safety as well as the efficient completion of the repair operation. Unless you are highly-skilled and have a thorough understanding of the procedures described, never attempt to bypass the use of any special tool when the procedure described specifies its use. Not only is there a very great risk of personal injury, but expensive damage could be caused to the components involved.

Environmental considerations

When disposing of used engine oil, brake fluid, antifreeze, etc, give due consideration to any detrimental environmental effects. Do not, for instance, pour any of the above liquids down drains into the general sewage system, or onto the ground to soak away. Many local council refuse tips provide a facility for waste oil disposal, as do some garages. If none of these facilities are available, consult your local Environmental Health Department, or the National Rivers Authority, for further advice.

With the universal tightening-up of legislation regarding the emission of environmentally-harmful substances from motor vehicles, most vehicles have tamperproof devices fitted to the main adjustment points of the fuel system. These devices are primarily designed to prevent unqualified persons from adjusting the fuel/air mixture, with the chance of a consequent increase in toxic emissions. If such devices are found during servicing or overhaul, they should, wherever possible, be renewed or refitted in accordance with the manufacturer's requirements or current legislation.

OIL CARE
FOLLOW THE CODE

OIL BANK LINE
0800 66 33 66
www.oilbankline.org.uk

Note: It is antisocial and illegal to dump oil down the drain. To find the location of your local oil recycling bank, call this number free.

The jack supplied with the vehicle tool kit should only be used for changing the roadwheels in an emergency – see *Wheel changing* at the front of this manual. When carrying out any other kind of work, raise the vehicle using a hydraulic (or 'trolley') jack, and always supplement the jack with axle stands positioned under the vehicle jacking points.

When using a hydraulic jack or axle stands, always position the jack head or axle stand head under one of the relevant jacking points (note that the jacking points for use with the vehicle jack are different from those for a hydraulic trolley jack). Nissan recommend the use of adapters when supporting the vehicle with axle stands – the adapters are grooved, and fit over the sill edge to prevent the vehicle weight damaging the sill **(see illustrations)**.

Do not jack the vehicle under the sump or any of the steering or suspension components other than those indicated.

⚠️ *Warning: Never work under, around, or near a raised vehicle, unless it is adequately supported in at least two places.*

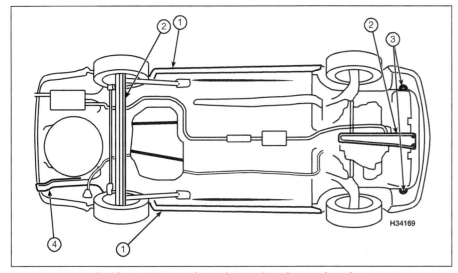

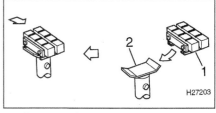

Nissan special adapters for use when supporting vehicle on axle stands. Fit the adapter to the stand, and position it underneath the vehicle so the sill edge is located in the adapter groove (arrowed)

1 Adapter   *2 Axle stand*

Jacking and supporting points and towing eye locations

1 Vehicle jack and axle stand location points (for use with adapters)

2 Hydraulic jack locating points
3 Front towing eye
4 Rear towing eye

Disconnecting the battery

Several systems fitted to the vehicle require battery power to be available at all times, either to ensure their continued operation (such as the clock) or to maintain control unit memories which could be erased if the battery were to be disconnected. Whenever the battery is to be disconnected therefore, first note the following, to ensure that there are no unforeseen consequences of this action:

a) *On any vehicle with central locking, it is a wise precaution to remove the key from the ignition, and to keep it with you, so that it does not get locked in if the central locking should engage accidentally when the battery is reconnected.*

b) *If a security-coded audio unit is fitted, and the unit and/or the battery is disconnected, the unit will not function again on reconnection until the correct security code is entered. Details of this procedure, which varies according to the unit fitted and vehicle model, are given in the vehicle owner's handbook. Where necessary, ensure you have the correct code before you disconnect the battery. If*

you do not have the code or details of the correct procedure, but can supply proof of ownership and a legitimate reason for wanting this information, a Nissan dealer may be able to help.

c) *The engine management system electronic control unit is of the 'self-learning' type, meaning that as it operates, it also monitors and stores the settings which give optimum engine performance under all operating conditions. When the battery is disconnected, these settings are lost and the ECU reverts to the base settings programmed into its memory at the factory. On restarting, this may lead to the engine running/idling roughly for a short while, until the ECU has re-learned the optimum settings. This process is best accomplished by taking the vehicle on a road test (for approximately 15 minutes), covering all engine speeds and loads, concentrating mainly in the 2500 to 3500 rpm region.*

d) *When reconnecting the battery after*

disconnection, switch on the ignition and wait 10 seconds to allow the electronic vehicle systems to stabilise and re-initialise.

Devices known as 'memory-savers' (or 'code-savers') can be used to maintain an electrical supply to various circuits. Precise details vary according to the device used. Typically, it is plugged into the cigarette lighter, and is connected by its own wires to a spare battery; the vehicle's own battery is then disconnected from the electrical system, leaving the 'memory-saver' to pass sufficient current to maintain audio unit security codes and any other memory values, and also to run permanently-live circuits such as the clock.

⚠️ *Warning: Some of these devices allow a considerable amount of current to pass, which can mean that many of the vehicle's systems are still operational when the main battery is disconnected. If a 'memory saver' is used, ensure that the circuit concerned is actually 'dead' before carrying out any work on it!*

Introduction

A selection of good tools is a fundamental requirement for anyone contemplating the maintenance and repair of a motor vehicle. For the owner who does not possess any, their purchase will prove a considerable expense, offsetting some of the savings made by doing-it-yourself. However, provided that the tools purchased meet the relevant national safety standards and are of good quality, they will last for many years and prove an extremely worthwhile investment.

To help the average owner to decide which tools are needed to carry out the various tasks detailed in this manual, we have compiled three lists of tools under the following headings: *Maintenance and minor repair, Repair and overhaul*, and *Special*. Newcomers to practical mechanics should start off with the *Maintenance and minor repair* tool kit, and confine themselves to the simpler jobs around the vehicle. Then, as confidence and experience grow, more difficult tasks can be undertaken, with extra tools being purchased as, and when, they are needed. In this way, a *Maintenance and minor repair* tool kit can be built up into a *Repair and overhaul* tool kit over a considerable period of time, without any major cash outlays. The experienced do-it-yourselfer will have a tool kit good enough for most repair and overhaul procedures, and will add tools from the *Special* category when it is felt that the expense is justified by the amount of use to which these tools will be put.

Maintenance and minor repair tool kit

The tools given in this list should be considered as a minimum requirement if routine maintenance, servicing and minor repair operations are to be undertaken. We recommend the purchase of combination spanners (ring one end, open-ended the other); although more expensive than open-ended ones, they do give the advantages of both types of spanner.

- ☐ *Combination spanners:*
  Metric - 8 to 19 mm inclusive
- ☐ *Adjustable spanner - 35 mm jaw (approx.)*
- ☐ *Spark plug spanner (with rubber insert) - petrol models*
- ☐ *Spark plug gap adjustment tool - petrol models*
- ☐ *Set of feeler gauges*
- ☐ *Brake bleed nipple spanner*
- ☐ *Screwdrivers:*
  Flat blade - 100 mm long x 6 mm dia
  Cross blade - 100 mm long x 6 mm dia
  Torx - various sizes (not all vehicles)
- ☐ *Combination pliers*
- ☐ *Hacksaw (junior)*
- ☐ *Tyre pump*
- ☐ *Tyre pressure gauge*
- ☐ *Oil can*
- ☐ *Oil filter removal tool*
- ☐ *Fine emery cloth*
- ☐ *Wire brush (small)*
- ☐ *Funnel (medium size)*
- ☐ *Sump drain plug key (not all vehicles)*

Repair and overhaul tool kit

These tools are virtually essential for anyone undertaking any major repairs to a motor vehicle, and are additional to those given in the *Maintenance and minor repair* list. Included in this list is a comprehensive set of sockets. Although these are expensive, they will be found invaluable as they are so versatile - particularly if various drives are included in the set. We recommend the half-inch square-drive type, as this can be used with most proprietary torque wrenches.

The tools in this list will sometimes need to be supplemented by tools from the *Special* list:

- ☐ *Sockets (or box spanners) to cover range in previous list (including Torx sockets)*
- ☐ *Reversible ratchet drive (for use with sockets)*
- ☐ *Extension piece, 250 mm (for use with sockets)*
- ☐ *Universal joint (for use with sockets)*
- ☐ *Flexible handle or sliding T "breaker bar" (for use with sockets)*
- ☐ *Torque wrench (for use with sockets)*
- ☐ *Self-locking grips*
- ☐ *Ball pein hammer*
- ☐ *Soft-faced mallet (plastic or rubber)*
- ☐ *Screwdrivers:*
  Flat blade - long & sturdy, short (chubby), and narrow (electrician's) types
  Cross blade - long & sturdy, and short (chubby) types
- ☐ *Pliers:*
  Long-nosed
  Side cutters (electrician's)
  Circlip (internal and external)
- ☐ *Cold chisel - 25 mm*
- ☐ *Scriber*
- ☐ *Scraper*
- ☐ *Centre-punch*
- ☐ *Pin punch*
- ☐ *Hacksaw*
- ☐ *Brake hose clamp*
- ☐ *Brake/clutch bleeding kit*
- ☐ *Selection of twist drills*
- ☐ *Steel rule/straight-edge*
- ☐ *Allen keys (inc. splined/Torx type)*
- ☐ *Selection of files*
- ☐ *Wire brush*
- ☐ *Axle stands*
- ☐ *Jack (strong trolley or hydraulic type)*
- ☐ *Light with extension lead*
- ☐ *Universal electrical multi-meter*

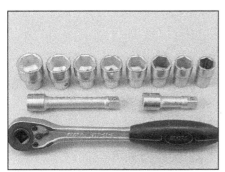

Sockets and reversible ratchet drive

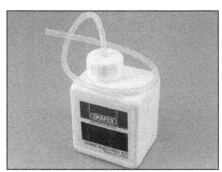

Brake bleeding kit

Torx key, socket and bit

Angular-tightening gauge

Special tools

The tools in this list are those which are not used regularly, are expensive to buy, or which need to be used in accordance with their manufacturers' instructions. Unless relatively difficult mechanical jobs are undertaken frequently, it will not be economic to buy many of these tools. Where this is the case, you could consider clubbing together with friends (or joining a motorists' club) to make a joint purchase, or borrowing the tools against a deposit from a local garage or tool hire specialist. It is worth noting that many of the larger DIY superstores now carry a large range of special tools for hire at modest rates.

The following list contains only those tools and instruments freely available to the public, and not those special tools produced by the vehicle manufacturer specifically for its dealer network. You will find occasional references to these manufacturers' special tools in the text of this manual. Generally, an alternative method of doing the job without the vehicle manufacturers' special tool is given. However, sometimes there is no alternative to using them. Where this is the case and the relevant tool cannot be bought or borrowed, you will have to entrust the work to a dealer.

- ☐ Angular-tightening gauge
- ☐ Valve spring compressor
- ☐ Valve grinding tool
- ☐ Piston ring compressor
- ☐ Piston ring removal/installation tool
- ☐ Cylinder bore hone
- ☐ Balljoint separator
- ☐ Coil spring compressors (where applicable)
- ☐ Two/three-legged hub and bearing puller
- ☐ Impact screwdriver
- ☐ Micrometer and/or vernier calipers
- ☐ Dial gauge
- ☐ Stroboscopic timing light
- ☐ Dwell angle meter/tachometer
- ☐ Fault code reader
- ☐ Cylinder compression gauge
- ☐ Hand-operated vacuum pump and gauge
- ☐ Clutch plate alignment set
- ☐ Brake shoe steady spring cup removal tool
- ☐ Bush and bearing removal/installation set
- ☐ Stud extractors
- ☐ Tap and die set
- ☐ Lifting tackle
- ☐ Trolley jack

Buying tools

Reputable motor accessory shops and superstores often offer excellent quality tools at discount prices, so it pays to shop around.

Remember, you don't have to buy the most expensive items on the shelf, but it is always advisable to steer clear of the very cheap tools. Beware of 'bargains' offered on market stalls or at car boot sales. There are plenty of good tools around at reasonable prices, but always aim to purchase items which meet the relevant national safety standards. If in doubt, ask the proprietor or manager of the shop for advice before making a purchase.

Care and maintenance of tools

Having purchased a reasonable tool kit, it is necessary to keep the tools in a clean and serviceable condition. After use, always wipe off any dirt, grease and metal particles using a clean, dry cloth, before putting the tools away. Never leave them lying around after they have been used. A simple tool rack on the garage or workshop wall for items such as screwdrivers and pliers is a good idea. Store all normal spanners and sockets in a metal box. Any measuring instruments, gauges, meters, etc, must be carefully stored where they cannot be damaged or become rusty.

Take a little care when tools are used. Hammer heads inevitably become marked, and screwdrivers lose the keen edge on their blades from time to time. A little timely attention with emery cloth or a file will soon restore items like this to a good finish.

Working facilities

Not to be forgotten when discussing tools is the workshop itself. If anything more than routine maintenance is to be carried out, a suitable working area becomes essential.

It is appreciated that many an owner-mechanic is forced by circumstances to remove an engine or similar item without the benefit of a garage or workshop. Having done this, any repairs should always be done under the cover of a roof.

Wherever possible, any dismantling should be done on a clean, flat workbench or table at a suitable working height.

Any workbench needs a vice; one with a jaw opening of 100 mm is suitable for most jobs. As mentioned previously, some clean dry storage space is also required for tools, as well as for any lubricants, cleaning fluids, touch-up paints etc, which become necessary.

Another item which may be required, and which has a much more general usage, is an electric drill with a chuck capacity of at least 8 mm. This, together with a good range of twist drills, is virtually essential for fitting accessories.

Last, but not least, always keep a supply of old newspapers and clean, lint-free rags available, and try to keep any working area as clean as possible.

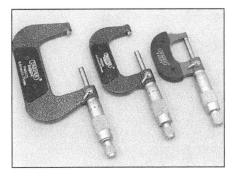

Micrometers

Dial test indicator ("dial gauge")

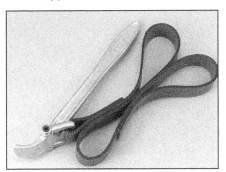

Strap wrench

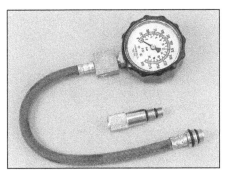

Compression tester

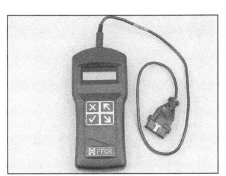

Fault code reader

This is a guide to getting your vehicle through the MOT test. Obviously it will not be possible to examine the vehicle to the same standard as the professional MOT tester. However, working through the following checks will enable you to identify any problem areas before submitting the vehicle for the test.

Where a testable component is in borderline condition, the tester has discretion in deciding whether to pass or fail it. The basis of such discretion is whether the tester would be happy for a close relative or friend to use the vehicle with the component in that condition. If the vehicle presented is clean and evidently well cared for, the tester may be more inclined to pass a borderline component than if the vehicle is scruffy and apparently neglected.

It has only been possible to summarise the test requirements here, based on the regulations in force at the time of printing. Test standards are becoming increasingly stringent, although there are some exemptions for older vehicles.

An assistant will be needed to help carry out some of these checks.

The checks have been sub-divided into four categories, as follows:

1 Checks carried out **FROM THE DRIVER'S SEAT**

2 Checks carried out **WITH THE VEHICLE ON THE GROUND**

3 Checks carried out **WITH THE VEHICLE RAISED AND THE WHEELS FREE TO TURN**

4 Checks carried out on **YOUR VEHICLE'S EXHAUST EMISSION SYSTEM**

---

1 Checks carried out **FROM THE DRIVER'S SEAT**

Handbrake

☐ Test the operation of the handbrake. Excessive travel (too many clicks) indicates incorrect brake or cable adjustment.

☐ Check that the handbrake cannot be released by tapping the lever sideways. Check the security of the lever mountings.

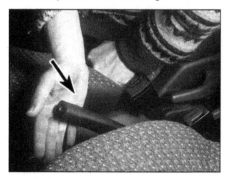

Footbrake

☐ Depress the brake pedal and check that it does not creep down to the floor, indicating a master cylinder fault. Release the pedal, wait a few seconds, then depress it again. If the pedal travels nearly to the floor before firm resistance is felt, brake adjustment or repair is necessary. If the pedal feels spongy, there is air in the hydraulic system which must be removed by bleeding.

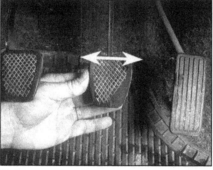

☐ Check that the brake pedal is secure and in good condition. Check also for signs of fluid leaks on the pedal, floor or carpets, which would indicate failed seals in the brake master cylinder.

☐ Check the servo unit (when applicable) by operating the brake pedal several times, then keeping the pedal depressed and starting the engine. As the engine starts, the pedal will move down slightly. If not, the vacuum hose or the servo itself may be faulty.

Steering wheel and column

☐ Examine the steering wheel for fractures or looseness of the hub, spokes or rim.

☐ Move the steering wheel from side to side and then up and down. Check that the steering wheel is not loose on the column, indicating wear or a loose retaining nut. Continue moving the steering wheel as before, but also turn it slightly from left to right.

☐ Check that the steering wheel is not loose on the column, and that there is no abnormal

movement of the steering wheel, indicating wear in the column support bearings or couplings.

Windscreen, mirrors and sunvisor

☐ The windscreen must be free of cracks or other significant damage within the driver's field of view. (Small stone chips are acceptable.) Rear view mirrors must be secure, intact, and capable of being adjusted.

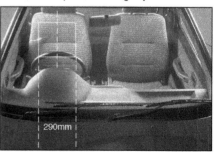

☐ The driver's sunvisor must be capable of being stored in the "up" position.

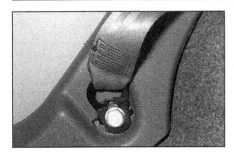

Seat belts and seats

Note: *The following checks are applicable to all seat belts, front and rear.*

☐ Examine the webbing of all the belts (including rear belts if fitted) for cuts, serious fraying or deterioration. Fasten and unfasten each belt to check the buckles. If applicable, check the retracting mechanism. Check the security of all seat belt mountings accessible from inside the vehicle.

☐ Seat belts with pre-tensioners, once activated, have a "flag" or similar showing on the seat belt stalk. This, in itself, is not a reason for test failure.

☐ The front seats themselves must be securely attached and the backrests must lock in the upright position.

Doors

☐ Both front doors must be able to be opened and closed from outside and inside, and must latch securely when closed.

2 Checks carried out WITH THE VEHICLE ON THE GROUND

Vehicle identification

☐ Number plates must be in good condition, secure and legible, with letters and numbers correctly spaced – spacing at (A) should be at least twice that at (B).

☐ The VIN plate and/or homologation plate must be legible.

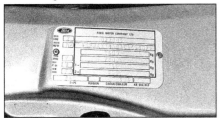

Electrical equipment

☐ Switch on the ignition and check the operation of the horn.

☐ Check the windscreen washers and wipers, examining the wiper blades; renew damaged or perished blades. Also check the operation of the stop-lights.

☐ Check the operation of the sidelights and number plate lights. The lenses and reflectors must be secure, clean and undamaged.

☐ Check the operation and alignment of the headlights. The headlight reflectors must not be tarnished and the lenses must be undamaged.

☐ Switch on the ignition and check the operation of the direction indicators (including the instrument panel tell-tale) and the hazard warning lights. Operation of the sidelights and stop-lights must not affect the indicators - if it does, the cause is usually a bad earth at the rear light cluster.

☐ Check the operation of the rear foglight(s), including the warning light on the instrument panel or in the switch.

☐ The ABS warning light must illuminate in accordance with the manufacturers' design. For most vehicles, the ABS warning light should illuminate when the ignition is switched on, and (if the system is operating properly) extinguish after a few seconds. Refer to the owner's handbook.

Footbrake

☐ Examine the master cylinder, brake pipes and servo unit for leaks, loose mountings, corrosion or other damage.

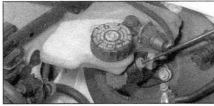

☐ The fluid reservoir must be secure and the fluid level must be between the upper (**A**) and lower (**B**) markings.

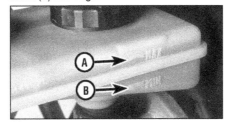

☐ Inspect both front brake flexible hoses for cracks or deterioration of the rubber. Turn the steering from lock to lock, and ensure that the hoses do not contact the wheel, tyre, or any part of the steering or suspension mechanism. With the brake pedal firmly depressed, check the hoses for bulges or leaks under pressure.

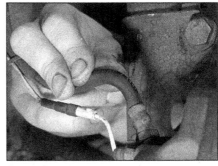

Steering and suspension

☐ Have your assistant turn the steering wheel from side to side slightly, up to the point where the steering gear just begins to transmit this movement to the roadwheels. Check for excessive free play between the steering wheel and the steering gear, indicating wear or insecurity of the steering column joints, the column-to-steering gear coupling, or the steering gear itself.

☐ Have your assistant turn the steering wheel more vigorously in each direction, so that the roadwheels just begin to turn. As this is done, examine all the steering joints, linkages, fittings and attachments. Renew any component that shows signs of wear or damage. On vehicles with power steering, check the security and condition of the steering pump, drivebelt and hoses.

☐ Check that the vehicle is standing level, and at approximately the correct ride height.

Shock absorbers

☐ Depress each corner of the vehicle in turn, then release it. The vehicle should rise and then settle in its normal position. If the vehicle continues to rise and fall, the shock absorber is defective. A shock absorber which has seized will also cause the vehicle to fail.

Exhaust system

☐ Start the engine. With your assistant holding a rag over the tailpipe, check the entire system for leaks. Repair or renew leaking sections.

3 Checks carried out
WITH THE VEHICLE RAISED AND THE WHEELS FREE TO TURN

Jack up the front and rear of the vehicle, and securely support it on axle stands. Position the stands clear of the suspension assemblies. Ensure that the wheels are clear of the ground and that the steering can be turned from lock to lock.

Steering mechanism

☐ Have your assistant turn the steering from lock to lock. Check that the steering turns smoothly, and that no part of the steering mechanism, including a wheel or tyre, fouls any brake hose or pipe or any part of the body structure.
☐ Examine the steering rack rubber gaiters for damage or insecurity of the retaining clips. If power steering is fitted, check for signs of damage or leakage of the fluid hoses, pipes or connections. Also check for excessive stiffness or binding of the steering, a missing split pin or locking device, or severe corrosion of the body structure within 30 cm of any steering component attachment point.

Front and rear suspension and wheel bearings

☐ Starting at the front right-hand side, grasp the roadwheel at the 3 o'clock and 9 o'clock positions and rock gently but firmly. Check for free play or insecurity at the wheel bearings, suspension balljoints, or suspension mountings, pivots and attachments.
☐ Now grasp the wheel at the 12 o'clock and 6 o'clock positions and repeat the previous inspection. Spin the wheel, and check for roughness or tightness of the front wheel bearing.

☐ If excess free play is suspected at a component pivot point, this can be confirmed by using a large screwdriver or similar tool and levering between the mounting and the component attachment. This will confirm whether the wear is in the pivot bush, its retaining bolt, or in the mounting itself (the bolt holes can often become elongated).

☐ Carry out all the above checks at the other front wheel, and then at both rear wheels.

Springs and shock absorbers

☐ Examine the suspension struts (when applicable) for serious fluid leakage, corrosion, or damage to the casing. Also check the security of the mounting points.
☐ If coil springs are fitted, check that the spring ends locate in their seats, and that the spring is not corroded, cracked or broken.
☐ If leaf springs are fitted, check that all leaves are intact, that the axle is securely attached to each spring, and that there is no deterioration of the spring eye mountings, bushes, and shackles.

☐ The same general checks apply to vehicles fitted with other suspension types, such as torsion bars, hydraulic displacer units, etc. Ensure that all mountings and attachments are secure, that there are no signs of excessive wear, corrosion or damage, and (on hydraulic types) that there are no fluid leaks or damaged pipes.
☐ Inspect the shock absorbers for signs of serious fluid leakage. Check for wear of the mounting bushes or attachments, or damage to the body of the unit.

Driveshafts
(fwd vehicles only)

☐ Rotate each front wheel in turn and inspect the constant velocity joint gaiters for splits or damage. Also check that each driveshaft is straight and undamaged.

Braking system

☐ If possible without dismantling, check brake pad wear and disc condition. Ensure that the friction lining material has not worn excessively, (A) and that the discs are not fractured, pitted, scored or badly worn (B).

☐ Examine all the rigid brake pipes underneath the vehicle, and the flexible hose(s) at the rear. Look for corrosion, chafing or insecurity of the pipes, and for signs of bulging under pressure, chafing, splits or deterioration of the flexible hoses.
☐ Look for signs of fluid leaks at the brake calipers or on the brake backplates. Repair or renew leaking components.
☐ Slowly spin each wheel, while your assistant depresses and releases the footbrake. Ensure that each brake is operating and does not bind when the pedal is released.

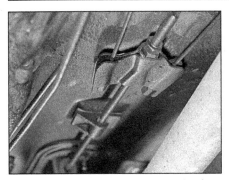

□ Examine the handbrake mechanism, checking for frayed or broken cables, excessive corrosion, or wear or insecurity of the linkage. Check that the mechanism works on each relevant wheel, and releases fully, without binding.

□ It is not possible to test brake efficiency without special equipment, but a road test can be carried out later to check that the vehicle pulls up in a straight line.

Fuel and exhaust systems

□ Inspect the fuel tank (including the filler cap), fuel pipes, hoses and unions. All components must be secure and free from leaks.

□ Examine the exhaust system over its entire length, checking for any damaged, broken or missing mountings, security of the retaining clamps and rust or corrosion.

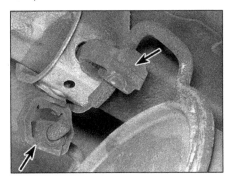

Wheels and tyres

□ Examine the sidewalls and tread area of each tyre in turn. Check for cuts, tears, lumps, bulges, separation of the tread, and exposure of the ply or cord due to wear or damage. Check that the tyre bead is correctly seated on the wheel rim, that the valve is sound and properly seated, and that the wheel is not distorted or damaged.

□ Check that the tyres are of the correct size for the vehicle, that they are of the same size and type on each axle, and that the pressures are correct.

□ Check the tyre tread depth. The legal minimum at the time of writing is 1.6 mm over at least three-quarters of the tread width. Abnormal tread wear may indicate incorrect front wheel alignment.

Body corrosion

□ Check the condition of the entire vehicle structure for signs of corrosion in load-bearing areas. (These include chassis box sections, side sills, cross-members, pillars, and all suspension, steering, braking system and seat belt mountings and anchorages.) Any corrosion which has seriously reduced the thickness of a load-bearing area is likely to cause the vehicle to fail. In this case professional repairs are likely to be needed.

□ Damage or corrosion which causes sharp or otherwise dangerous edges to be exposed will also cause the vehicle to fail.

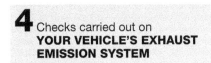

4 Checks carried out on **YOUR VEHICLE'S EXHAUST EMISSION SYSTEM**

Petrol models

□ Have the engine at normal operating temperature, and make sure that it is in good tune (ignition system in good order, air filter element clean, etc).

□ Before any measurements are carried out, raise the engine speed to around 2500 rpm, and hold it at this speed for 20 seconds. Allow the engine speed to return to idle, and watch for smoke emissions from the exhaust tailpipe. If the idle speed is obviously much too high, or if dense blue or clearly-visible black smoke comes from the tailpipe for more than 5 seconds, the vehicle will fail. As a rule of thumb, blue smoke signifies oil being burnt (engine wear) while black smoke signifies unburnt fuel (dirty air cleaner element, or other carburettor or fuel system fault).

□ An exhaust gas analyser capable of measuring carbon monoxide (CO) and hydrocarbons (HC) is now needed. If such an instrument cannot be hired or borrowed, a local garage may agree to perform the check for a small fee.

CO emissions (mixture)

□ At the time of writing, for vehicles first used between 1st August 1975 and 31st July 1986 (P to C registration), the CO level must not exceed 4.5% by volume. For vehicles first used between 1st August 1986 and 31st July 1992 (D to J registration), the CO level must not exceed 3.5% by volume. Vehicles first

used after 1st August 1992 (K registration) must conform to the manufacturer's specification. The MOT tester has access to a DOT database or emissions handbook, which lists the CO and HC limits for each make and model of vehicle. The CO level is measured with the engine at idle speed, and at "fast idle". The following limits are given as a general guide:

At idle speed -
  CO level no more than 0.5%
At "fast idle" (2500 to 3000 rpm) -
  CO level no more than 0.3%
  (Minimum oil temperature 60ºC)

□ If the CO level cannot be reduced far enough to pass the test (and the fuel and ignition systems are otherwise in good condition) then the carburettor is badly worn, or there is some problem in the fuel injection system or catalytic converter (as applicable).

HC emissions

□ With the CO within limits, HC emissions for vehicles first used between 1st August 1975 and 31st July 1992 (P to J registration) must not exceed 1200 ppm. Vehicles first used after 1st August 1992 (K registration) must conform to the manufacturer's specification. The MOT tester has access to a DOT database or emissions handbook, which lists the CO and HC limits for each make and model of vehicle. The HC level is measured with the engine at "fast idle". The following is given as a general guide:

At "fast idle" (2500 to 3000 rpm) -
  HC level no more than 200 ppm
  (Minimum oil temperature 60ºC)

□ Excessive HC emissions are caused by incomplete combustion, the causes of which can include oil being burnt, mechanical wear and ignition/fuel system malfunction.

Diesel models

□ The only emission test applicable to Diesel engines is the measuring of exhaust smoke density. The test involves accelerating the engine several times to its maximum unloaded speed.

Note: *It is of the utmost importance that the engine timing belt is in good condition before the test is carried out.*

□ The limits for Diesel engine exhaust smoke, introduced in September 1995 are:
Vehicles first used before 1st August 1979:
  Exempt from metered smoke testing, but must not emit "dense blue or clearly visible black smoke for a period of more than 5 seconds at idle" or "dense blue or clearly visible black smoke during acceleration which would obscure the view of other road users".
Non-turbocharged vehicles first used after 1st August 1979: 2.5m-1
Turbocharged vehicles first used after 1st August 1979: 3.0m-1
□ Excessive smoke can be caused by a dirty air cleaner element. Otherwise, professional advice may be needed to find the cause.

Engine

☐ Engine fails to rotate when attempting to start
☐ Engine rotates, but will not start
☐ Engine difficult to start when cold
☐ Engine difficult to start when hot
☐ Starter motor noisy or excessively-rough in engagement
☐ Engine starts, but stops immediately
☐ Engine idles erratically
☐ Engine misfires at idle speed
☐ Engine misfires throughout the driving speed range
☐ Engine hesitates on acceleration
☐ Engine stalls
☐ Engine lacks power
☐ Engine backfires
☐ Oil pressure warning light illuminated with engine running
☐ Engine runs-on after switching off
☐ Engine noises

Cooling system

☐ Overheating
☐ Overcooling
☐ External coolant leakage
☐ Internal coolant leakage
☐ Corrosion

Fuel and exhaust systems

☐ Excessive fuel consumption
☐ Fuel leakage and/or fuel odour
☐ Excessive noise or fumes from exhaust system

Clutch

☐ Pedal travels to floor – no pressure or very little resistance
☐ Clutch fails to disengage (unable to select gears)
☐ Clutch slips (engine speed increases, with no increase in vehicle speed)
☐ Judder as clutch is engaged
☐ Noise when depressing or releasing clutch pedal

Manual transmission

☐ Noisy in neutral with engine running
☐ Noisy in one particular gear
☐ Difficulty engaging gears
☐ Jumps out of gear
☐ Vibration
☐ Lubricant leaks

Automatic transmission

☐ Fluid leakage
☐ Transmission fluid brown, or has burned smell
☐ General gear selection problems
☐ Transmission will not downshift (kickdown) with accelerator fully depressed
☐ Engine will not start in any gear, or starts in gears other than Park or Neutral
☐ Transmission slips, shifts roughly, is noisy, or has no drive in forward or reverse gears

Driveshafts

☐ Clicking or knocking noise on turns (at slow speed on full-lock)
☐ Vibration when accelerating or decelerating

Braking system

☐ Vehicle pulls to one side under braking
☐ Noise (grinding or high-pitched squeal) when brakes applied
☐ Excessive brake pedal travel
☐ Brake pedal feels spongy when depressed
☐ Excessive brake pedal effort required to stop vehicle
☐ Judder felt through brake pedal or steering wheel when braking
☐ Brakes binding
☐ Rear wheels locking under normal braking

Suspension and steering systems

☐ Vehicle pulls to one side
☐ Wheel wobble and vibration
☐ Excessive pitching and/or rolling around corners, or during braking
☐ Wandering or general instability
☐ Excessively-stiff steering
☐ Excessive play in steering
☐ Lack of power assistance
☐ Tyre wear excessive

Electrical system

☐ Battery will not hold a charge for more than a few days
☐ Ignition/no-charge warning light remains illuminated with engine running
☐ Ignition/no-charge warning light fails to come on
☐ Lights inoperative
☐ Instrument readings inaccurate or erratic
☐ Horn inoperative, or unsatisfactory in operation
☐ Windscreen/tailgate wipers inoperative, or unsatisfactory in operation
☐ Windscreen/tailgate washers inoperative, or unsatisfactory in operation
☐ Electric windows inoperative, or unsatisfactory in operation
☐ Central locking system inoperative, or unsatisfactory in operation

Introduction

The vehicle owner who does his or her own maintenance according to the recommended service schedules should not have to use this section of the manual very often. Modern component reliability is such that, provided those items subject to wear or deterioration are inspected or renewed at the specified intervals, sudden failure is comparatively rare. Faults do not usually just happen as a result of sudden failure, but develop over a period of time. Major mechanical failures in particular are usually preceded by characteristic symptoms over hundreds or even thousands of miles. Those components which do occasionally fail without warning are often small and easily carried in the vehicle.

With any fault-finding, the first step is to decide where to begin investigations. Sometimes this is obvious, but on other occasions, a little detective work will be necessary. The owner who makes half a dozen haphazard adjustments or replacements may be successful in curing a fault (or its symptoms), but will be none the wiser if the fault recurs, and ultimately may have spent more time and money than was necessary. A calm and logical approach will be found to be more satisfactory in the long run. Always take into account any warning signs or abnormalities that may have been noticed in the period preceding the fault – power loss, high or low gauge readings, unusual smells, etc – and remember that failure of components such as fuses or spark plugs may only be pointers to some underlying fault.

The pages which follow provide an easy-reference guide to the more common problems which may occur during the operation of the vehicle. These problems and their possible causes are grouped under headings denoting various components or systems, such as Engine, Cooling system, etc. The Chapter and/or Section which deals with the problem is also shown in brackets. Whatever the fault, certain basic principles apply. These are as follows:

Verify the fault. This is simply a matter of being sure that you know what the symptoms are before starting work. This is particularly important if you are investigating a fault for someone else, who may not have described it very accurately.

Don't overlook the obvious. For example, if the vehicle won't start, is there petrol in the tank? (Don't take anyone else's word on this particular point, and don't trust the fuel gauge either!) If an electrical fault is indicated, look for loose or broken wires before digging out the test gear.

Cure the disease, not the symptom. Substituting a flat battery with a fully-charged one will get you off the hard shoulder, but if the underlying cause is not attended to, the new battery will go the same way. Similarly, changing oil-fouled spark plugs for a new set will get you moving again, but remember that the reason for the fouling (if it wasn't simply an incorrect grade of plug) will have to be established and corrected.

Don't take anything for granted. Particularly, don't forget that a 'new' component may itself be defective (especially if it's been rattling around in the boot for months), and don't leave components out of a fault diagnosis sequence just because they are new or recently-fitted. When you do finally diagnose a difficult fault, you'll probably realise that all the evidence was there from the start.

Engine

Engine fails to rotate when attempting to start

- [ ] Battery terminal connections loose or corroded (*Weekly Checks*).
- [ ] Battery discharged or faulty (Chapter 5A).
- [ ] Broken, loose or disconnected wiring in the starting circuit (Chapter 5A).
- [ ] Defective starter solenoid or switch (Chapter 5A).
- [ ] Defective starter motor (Chapter 5A).
- [ ] Starter pinion or flywheel ring gear teeth loose or broken (Chapter 2A).
- [ ] Engine earth strap broken or disconnected (Chapter 5A).
- [ ] Automatic transmission not in Park/Neutral position or starter inhibitor switch faulty (Chapter 7B).

Engine rotates, but will not start

- [ ] Fuel tank empty.
- [ ] Battery discharged (engine rotates slowly) (Chapter 5A).
- [ ] Battery terminal connections loose or corroded (*Weekly checks*).
- [ ] Ignition components damp or damaged (Chapters 1 and 5B).
- [ ] Broken, loose or disconnected wiring in the ignition circuit (Chapters 1 and 5B).
- [ ] Worn, faulty or incorrectly-gapped spark plugs (Chapter 1).
- [ ] Fuel injection/engine management system fault (Chapter 4A or 4B).
- [ ] Major mechanical failure (eg camshaft drive) (Chapter 2A or 2B).

Engine difficult to start when cold

- [ ] Battery discharged (Chapter 5A).
- [ ] Battery terminal connections loose or corroded (*Weekly checks*).
- [ ] Worn, faulty or incorrectly-gapped spark plugs (Chapter 1).
- [ ] Fuel injection/engine management system fault (Chapter 4A or 4B).
- [ ] Other ignition system fault (Chapters 1 and 5A).
- [ ] Low cylinder compressions (Chapter 2A).

Engine difficult to start when hot

- [ ] Air filter element dirty or clogged (Chapter 1).
- [ ] Fuel injection/engine management system fault (Chapter 4A or 4B).
- [ ] Other ignition system fault (Chapters 1 and 5A).
- [ ] Low cylinder compressions (Chapter 2A).

Starter motor noisy or excessively-rough in engagement

- [ ] Starter pinion or flywheel ring gear teeth loose or broken (Chapters 2A and 5A).
- [ ] Starter motor mounting bolts loose or missing (Chapter 5A).
- [ ] Starter motor internal components worn or damaged (Chapter 5A).

Engine starts, but stops immediately

- [ ] Loose or faulty electrical connections in the ignition circuit (Chapters 1 and 5B).
- [ ] Vacuum leak at the throttle housing or inlet manifold (Chapter 4A or 4B).
- [ ] Fuel injection/engine management system fault (Chapter 4A or 4B).

Engine idles erratically

- [ ] Air filter element clogged (Chapter 1).
- [ ] Vacuum leak at the throttle housing, inlet manifold or associated hoses (Chapter 4A or 4B).
- [ ] Worn, faulty or incorrectly-gapped spark plugs (Chapter 1).
- [ ] Uneven or low cylinder compressions (Chapter 2A).
- [ ] Camshaft lobes worn (Chapter 2A).
- [ ] Timing chain(s) incorrectly fitted (Chapter 2A).
- [ ] Fuel injection/engine management system fault (Chapter 4A or 4B).

Engine misfires at idle speed

- [ ] Worn, faulty or incorrectly-gapped spark plugs (Chapter 1).
- [ ] Faulty spark plug HT leads (Chapter 1).
- [ ] Vacuum leak at the throttle housing, inlet manifold or associated hoses (Chapter 4A or 4B).
- [ ] Fuel injection/engine management system fault (Chapter 4A or 4B).
- [ ] Distributor cap cracked or tracking internally (Chapter 1).
- [ ] Uneven or low cylinder compressions (Chapter 2A).
- [ ] Disconnected, leaking, or perished crankcase ventilation hoses (Chapter 4B).

Engine (continued)

Engine misfires throughout the driving speed range

☐ Fuel filter choked (Chapter 1).
☐ Fuel pump faulty, or delivery pressure low (Chapter 4A).
☐ Fuel tank vent blocked, or fuel pipes restricted (Chapter 4A).
☐ Vacuum leak at the throttle housing, inlet manifold or associated hoses (Chapter 4A or 4B).
☐ Worn, faulty or incorrectly-gapped spark plugs (Chapter 1).
☐ Faulty spark plug HT (Chapter 1).
☐ Distributor cap cracked or tracking internally (Chapter 1).
☐ Faulty ignition coil (Chapter 5B).
☐ Uneven or low cylinder compressions (Chapter 2A).
☐ Fuel injection/engine management system fault (Chapter 4A or 4B).

Engine hesitates on acceleration

☐ Worn, faulty or incorrectly-gapped spark plugs (Chapter 1).
☐ Vacuum leak at the throttle housing, inlet manifold or associated hoses (Chapter 4A or 4B).
☐ Fuel injection/engine management system fault (Chapter 4A or 4B).

Engine stalls

☐ Vacuum leak at the throttle housing, inlet manifold or associated hoses (Chapter 4A or 4B).
☐ Fuel filter choked (Chapter 1).
☐ Fuel pump faulty, or delivery pressure low (Chapter 4A).
☐ Fuel tank vent blocked, or fuel pipes restricted (Chapter 4A).
☐ Fuel injection/engine management system fault (Chapter 4A or 4B).

Engine lacks power

☐ Timing chain(s) incorrectly fitted (Chapter 2A).
☐ Fuel filter choked (Chapter 1).
☐ Fuel pump faulty, or delivery pressure low (Chapter 4A).
☐ Uneven or low cylinder compressions (Chapter 2A).
☐ Worn, faulty or incorrectly-gapped spark plugs (Chapter 1).
☐ Vacuum leak at the throttle housing, inlet manifold or associated hoses (Chapter 4A or 4B).
☐ Fuel injection/engine management system fault (Chapter 4A or 4B).
☐ Brakes binding (Chapters 1 and 9).
☐ Clutch slipping – manual transmission models (Chapter 6).

Engine backfires

☐ Timing chain(s) incorrectly fitted (Chapter 2A).
☐ Vacuum leak at the throttle housing, inlet manifold or associated hoses (Chapter 4A or 4B).
☐ Fuel injection/engine management system fault (Chapter 4A or 4B).

Oil pressure warning light illuminated with engine running

☐ Low oil level, or incorrect oil grade (*Weekly checks*).
☐ Faulty oil pressure warning light switch (Chapter 5A).
☐ Worn engine bearings and/or oil pump (Chapter 2A or 2B).
☐ High engine operating temperature (Chapter 3).
☐ Oil pressure relief valve defective (Chapter 2A).
☐ Oil pick-up strainer clogged (Chapter 2A).

Engine runs-on after switching off

☐ Excessive carbon build-up in engine (Chapter 2A).
☐ High engine operating temperature (Chapter 3).
☐ Fuel injection/engine management system fault (Chapter 4A or 4B).

Engine noises

Pre-ignition (pinking) or knocking during acceleration or under load

☐ Ignition timing incorrect/ignition system fault (Chapters 1 and 5B).
☐ Incorrect grade of spark plug (Chapter 1).
☐ Incorrect grade of fuel (Chapter 4A).
☐ Vacuum leak at the throttle housing, inlet manifold or associated hoses (Chapter 4A or 4B).
☐ Excessive carbon build-up in engine (Chapter 2A).
☐ Fuel injection/engine management system fault (Chapter 4A or 4B).

Whistling or wheezing noises

☐ Leaking inlet manifold or throttle housing gasket (Chapter 4A).
☐ Leaking exhaust manifold gasket or pipe-to-manifold joint (Chapter 4A).
☐ Leaking vacuum hose (Chapters 4A, 4B, 5B and 9).
☐ Blowing cylinder head gasket (Chapter 2A).

Tapping or rattling noises

☐ Worn valve gear or camshaft (Chapter 2A).
☐ Ancillary component fault (coolant pump, alternator, etc) (Chapters 3, 5A, etc).

Knocking or thumping noises

☐ Worn big-end bearings (regular heavy knocking, perhaps less under load) (Chapter 2B).
☐ Worn main bearings (rumbling and knocking, perhaps worsening under load) (Chapter 2B).
☐ Piston slap (most noticeable when cold) (Chapter 2B).
☐ Ancillary component fault (coolant pump, alternator, etc) (Chapters 3, 5A, etc).

Cooling system

Overheating

- [ ] Insufficient coolant in system (*Weekly checks*).
- [ ] Thermostat faulty (Chapter 3).
- [ ] Radiator core blocked, or grille restricted (Chapter 3).
- [ ] Electric cooling fan faulty (Chapter 3).
- [ ] Pressure cap faulty (Chapter 3).
- [ ] Ignition timing incorrect/ignition system fault (Chapters 1 and 5B).
- [ ] Inaccurate temperature gauge sender unit (Chapter 3).
- [ ] Airlock in cooling system (Chapter 1).

Overcooling

- [ ] Thermostat faulty (Chapter 3).
- [ ] Inaccurate temperature gauge sender unit (Chapter 3).

External coolant leakage

- [ ] Deteriorated or damaged hoses or hose clips (Chapter 1).
- [ ] Radiator core or heater matrix leaking (Chapter 3).
- [ ] Pressure cap faulty (Chapter 3).
- [ ] Coolant pump seal leaking (Chapter 3).
- [ ] Boiling due to overheating (Chapter 3).
- [ ] Cylinder block core plug leaking (Chapter 2B).

Internal coolant leakage

- [ ] Leaking cylinder head gasket (Chapter 2A).
- [ ] Cracked cylinder head or cylinder bore (Chapter 2A or 2B).

Corrosion

- [ ] Infrequent draining and flushing (Chapter 1).
- [ ] Incorrect coolant mixture or inappropriate coolant type (Chapter 1).

Fuel and exhaust systems

Excessive fuel consumption

- [ ] Air filter element dirty or clogged (Chapter 1).
- [ ] Fuel injection/engine management system fault (Chapter 4A or 4B).
- [ ] Ignition timing incorrect/ignition system fault (Chapters 1 and 5B).
- [ ] Tyres under-inflated (*Weekly checks*).

Fuel leakage and/or fuel odour

- [ ] Damaged or corroded fuel tank, pipes or connections (Chapter 4A or 4B).

Excessive noise or fumes from exhaust system

- [ ] Leaking exhaust system or manifold joints (Chapters 1 or 4B).
- [ ] Leaking, corroded or damaged silencers or pipe (Chapters 1 or 4B).
- [ ] Broken mountings causing body or suspension contact (Chapter 1).

Clutch

Pedal travels to floor – no pressure or very little resistance

- [ ] Broken clutch cable (Chapter 6).
- [ ] Incorrect clutch cable adjustment (Chapter 6).
- [ ] Broken clutch release bearing or fork (Chapter 6).
- [ ] Broken diaphragm spring in clutch pressure plate (Chapter 6).

Clutch fails to disengage (unable to select gears).

- [ ] Incorrect clutch cable adjustment (Chapter 6).
- [ ] Clutch friction plate sticking on gearbox input shaft splines (Chapter 6).
- [ ] Clutch friction plate sticking to flywheel or pressure plate (Chapter 6).
- [ ] Faulty pressure plate assembly (Chapter 6).
- [ ] Clutch release mechanism worn or incorrectly assembled (Chapter 6).

Clutch slips (engine speed increases, with no increase in vehicle speed).

- [ ] Incorrect clutch cable adjustment (Chapter 6).
- [ ] Clutch friction plate linings excessively worn (Chapter 6).
- [ ] Clutch friction plate linings contaminated with oil or grease (Chapter 6).
- [ ] Faulty pressure plate or weak diaphragm spring (Chapter 6).

Judder as clutch is engaged

- [ ] Clutch friction plate linings contaminated with oil or grease (Chapter 6).
- [ ] Clutch friction plate linings excessively worn (Chapter 6).
- [ ] Clutch cable sticking or frayed (Chapter 6).
- [ ] Faulty or distorted pressure plate or diaphragm spring (Chapter 6).
- [ ] Worn or loose engine or gearbox mountings (Chapter 2A).
- [ ] Clutch friction plate hub or gearbox input shaft splines worn (Chapter 6).

Noise when depressing or releasing clutch pedal

- [ ] Worn clutch release bearing (Chapter 6).
- [ ] Worn or dry clutch pedal bushes (Chapter 6).
- [ ] Faulty pressure plate assembly (Chapter 6).
- [ ] Pressure plate diaphragm spring broken (Chapter 6).
- [ ] Broken clutch friction plate cushioning springs (Chapter 6).

Manual transmission

Noisy in neutral with engine running

☐ Input shaft bearings worn (noise apparent with clutch pedal released, but not when depressed) (Chapter 7A).*
☐ Clutch release bearing worn (noise apparent with clutch pedal depressed, possibly less when released) (Chapter 6).

Noisy in one particular gear

☐ Worn, damaged or chipped gear teeth (Chapter 7A).*

Difficulty engaging gears

☐ Clutch fault (Chapter 6).
☐ Oil level low (Chapter 1).
☐ Worn or damaged gearchange linkage (Chapter 7A).
☐ Worn synchroniser units (Chapter 7A).*

Jumps out of gear

☐ Worn or damaged gearchange linkage (Chapter 7A).
☐ Worn synchroniser units (Chapter 7A).*
☐ Worn selector forks (Chapter 7A).*

Vibration

☐ Lack of oil (Chapter 1).
☐ Worn bearings (Chapter 7A).*

Lubricant leaks

☐ Leaking driveshaft oil seal (Chapter 7A).
☐ Leaking housing joint (Chapter 7A).*
☐ Leaking input shaft oil seal (Chapter 7A).*
☐ Leaking selector shaft oil seal (Chapter 7A).
Although the corrective action necessary to remedy the symptoms described is beyond the scope of the home mechanic, the above information should be helpful in isolating the cause of the condition, so that the owner can communicate clearly with a professional mechanic.

Automatic transmission

Note: *Due to the complexity of the automatic transmission, it is difficult for the home mechanic to properly diagnose and service this unit. For problems other than the following, the vehicle should be taken to a dealer service department or automatic transmission specialist. Do not be too hasty in removing the transmission if a fault is suspected, as most of the testing is carried out with the unit still fitted.*

Fluid leakage

☐ Automatic transmission fluid is usually dark in colour. Fluid leaks should not be confused with engine oil, which can easily be blown onto the transmission by airflow.
☐ To determine the source of a leak, first remove all built-up dirt and grime from the transmission housing and surrounding areas using a degreasing agent, or by steam-cleaning. Drive the vehicle at low speed, so airflow will not blow the leak far from its source. Raise and support the vehicle, and determine where the leak is coming from. The following are common areas of leakage:
a) Oil pan (Chapter 1).
b) Dipstick tube (Chapter 1).
c) Transmission-to-fluid cooler pipes/unions (Chapter 7B).

Transmission fluid brown, or has burned smell

☐ Transmission fluid level low, or fluid in need of renewal (Chapter 1).

General gear selection problems

☐ Chapter 7B deals with checking and adjusting the selector cable on automatic transmissions. The following are common problems which may be caused by a poorly-adjusted cable:

a) Engine starting in gears other than Park or Neutral.
b) Indicator panel indicating a gear other than the one actually being used.
c) Vehicle moves when in Park or Neutral.
d) Poor gearshift quality or erratic gearchanges
☐ Refer to Chapter 7B for the selector cable adjustment procedure.

Transmission will not downshift (kickdown) with accelerator pedal fully depressed

☐ Low transmission fluid level (Chapter 1).
☐ Incorrect selector cable adjustment (Chapter 7B).
☐ Incorrect kickdown cable adjustment (Chapter 7B).

Engine will not start in any gear, or starts in gears other than Park or Neutral

☐ Incorrect selector cable adjustment (Chapter 7B).
☐ Incorrect starter inhibitor switch adjustment (Chapter 7B).

Transmission slips, is noisy, or has no drive in forward or reverse gears

☐ There are many probable causes for the above problems, but the home mechanic should be concerned with only one possibility – fluid level. Before taking the vehicle to a dealer or transmission specialist, check the fluid level and condition of the fluid as described in Chapter 1. Correct the fluid level as necessary, or change the fluid and filter if needed. If the problem persists, professional help will be necessary.

Driveshafts

Clicking or knocking noise on turns (at slow speed on full-lock).

☐ Lack of constant velocity joint lubricant, possibly due to damaged gaiter (Chapter 8).
☐ Worn outer constant velocity joint (Chapter 8).

Vibration when accelerating or decelerating

☐ Worn inner constant velocity joint (Chapter 8).
☐ Bent or distorted driveshaft (Chapter 8).

Braking system

Note: *Before assuming that a brake problem exists, make sure that the tyres are in good condition and correctly inflated, that the front wheel alignment is correct, and that the vehicle is not loaded with weight in an unequal manner. Apart from checking the condition of all pipe and hose connections, any faults occurring on the anti-lock braking system should be referred to a Nissan dealer for diagnosis.*

Vehicle pulls to one side under braking

☐ Worn, defective, damaged or contaminated brake pads/shoes on one side (Chapters 1 and 9).
☐ Seized or partially-seized front brake caliper or rear wheel cylinder/caliper piston (Chapters 1 and 9).
☐ A mixture of brake pad/shoe lining materials fitted between sides (Chapters 1 and 9).
☐ Brake caliper or backplate mounting bolts loose (Chapter 9).
☐ Worn or damaged steering or suspension components (Chapters 1 and 10).

Noise (grinding or high-pitched squeal) when brakes applied

☐ Brake pad or shoe friction lining material worn down to metal backing (Chapters 1 and 9).
☐ Excessive corrosion of brake disc or drum. (May be apparent after the vehicle has been standing for some time (Chapters 1 and 9).
☐ Foreign object (stone chipping, etc) trapped between brake disc and shield (Chapters 1 and 9).

Excessive brake pedal travel

☐ Inoperative rear brake self-adjust mechanism – drum brake models (Chapters 1 and 9).
☐ Faulty master cylinder (Chapter 9).
☐ Air in hydraulic system (Chapter 9).
☐ Faulty vacuum servo unit (Chapters 1 and 9).

Brake pedal feels spongy when depressed

☐ Air in hydraulic system (Chapter 9).
☐ Deteriorated flexible rubber brake hoses (Chapters 1 and 9).
☐ Master cylinder mounting nuts loose (Chapter 9).
☐ Faulty master cylinder (Chapter 9).

Excessive brake pedal effort required to stop vehicle

☐ Faulty vacuum servo unit (Chapters 1 and 9).
☐ Disconnected, damaged or insecure brake servo vacuum hose (Chapter 9).
☐ Primary or secondary hydraulic circuit failure (Chapter 9).
☐ Seized brake caliper or wheel cylinder piston(s) (Chapter 9).
☐ Brake pads or brake shoes incorrectly fitted (Chapter 9).
☐ Incorrect grade of brake pads or brake shoes fitted (Chapter 9).
☐ Brake pads or brake shoe linings contaminated (Chapter 9).

Judder felt through brake pedal or steering wheel when braking

☐ Excessive run-out or distortion of discs/drums (Chapter 9).
☐ Brake pad or brake shoe linings worn (Chapters 1 and 9).
☐ Brake caliper or brake backplate mounting bolts loose (Chapter 9).
☐ Wear in suspension or steering components or mountings (Chapters 1 and 10).

Brakes binding

☐ Seized brake caliper or wheel cylinder piston(s) (Chapter 9).
☐ Incorrectly-adjusted handbrake mechanism (Chapter 1).
☐ Faulty master cylinder (Chapter 9).

Rear wheels locking under normal braking

☐ Rear brake pad/shoe linings contaminated (Chapters 1 and 9).
☐ Faulty brake pressure regulator (Chapter 9).

Suspension and steering

Note: *Before diagnosing suspension or steering faults, be sure that the trouble is not due to incorrect tyre pressures, mixtures of tyre types, or binding brakes.*

Vehicle pulls to one side

☐ Defective tyre (*Weekly checks*).
☐ Excessive wear in suspension or steering components (Chapters 1 and 10).
☐ Incorrect front wheel alignment (Chapter 1).
☐ Accident damage to steering or suspension components (Chapter 1).

Wheel wobble and vibration

☐ Front roadwheels out of balance (vibration felt mainly through the steering wheel) (Chapters 1 and 10).
☐ Rear roadwheels out of balance (vibration felt throughout the vehicle) (Chapters 1 and 10).
☐ Roadwheels damaged or distorted (Chapters 1 and 10).
☐ Faulty or damaged tyre (*Weekly checks*).
☐ Worn steering or suspension joints, bushes or components (Chapters 1 and 10).
☐ Wheel nuts loose (Chapters 1 and 10).

Excessive pitching and/or rolling around corners, or during braking

☐ Defective shock absorbers (Chapters 1 and 10).
☐ Broken or weak spring and/or suspension component (Chapters 1 and 10).
☐ Worn or damaged anti-roll bar or mountings (Chapter 10).

Wandering or general instability

☐ Incorrect front wheel alignment (Chapter 1).
☐ Worn steering or suspension joints, bushes or components (Chapters 1 and 10).
☐ Roadwheels out of balance (Chapters 1 and 10).
☐ Faulty or damaged tyre (*Weekly checks*).
☐ Wheel nuts loose (Chapters 1 and 10).
☐ Defective shock absorbers (Chapters 1 and 10).

Excessively-stiff steering

☐ Lack of steering gear lubricant (Chapter 10).
☐ Seized track rod end balljoint or suspension balljoint (Chapters 1 and 10).
☐ Broken or incorrectly-adjusted auxiliary drivebelt – power steering (Chapter 1).
☐ Incorrect front wheel alignment (Chapter 1).
☐ Steering rack or column bent or damaged (Chapter 10).

Suspension and steering (continued)

Excessive play in steering

- ☐ Worn steering track rod end balljoints (Chapters 1 and 10).
- ☐ Worn rack-and-pinion steering gear (Chapter 10).
- ☐ Worn steering or suspension joints, bushes or components (Chapters 1 and 10).

Lack of power assistance

- ☐ Broken or incorrectly-adjusted auxiliary drivebelt (Chapter 1).
- ☐ Incorrect power steering fluid level (Weekly checks).
- ☐ Restriction in power steering fluid hoses (Chapter 1).
- ☐ Faulty power steering pump (Chapter 10).
- ☐ Faulty rack-and-pinion steering gear (Chapter 10).

Tyre wear excessive

Tyres worn on inside or outside edges

- ☐ Tyres under-inflated (wear on both edges) (Weekly checks).
- ☐ Incorrect camber or castor angles (wear on one edge only) (Chapter 1).
- ☐ Worn steering or suspension joints, bushes or components (Chapters 1 and 10).
- ☐ Excessively-hard cornering.
- ☐ Accident damage.

Tyre treads exhibit feathered edges

- ☐ Incorrect toe setting (Chapter 1).

Tyres worn in centre of tread

- ☐ Tyres over-inflated (Weekly checks).

Tyres worn on inside and outside edges

- ☐ Tyres under-inflated (Weekly checks).

Tyres worn unevenly

- ☐ Tyres/wheels out of balance (Chapter 1).
- ☐ Excessive wheel or tyre run-out (Chapter 1).
- ☐ Worn shock absorbers (Chapters 1 and 10).
- ☐ Faulty tyre (Weekly checks).

Electrical system

Note: For problems associated with the starting system, refer to the faults listed under 'Engine' earlier in this Section.

Battery will not hold a charge for more than a few days

- ☐ Battery defective internally (Chapter 5A).
- ☐ Battery terminal connections loose or corroded (Weekly checks).
- ☐ Auxiliary drivebelt worn or incorrectly adjusted (Chapter 1).
- ☐ Alternator not charging at correct output (Chapter 5A).
- ☐ Alternator or voltage regulator faulty (Chapter 5A).
- ☐ Short-circuit causing continual battery drain (Chapters 5A and 12).

Ignition/no-charge warning light remains illuminated with engine running

- ☐ Auxiliary drivebelt broken, worn, or incorrectly adjusted (Chapter 1).
- ☐ Alternator brushes worn, sticking, or dirty (Chapter 5A).
- ☐ Alternator brush springs weak or broken (Chapter 5A).
- ☐ Internal fault in alternator or voltage regulator (Chapter 5A).
- ☐ Broken, disconnected, or loose wiring in charging circuit (Chapter 5A).

Ignition/no-charge warning light fails to come on

- ☐ Warning light bulb blown (Chapter 12).
- ☐ Broken, disconnected, or loose wiring in warning light circuit (Chapter 12).
- ☐ Alternator faulty (Chapter 5A).

Lights inoperative

- ☐ Bulb blown (Chapter 12).
- ☐ Corrosion of bulb or bulbholder contacts (Chapter 12).
- ☐ Blown fuse (Chapter 12).
- ☐ Faulty relay (Chapter 12).
- ☐ Broken, loose, or disconnected wiring (Chapter 12).
- ☐ Faulty switch (Chapter 12).

Instrument readings inaccurate or erratic

Fuel or temperature gauges give no reading

- ☐ Faulty gauge sender unit (Chapters 3 or 4A).
- ☐ Wiring open-circuit (Chapter 12).
- ☐ Faulty gauge (Chapter 12).

Fuel or temperature gauges give continuous maximum reading

- ☐ Faulty gauge sender unit (Chapters 3 or 4A).
- ☐ Wiring short-circuit (Chapter 12).
- ☐ Faulty gauge (Chapter 12).

Horn inoperative, or unsatisfactory in operation

Horn operates all the time

- ☐ Horn push either earthed or stuck down (Chapter 12).
- ☐ Horn cable-to-horn push earthed (Chapter 12).

Horn fails to operate

- ☐ Blown fuse (Chapter 12).
- ☐ Cable or cable connections loose, broken or disconnected (Chapter 12).
- ☐ Faulty horn (Chapter 12).

Horn emits intermittent or unsatisfactory sound

- ☐ Cable connections loose (Chapter 12).
- ☐ Horn mountings loose (Chapter 12).
- ☐ Faulty horn (Chapter 12).

Electrical system (continued)

Windscreen/tailgate wipers inoperative, or unsatisfactory in operation

Wipers fail to operate, or operate very slowly

- ☐ Wiper blades stuck to screen, or linkage seized or binding (Chapters 1 and 12).
- ☐ Blown fuse (Chapter 12).
- ☐ Cable or cable connections loose, broken or disconnected (Chapter 12).
- ☐ Faulty relay (Chapter 12).
- ☐ Faulty wiper motor (Chapter 12).

Wiper blades sweep over too large or too small an area of the glass

- ☐ Wiper arms incorrectly positioned on spindles (Chapter 1).
- ☐ Excessive wear of wiper linkage (Chapter 12).
- ☐ Wiper motor or linkage mountings loose or insecure (Chapter 12).

Wiper blades fail to clean the glass effectively

- ☐ Wiper blade rubbers worn or perished (*Weekly checks*).
- ☐ Wiper arm tension springs broken, or arm pivots seized (Chapter 12).
- ☐ Insufficient windscreen washer additive to adequately remove road film (*Weekly checks*).

Windscreen/tailgate washers inoperative, or unsatisfactory in operation

One or more washer jets inoperative

- ☐ Blocked washer jet (Chapter 1 or 12).
- ☐ Disconnected, kinked or restricted fluid hose (Chapter 12).
- ☐ Insufficient fluid in washer reservoir (*Weekly checks*).

Washer pump fails to operate

- ☐ Broken or disconnected wiring or connections (Chapter 12).
- ☐ Blown fuse (Chapter 12).
- ☐ Faulty washer switch (Chapter 12).
- ☐ Faulty washer pump (Chapter 12).

Washer pump runs for some time before fluid is emitted from jets

- ☐ Faulty one-way valve in fluid supply hose (Chapter 12).

Electric windows inoperative, or unsatisfactory in operation

Window glass will only move in one direction

- ☐ Faulty switch (Chapter 12).

Window glass slow to move

- ☐ Incorrectly-adjusted door glass guide channels (Chapter 11).
- ☐ Regulator seized or damaged, or in need of lubrication (Chapter 11).
- ☐ Door internal components or trim fouling regulator (Chapter 11).
- ☐ Faulty motor (Chapter 11).

Window glass fails to move

- ☐ Incorrectly-adjusted door glass guide channels (Chapter 11).
- ☐ Blown fuse (Chapter 12).
- ☐ Faulty relay (Chapter 12).
- ☐ Broken or disconnected wiring or connections (Chapter 12).
- ☐ Faulty motor (Chapter 11).

Central locking system inoperative, or unsatisfactory in operation

Complete system failure

- ☐ Blown fuse (Chapter 12).
- ☐ Faulty relay (Chapter 12).
- ☐ Broken or disconnected wiring or connections (Chapter 12).
- ☐ Faulty control unit (Chapter 11).

Latch locks but will not unlock, or unlocks but will not lock

- ☐ Faulty master switch (Chapter 12).
- ☐ Broken or disconnected latch operating rods or levers (Chapter 11).
- ☐ Faulty relay (Chapter 12).
- ☐ Faulty control unit (Chapter 11).

One solenoid/motor fails to operate

- ☐ Broken or disconnected wiring or connections (Chapter 12).
- ☐ Faulty solenoid/motor (Chapter 11).
- ☐ Broken, binding or disconnected latch operating rods or levers (Chapter 11).
- ☐ Fault in door latch (Chapter 11).

A

ABS (Anti-lock brake system) A system, usually electronically controlled, that senses incipient wheel lockup during braking and relieves hydraulic pressure at wheels that are about to skid.

Air bag An inflatable bag hidden in the steering wheel (driver's side) or the dash or glovebox (passenger side). In a head-on collision, the bags inflate, preventing the driver and front passenger from being thrown forward into the steering wheel or windscreen.

Air cleaner A metal or plastic housing, containing a filter element, which removes dust and dirt from the air being drawn into the engine.

Air filter element The actual filter in an air cleaner system, usually manufactured from pleated paper and requiring renewal at regular intervals.

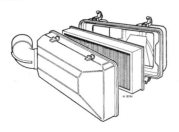

Air filter

Allen key A hexagonal wrench which fits into a recessed hexagonal hole.

Alligator clip A long-nosed spring-loaded metal clip with meshing teeth. Used to make temporary electrical connections.

Alternator A component in the electrical system which converts mechanical energy from a drivebelt into electrical energy to charge the battery and to operate the starting system, ignition system and electrical accessories.

Alternator (exploded view)

Ampere (amp) A unit of measurement for the flow of electric current. One amp is the amount of current produced by one volt acting through a resistance of one ohm.

Anaerobic sealer A substance used to prevent bolts and screws from loosening. Anaerobic means that it does not require oxygen for activation. The Loctite brand is widely used.

Antifreeze A substance (usually ethylene glycol) mixed with water, and added to a vehicle's cooling system, to prevent freezing of the coolant in winter. Antifreeze also contains chemicals to inhibit corrosion and the formation of rust and other deposits that would tend to clog the radiator and coolant passages and reduce cooling efficiency.

Anti-seize compound A coating that reduces the risk of seizing on fasteners that are subjected to high temperatures, such as exhaust manifold bolts and nuts.

Anti-seize compound

Asbestos A natural fibrous mineral with great heat resistance, commonly used in the composition of brake friction materials. Asbestos is a health hazard and the dust created by brake systems should never be inhaled or ingested.

Axle A shaft on which a wheel revolves, or which revolves with a wheel. Also, a solid beam that connects the two wheels at one end of the vehicle. An axle which also transmits power to the wheels is known as a live axle.

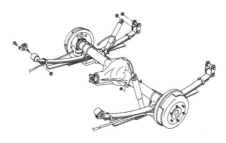

Axle assembly

Axleshaft A single rotating shaft, on either side of the differential, which delivers power from the final drive assembly to the drive wheels. Also called a driveshaft or a halfshaft.

B

Ball bearing An anti-friction bearing consisting of a hardened inner and outer race with hardened steel balls between two races.

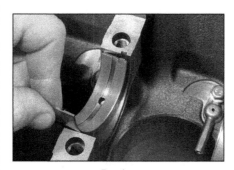

Bearing

Bearing The curved surface on a shaft or in a bore, or the part assembled into either, that permits relative motion between them with minimum wear and friction.

Big-end bearing The bearing in the end of the connecting rod that's attached to the crankshaft.

Bleed nipple A valve on a brake wheel cylinder, caliper or other hydraulic component that is opened to purge the hydraulic system of air. Also called a bleed screw.

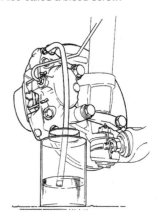

Brake bleeding

Brake bleeding Procedure for removing air from lines of a hydraulic brake system.

Brake disc The component of a disc brake that rotates with the wheels.

Brake drum The component of a drum brake that rotates with the wheels.

Brake linings The friction material which contacts the brake disc or drum to retard the vehicle's speed. The linings are bonded or riveted to the brake pads or shoes.

Brake pads The replaceable friction pads that pinch the brake disc when the brakes are applied. Brake pads consist of a friction material bonded or riveted to a rigid backing plate.

Brake shoe The crescent-shaped carrier to which the brake linings are mounted and which forces the lining against the rotating drum during braking.

Braking systems For more information on braking systems, consult the *Haynes Automotive Brake Manual*.

Breaker bar A long socket wrench handle providing greater leverage.

Bulkhead The insulated partition between the engine and the passenger compartment.

C

Caliper The non-rotating part of a disc-brake assembly that straddles the disc and carries the brake pads. The caliper also contains the hydraulic components that cause the pads to pinch the disc when the brakes are applied. A caliper is also a measuring tool that can be set to measure inside or outside dimensions of an object.

Camshaft A rotating shaft on which a series of cam lobes operate the valve mechanisms. The camshaft may be driven by gears, by sprockets and chain or by sprockets and a belt.

Canister A container in an evaporative emission control system; contains activated charcoal granules to trap vapours from the fuel system.

Canister

Carburettor A device which mixes fuel with air in the proper proportions to provide a desired power output from a spark ignition internal combustion engine.

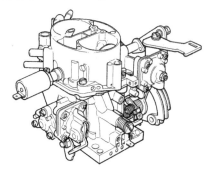

Carburettor

Castellated Resembling the parapets along the top of a castle wall. For example, a castellated balljoint stud nut.

Castellated nut

Castor In wheel alignment, the backward or forward tilt of the steering axis. Castor is positive when the steering axis is inclined rearward at the top.

Catalytic converter A silencer-like device in the exhaust system which converts certain pollutants in the exhaust gases into less harmful substances.

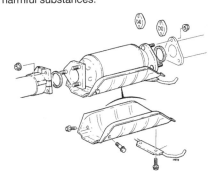

Catalytic converter

Circlip A ring-shaped clip used to prevent endwise movement of cylindrical parts and shafts. An internal circlip is installed in a groove in a housing; an external circlip fits into a groove on the outside of a cylindrical piece such as a shaft.

Clearance The amount of space between two parts. For example, between a piston and a cylinder, between a bearing and a journal, etc.

Coil spring A spiral of elastic steel found in various sizes throughout a vehicle, for example as a springing medium in the suspension and in the valve train.

Compression Reduction in volume, and increase in pressure and temperature, of a gas, caused by squeezing it into a smaller space.

Compression ratio The relationship between cylinder volume when the piston is at top dead centre and cylinder volume when the piston is at bottom dead centre.

Constant velocity (CV) joint A type of universal joint that cancels out vibrations caused by driving power being transmitted through an angle.

Core plug A disc or cup-shaped metal device inserted in a hole in a casting through which core was removed when the casting was formed. Also known as a freeze plug or expansion plug.

Crankcase The lower part of the engine block in which the crankshaft rotates.

Crankshaft The main rotating member, or shaft, running the length of the crankcase, with offset "throws" to which the connecting rods are attached.

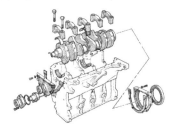

Crankshaft assembly

Crocodile clip See Alligator clip

D

Diagnostic code Code numbers obtained by accessing the diagnostic mode of an engine management computer. This code can be used to determine the area in the system where a malfunction may be located.

Disc brake A brake design incorporating a rotating disc onto which brake pads are squeezed. The resulting friction converts the energy of a moving vehicle into heat.

Double-overhead cam (DOHC) An engine that uses two overhead camshafts, usually one for the intake valves and one for the exhaust valves.

Drivebelt(s) The belt(s) used to drive accessories such as the alternator, water pump, power steering pump, air conditioning compressor, etc. off the crankshaft pulley.

Accessory drivebelts

Driveshaft Any shaft used to transmit motion. Commonly used when referring to the axleshafts on a front wheel drive vehicle.

Driveshaft

Drum brake A type of brake using a drum-shaped metal cylinder attached to the inner surface of the wheel. When the brake pedal is pressed, curved brake shoes with friction linings press against the inside of the drum to slow or stop the vehicle.

Drum brake assembly

E

EGR valve A valve used to introduce exhaust gases into the intake air stream.

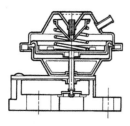

EGR valve

Electronic control unit (ECU) A computer which controls (for instance) ignition and fuel injection systems, or an anti-lock braking system. For more information refer to the *Haynes Automotive Electrical and Electronic Systems Manual.*

Electronic Fuel Injection (EFI) A computer controlled fuel system that distributes fuel through an injector located in each intake port of the engine.

Emergency brake A braking system, independent of the main hydraulic system, that can be used to slow or stop the vehicle if the primary brakes fail, or to hold the vehicle stationary even though the brake pedal isn't depressed. It usually consists of a hand lever that actuates either front or rear brakes mechanically through a series of cables and linkages. Also known as a handbrake or parking brake.

Endfloat The amount of lengthwise movement between two parts. As applied to a crankshaft, the distance that the crankshaft can move forward and back in the cylinder block.

Engine management system (EMS) A computer controlled system which manages the fuel injection and the ignition systems in an integrated fashion.

Exhaust manifold A part with several passages through which exhaust gases leave the engine combustion chambers and enter the exhaust pipe.

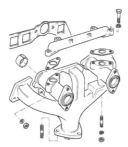

Exhaust manifold

F

Fan clutch A viscous (fluid) drive coupling device which permits variable engine fan speeds in relation to engine speeds.

Feeler blade A thin strip or blade of hardened steel, ground to an exact thickness, used to check or measure clearances between parts.

Feeler blade

Firing order The order in which the engine cylinders fire, or deliver their power strokes, beginning with the number one cylinder.

Flywheel A heavy spinning wheel in which energy is absorbed and stored by means of momentum. On cars, the flywheel is attached to the crankshaft to smooth out firing impulses.

Free play The amount of travel before any action takes place. The "looseness" in a linkage, or an assembly of parts, between the initial application of force and actual movement. For example, the distance the brake pedal moves before the pistons in the master cylinder are actuated.

Fuse An electrical device which protects a circuit against accidental overload. The typical fuse contains a soft piece of metal which is calibrated to melt at a predetermined current flow (expressed as amps) and break the circuit.

Fusible link A circuit protection device consisting of a conductor surrounded by heat-resistant insulation. The conductor is smaller than the wire it protects, so it acts as the weakest link in the circuit. Unlike a blown fuse, a failed fusible link must frequently be cut from the wire for replacement.

G

Gap The distance the spark must travel in jumping from the centre electrode to the side

Adjusting spark plug gap

electrode in a spark plug. Also refers to the spacing between the points in a contact breaker assembly in a conventional points-type ignition, or to the distance between the reluctor or rotor and the pickup coil in an electronic ignition.

Gasket Any thin, soft material - usually cork, cardboard, asbestos or soft metal - installed between two metal surfaces to ensure a good seal. For instance, the cylinder head gasket seals the joint between the block and the cylinder head.

Gasket

Gauge An instrument panel display used to monitor engine conditions. A gauge with a movable pointer on a dial or a fixed scale is an analogue gauge. A gauge with a numerical readout is called a digital gauge.

H

Halfshaft A rotating shaft that transmits power from the final drive unit to a drive wheel, usually when referring to a live rear axle.

Harmonic balancer A device designed to reduce torsion or twisting vibration in the crankshaft. May be incorporated in the crankshaft pulley. Also known as a vibration damper.

Hone An abrasive tool for correcting small irregularities or differences in diameter in an engine cylinder, brake cylinder, etc.

Hydraulic tappet A tappet that utilises hydraulic pressure from the engine's lubrication system to maintain zero clearance (constant contact with both camshaft and valve stem). Automatically adjusts to variation in valve stem length. Hydraulic tappets also reduce valve noise.

I

Ignition timing The moment at which the spark plug fires, usually expressed in the number of crankshaft degrees before the piston reaches the top of its stroke.

Inlet manifold A tube or housing with passages through which flows the air-fuel mixture (carburettor vehicles and vehicles with throttle body injection) or air only (port fuel-injected vehicles) to the port openings in the cylinder head.

J

Jump start Starting the engine of a vehicle with a discharged or weak battery by attaching jump leads from the weak battery to a charged or helper battery.

L

Load Sensing Proportioning Valve (LSPV) A brake hydraulic system control valve that works like a proportioning valve, but also takes into consideration the amount of weight carried by the rear axle.
Locknut A nut used to lock an adjustment nut, or other threaded component, in place. For example, a locknut is employed to keep the adjusting nut on the rocker arm in position.
Lockwasher A form of washer designed to prevent an attaching nut from working loose.

M

MacPherson strut A type of front suspension system devised by Earle MacPherson at Ford of England. In its original form, a simple lateral link with the anti-roll bar creates the lower control arm. A long strut - an integral coil spring and shock absorber - is mounted between the body and the steering knuckle. Many modern so-called MacPherson strut systems use a conventional lower A-arm and don't rely on the anti-roll bar for location.
Multimeter An electrical test instrument with the capability to measure voltage, current and resistance.

N

NOx Oxides of Nitrogen. A common toxic pollutant emitted by petrol and diesel engines at higher temperatures.

O

Ohm The unit of electrical resistance. One volt applied to a resistance of one ohm will produce a current of one amp.
Ohmmeter An instrument for measuring electrical resistance.
O-ring A type of sealing ring made of a special rubber-like material; in use, the O-ring is compressed into a groove to provide the sealing action.

O-ring

Overhead cam (ohc) engine An engine with the camshaft(s) located on top of the cylinder head(s).
Overhead valve (ohv) engine An engine with the valves located in the cylinder head, but with the camshaft located in the engine block.
Oxygen sensor A device installed in the engine exhaust manifold, which senses the oxygen content in the exhaust and converts this information into an electric current. Also called a Lambda sensor.

P

Phillips screw A type of screw head having a cross instead of a slot for a corresponding type of screwdriver.
Plastigage A thin strip of plastic thread, available in different sizes, used for measuring clearances. For example, a strip of Plastigage is laid across a bearing journal. The parts are assembled and dismantled; the width of the crushed strip indicates the clearance between journal and bearing.

Plastigage

Propeller shaft The long hollow tube with universal joints at both ends that carries power from the transmission to the differential on front-engined rear wheel drive vehicles.
Proportioning valve A hydraulic control valve which limits the amount of pressure to the rear brakes during panic stops to prevent wheel lock-up.

R

Rack-and-pinion steering A steering system with a pinion gear on the end of the steering shaft that mates with a rack (think of a geared wheel opened up and laid flat). When the steering wheel is turned, the pinion turns, moving the rack to the left or right. This movement is transmitted through the track rods to the steering arms at the wheels.
Radiator A liquid-to-air heat transfer device designed to reduce the temperature of the coolant in an internal combustion engine cooling system.
Refrigerant Any substance used as a heat transfer agent in an air-conditioning system. R-12 has been the principle refrigerant for many years; recently, however, manufacturers have begun using R-134a, a non-CFC substance that is considered less harmful to the ozone in the upper atmosphere.

Rocker arm A lever arm that rocks on a shaft or pivots on a stud. In an overhead valve engine, the rocker arm converts the upward movement of the pushrod into a downward movement to open a valve.
Rotor In a distributor, the rotating device inside the cap that connects the centre electrode and the outer terminals as it turns, distributing the high voltage from the coil secondary winding to the proper spark plug. Also, that part of an alternator which rotates inside the stator. Also, the rotating assembly of a turbocharger, including the compressor wheel, shaft and turbine wheel.
Runout The amount of wobble (in-and-out movement) of a gear or wheel as it's rotated. The amount a shaft rotates "out-of-true." The out-of-round condition of a rotating part.

S

Sealant A liquid or paste used to prevent leakage at a joint. Sometimes used in conjunction with a gasket.
Sealed beam lamp An older headlight design which integrates the reflector, lens and filaments into a hermetically-sealed one-piece unit. When a filament burns out or the lens cracks, the entire unit is simply replaced.
Serpentine drivebelt A single, long, wide accessory drivebelt that's used on some newer vehicles to drive all the accessories, instead of a series of smaller, shorter belts. Serpentine drivebelts are usually tensioned by an automatic tensioner.

Serpentine drivebelt

Shim Thin spacer, commonly used to adjust the clearance or relative positions between two parts. For example, shims inserted into or under bucket tappets control valve clearances. Clearance is adjusted by changing the thickness of the shim.
Slide hammer A special puller that screws into or hooks onto a component such as a shaft or bearing; a heavy sliding handle on the shaft bottoms against the end of the shaft to knock the component free.
Sprocket A tooth or projection on the periphery of a wheel, shaped to engage with a chain or drivebelt. Commonly used to refer to the sprocket wheel itself.

Starter inhibitor switch On vehicles with an automatic transmission, a switch that prevents starting if the vehicle is not in Neutral or Park.

Strut See MacPherson strut.

T

Tappet A cylindrical component which transmits motion from the cam to the valve stem, either directly or via a pushrod and rocker arm. Also called a cam follower.

Thermostat A heat-controlled valve that regulates the flow of coolant between the cylinder block and the radiator, so maintaining optimum engine operating temperature. A thermostat is also used in some air cleaners in which the temperature is regulated.

Thrust bearing The bearing in the clutch assembly that is moved in to the release levers by clutch pedal action to disengage the clutch. Also referred to as a release bearing.

Timing belt A toothed belt which drives the camshaft. Serious engine damage may result if it breaks in service.

Timing chain A chain which drives the camshaft.

Toe-in The amount the front wheels are closer together at the front than at the rear. On rear wheel drive vehicles, a slight amount of toe-in is usually specified to keep the front wheels running parallel on the road by offsetting other forces that tend to spread the wheels apart.

Toe-out The amount the front wheels are closer together at the rear than at the front. On front wheel drive vehicles, a slight amount of toe-out is usually specified.

Tools For full information on choosing and using tools, refer to the *Haynes Automotive Tools Manual*.

Tracer A stripe of a second colour applied to a wire insulator to distinguish that wire from another one with the same colour insulator.

Tune-up A process of accurate and careful adjustments and parts replacement to obtain the best possible engine performance.

Turbocharger A centrifugal device, driven by exhaust gases, that pressurises the intake air. Normally used to increase the power output from a given engine displacement, but can also be used primarily to reduce exhaust emissions (as on VW's "Umwelt" Diesel engine).

U

Universal joint or U-joint A double-pivoted connection for transmitting power from a driving to a driven shaft through an angle. A U-joint consists of two Y-shaped yokes and a cross-shaped member called the spider.

V

Valve A device through which the flow of liquid, gas, vacuum, or loose material in bulk may be started, stopped, or regulated by a movable part that opens, shuts, or partially obstructs one or more ports or passageways. A valve is also the movable part of such a device.

Valve clearance The clearance between the valve tip (the end of the valve stem) and the rocker arm or tappet. The valve clearance is measured when the valve is closed.

Vernier caliper A precision measuring instrument that measures inside and outside dimensions. Not quite as accurate as a micrometer, but more convenient.

Viscosity The thickness of a liquid or its resistance to flow.

Volt A unit for expressing electrical "pressure" in a circuit. One volt that will produce a current of one ampere through a resistance of one ohm.

W

Welding Various processes used to join metal items by heating the areas to be joined to a molten state and fusing them together. For more information refer to the *Haynes Automotive Welding Manual*.

Wiring diagram A drawing portraying the components and wires in a vehicle's electrical system, using standardised symbols. For more information refer to the *Haynes Automotive Electrical and Electronic Systems Manual*.

Note: *References throughout this index are in the form* **"Chapter number"** • **"Page number"**

U

V

W

Preserving Our Motoring Heritage

< *The Model J Duesenberg Derham Tourster. Only eight of these magnificent cars were ever built – this is the only example to be found outside the United States of America*

Almost every car you've ever loved, loathed or desired is gathered under one roof at the Haynes Motor Museum. Over 300 immaculately presented cars and motorbikes represent every aspect of our motoring heritage, from elegant reminders of bygone days, such as the superb Model J Duesenberg to curiosities like the bug-eyed BMW Isetta. There are also many old friends and flames. Perhaps you remember the 1959 Ford Popular that you did your courting in? The magnificent 'Red Collection' is a spectacle of classic sports cars including AC, Alfa Romeo, Austin Healey, Ferrari, Lamborghini, Maserati, MG, Riley, Porsche and Triumph.

A Perfect Day Out

Each and every vehicle at the Haynes Motor Museum has played its part in the history and culture of Motoring. Today, they make a wonderful spectacle and a great day out for all the family. Bring the kids, bring Mum and Dad, but above all bring your camera to capture those golden memories for ever. You will also find an impressive array of motoring memorabilia, a comfortable 70 seat video cinema and one of the most extensive transport book shops in Britain. The Pit Stop Cafe serves everything from a cup of tea to wholesome, home-made meals or, if you prefer, you can enjoy the large picnic area nestled in the beautiful rural surroundings of Somerset.

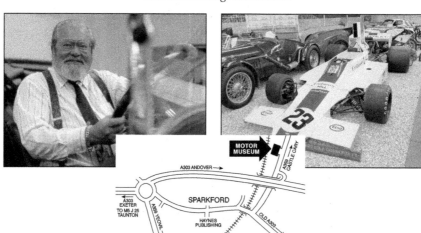

> *John Haynes O.B.E., Founder and Chairman of the museum at the wheel of a Haynes Light 12.*

< *Graham Hill's Lola Cosworth Formula 1 car next to a 1934 Riley Sports.*

The Museum is situated on the A359 Yeovil to Frome road at Sparkford, just off the A303 in Somerset. It is about 40 miles south of Bristol, and 25 minutes drive from the M5 intersection at Taunton.
Open 9.30am - 5.30pm (10.00am - 4.00pm Winter) 7 days a week, *except Christmas Day, Boxing Day and New Years Day*
Special rates available for schools, coach parties and outings Charitable Trust No. 292048